T0257775

Contents

Preface

The purpose of the book is to provide a glimpse into the dynamics and to present opinions and studies of some of the scientists engaged in the development of new ideas in the field from very different standpoints. This book will prove useful to students and researchers owing to its high content quality.

This is a wide-ranging book devoted to preinvasive lesions of the human body. Written with the aim of serving experts to not only diagnose but also understand the etiopathogenesis of precursor lesions, the book also endeavors to identify its molecular and genetic techniques. Chapters in this book consist of a significant quantity of novel information with an efficient categorization of intraepithelial lesions of the cervix and the vulva. This book has been modernized according to the most recent technological developments and can be described as a brief, educational and practical text at all levels, discussing the important role of the molecular analysis of intraepithelial lesions.

At the end, I would like to appreciate all the efforts made by the authors in completing their chapters professionally. I express my deepest gratitude to all of them for contributing to this book by sharing their valuable works. A special thanks to my family and friends for their constant support in this journey.

Editor

Part 1

Intraepithelial Neoplasia of Oral Cavity

Novel Markers for Diagnosis and Prognosis of Oral Intraepithelial Neoplasia

Angela Celetti[1], Francesco Merolla[1,2], Chiara Luise[1],
Maria Siano[2] and Stefania Staibano[2]
*[1]Istituto di Endocrinologia e Oncologia Sperimentale, CNR, c/o Dipartimento di
Biologia e Patologia Cellulare e Molecolare, Università Federico II, Napoli,
[2]Dipartimento di Scienze Biomorfologiche e Funzionali, Università Federico II, Napoli*
Italy

1. Introduction

Squamous carcinoma of the oral cavity is a slow multi-steps process, based on progressive accumulation of genetic events leading to the selection of clonal populations of transformed epithelial cells (Ha&Califano, 2002) The spectrum of histological changes occurring in this process ranges from atypical squamous hyperplasia to carcinoma in situ (CIS), and is grouped under the designation of oral intraepithelial lesions (OILs) (Gale et al, 2005; Gale et al 2006). In their evolution, most cases of OILs are self-limiting and reversible, whereas some persist and may progress to SCC in spite of careful follow-up and treatment (Kambic̆&Gale, 1986; Crissman et al, 1993).

As for the largest group of head and neck intraepithelial lesions, in the last years, various aspects of oral carcinogenesis have been investigated, including the aetiology, histological classification, treatment, frequency of malignant transformation and predictive factors. Particular attention has been directed to the analysis of the interrelationship between histological parameters and their biological behaviour (Gale et al 2005, Gale et al 2006; Kambic̆&Gale, 1986; Putney&O'Keefe 1953; Kambic̆ 1978; Crissman 1979; Henry 1979; Hellquist et al, 1982; Gillis et al, 1983; Grundmann 1983; Goodman 1984; Crissman&Fu 1986; Velasco et al 1987; Olde-Kalter et al 1987; Crissman&Zarbo 1989; Sllamniku et al 1989; Bouquot et al, 1991a; Kambic&Gale 1995; Hellquist et al 1999; Gale et al, 2000; Gallo et al 2001; Ricci et al 2003). These analyses have been recently further supplemented by molecular genetic investigations trying to include the molecular events involved in the pathogenesis of oral squamous cell carcinoma (OSCC) to improve the prognostic evaluation of OIN (Ha&Califano 2002; Somers et al 1992; Saglam et al 2007).

A precise and uniform terminology of squamous intraepithelia lesions is essential for successful collaboration among pathologists, as well as for proper communication with clinicians. The terminology used in clinical and pathological reports has changed significantly over the last six decades. Common agreement has recently been achieved for terms that are used only for the clinical appearance and do not have any histopathological and prognostic implications. The most frequently applied clinical diagnoses are oral leukoplakia and erythroplakia (Kambic̆&Gale 1995; Gale et al 2000; Gallo et al 2001). In contrast, keratosis remains a controversial term, since it is often wrongly applied interchangeably to macroscopic and microscopic features, whereas it really represents a histological term denoting the appearance of a keratin layer on the surface of the squamous epithelium.

Unfortunately, inconsistent terminology still exists for the histological classification of OIN. The spectrum of epithelial changes has been variously described as keratosis, dysplasia, squamous intraepithelial neoplasia (SIN), oral intraepithelial neoplasia (OIN), etc, to list only the most commonly used terms. Because of our inability to harmonize different views and establish a single classification of squamous intraepithelial lesions, there are three classification schemes in the most recent edition of the World Health Organization (WHO) classification of tumours, pathology of the head and neck tumours, as follows: (i) dysplasia system, (ii) SIN system, and (iii) Ljubljana classification (Gale et al 2005). These classifications differ conceptually and terminologically, and analogy between them can only be approximate.

Chronic inflammation, leukoplakia or, occasionally, erythroplakia, appear mainly in the buccal mucosa, labial commissure, gingiva/alveolar ridge, tongue, floor of the mouth. Lesions can be either sharply circumscribed and grow exophytically, or be predominantly flat and diffuse, related in part to the amount of keratin layer. Their surface is rough, may be muddy brown to red (erythroplakia), perhaps with increasingly visible vascularity, or coated with diffuse or dispersed circumscribed whitish plaques. A circumscribed whitish thickening of the mucosa may be observed, covered by irregularly exophytic warty plaques. A speckled appearance of lesions can also be present, caused by an unequal thickness of the keratin layer (Gale et al 2005; Kambic&Gale 1995). Some leukoplakic lesions are ulcerated (6.5%) or combined with erythroplakia (15%) (Bouquot et al 1991a). In general, leukoplakic lesions are thought to have a low risk of malignant transformation, mixed white and red lesions, or speckled leukoplakia, an intermediate risk, and pure erythroplakia (red lesions) the highest risk of cancer development. However, none of these features can be used as an indicator of the overlying changes of the epithelium, and histological analysis of these lesions is mandatory to determine their biological potential.

Symptoms depend on the location and severity of the disease and usually last a few months before clinical notice.

2. Clinical classification of leukoplakia and epithelial dysplasia

Leukoplakia, erythroplakia and palatal keratosis, associated with reverse smoking, are categorized as precancerous lesions (Axell et al 1996; Pindborg JJ et al 1997). Oral leukoplakia is the most common disease among precancerous lesions, whereas erythroplakia is relatively uncommon, and palatal keratosis associated with reverse smoking is rarely reported in Japan (Warnakulasuriya et al 2007). Pindborg et al (1963) confirmed that speckled leukoplakia, which is characterized by the presence of white nodular patches or white lesions interspersed with erythematous areas, was often associated with epithelial dysplasia or carcinoma. These findings were supported by subsequent reports showing the association of nonhomogeneous leukoplakia with epithelial dysplasia (Silverman et al 1976; Gupta et al 1980). The two-tiered clinical classification system, used to divide oral leukoplakia into homogeneous and nonhomogeneous leukoplakia, was created by an international symposium (Axell et al 1996; Pindborg et al 1997). Under this system, homogeneous leukoplakia is further divided into four subtypes: flat, corrugated, wrinkled, or pumice; and similarly nonhomogeneous leukoplakia is subdivided into four types: verrucous, nodular (speckled), ulcerated, or erythroleukoplakia. The adjective "nonhomogeneous" refers to the color (i.e., a mixture of white and red changes for erythroleukoplakia) and texture (i.e., exophytic, papillary, or verrucous) of the lesion (van der Waal et al 1997). However, with regard to nonhomologous lesions, there are no reproducible criteria under this system for the clinical differentiation of proliferative verrucous leukoplakia from verrucous hyperplasia or verrucous carcinoma (van der Waal et al 1997; Shear&Pindborg 1980).

Sugar& Banoczy (1969) reported that leukoplakia erosiva and leukoplakia verrucosa were more often associated with epithelial dysplasia than leukoplakia simplex. Furthermore, because the clinical features of oral leukoplakia in Japan did not correlate with the two aforementioned systems, Amagasa et al (1977) developed a clinical classification system of oral leukoplakia in Japan, which was subsequently further developed in 2006 (Amagasa et al 2006). Under this system, oral leukoplakia is classified into four clinical types: type I, a flat white patch or plaque without red components; type II, a flat white patch or plaque with red components; type III, a slightly raised or elevated white plaque; and type IV, a markedly raised or elevated white plaque. Using this classification system, it was found that type II leukoplakia was significantly associated with epithelial dysplasia.

3. Histopathological features of Squamous Intraepithelial Lesions (SILs)

Traditional light microscopic examination, in spite of a certain subjectivity in interpretation, remains the most reliable method for determining an accurate diagnosis of SILs. Jackson first defined chronic laryngitis and keratosis as precancerous lesions (Jackson C, 1923); later, numerous studies and classifications have attempted to correlate phenotypic and genetic changes with the biological behaviour of the lesions (Michaels&Hellquist 2001). Regrettably, neither generally accepted criteria nor unified terminology have to date been provided for a histological grading system of oral SILs. Evidence of the inability of pathologists to set up a single, unified classification of SILs was manifest in the WHO Classification of head and neck tumours, published in 2005, where the dysplasia system is presented as the 2005 WHO classification simultaneously with the classification of SIN and the Ljubljana classification (Gale et al 2005). The majority of current classifications, such as the traditional dysplasia system (Hellquist et al 1982; Blackwell et al 1995), keratosis without (KWA) and with atypia/ in situ carcinoma (CIS) (Crissman 1979; Crissman 1982), Squamous Intraepithelial Neoplasia (SIN) (Crissman et al 1993; Crissman&Zarbo 1989) and Laryngeal Intraepithelial Neoplasia (LIN), (Friedmann&Ferlito 1993; Resta et al 1992) follow criteria similar to those commonly used for epithelial lesions of the uterine cervix. However, the different aetiology of oral lesions and their particular clinical and histological features require a grading system more appropriate to this region (Hellquist et al 1999). One can object that grading SILs, in spite of the clear histological criteria, is an attempt to impose arbitrary distinct categories of a continually progressing process without naturally and sharply defined borders (Bosman 2001; Kujan et al 2011). However, this continuous process, which is of long duration, may eventually stop, regress or progress, depending above all on the influence of various detrimental factors causing genetic and, consequently, phenotypic epithelial changes. When a biopsy is performed with a representative tissue sample, the established histological changes still serve at present as the main guidance for clinicians on how to treat the patient, as well as being the most reliable prognostic factor of the biological behaviour of the disease.

4. A lesson from premalignant lesions of the uterine cervix

One of the most significant advances in oncology has been the realization that cervical carcinoma arises from precursor lesions. There is probably more known about cervical neoplasia and its natural history than about any other human epithelial neoplasm. Most medical authorities now agree that cervical cancer is the end stage of a continuum of progressively more atypical changes in which one stage merges imperceptibility with the next. The first and apparently earliest change is the appearance of atypical cells in the basal layers of the squamous epithelium, but this occurs alongside normal differentiation toward the prickle and keratinizing cell layers. As the lesion evolves, there is progressive involvement of more and more layers of the epithelium, until it is totally replaced by atypical cells, exhibiting no surface differentiation (Robbins&Cortran 1979).

The most widely used term for the various stages in the evolution of these precursor lesions is "dysplasia" (Reagan&Hamonic 1956), which literally means bad molding or, in more scientific terms, disordered development. In WHO's 1975 "Histological Typing of Female Genital Tract Tumours" (Poulsen et al 1975), dysplasia is subdivided into mild, moderate and severe, depending on the thickness of the squamous epithelium involved by atypical cells. When there is full-thickness involvement, we use the term "carcinoma in situ", which was coined by Broders in 1932 in relation to head and neck lesions (Bouquot et al 2006; Broders 1932).

A newer terminology, "cervical intraepithelial neoplasia" (CIN), was subsequently proposed in an attempt to emphasize that these dysplastic changes represent a spectrum of the same basic changes (Richart 1966; 1973). CIN involves one or more clones of transformed cells slowly replacing normal keratinocytes, starting from the basal and parabasal layers to progressively invading the entire epithelial height. Richart subdivided CIN into three grades, CIN I, CIN II and CIN III, corresponding to mild, moderate and severe dysplasia, respectively, which then progresses to CIS.

The classifications and concepts of premalignant lesions of the uterine cervix have been extended to all other mucosal sites covered by squamous epithelial as oral mucosa.

5. WHO classification

In 1973, WHO defined an oral premalignant lesion as "a morphologically altered tissue in which oral cancer was more likely to occur than in its apparently normal counterpart"; more recently, researchers have recommended use of the term "potentially malignant disorder" (Warnakulasuriya et al 2007). Under the WHO classification, atypical epithelium is divided into two pathological entities, one with progression to SCC and the other without progression. Although the former is a true premalignant lesion and the latter is a reactive atypical epithelium, the concept of epithelial dysplasia (mild, moderate or severe) includes both lesions and is a borderline category which can be placed in neither of the WHO's classifications.

As mentioned earlier, the dysplasia–carcinoma sequence theory as applied to the oral mucosa was adopted from the case of the uterine cervix, and the fundamental view of the WHO classification for oral cancer remained unchanged for more than three decades, from the first edition in 1971 (Wahi et al 1971, Napier&Speight 2008) to the latest version in 2005 (Gale et al 2005). WHO's "Histo-pathological Typing of Cancer and Precancer of the Oral Mucosa" (Pindborg et al 1997), is now used as a worldwide standard guide to diagnosis. The dysplastic features of oral mucosa are characterized by cellular atypia and loss of normal maturation and stratification, and the more severe the dysplasia, the greater the likelihood of malignant transformation. On the basis of the various criteria thought to be typical for the transformation of a dysplastic lesion to carcinoma, lesions are most frequently graded into one of four different groups: mild, moderate or severe dysplasia, or CIS, with the latter considered to be pre-invasive malignancy at the extreme end of epithelial dysplasia. Several histopathological changes may occur in epithelial dysplasia (Pindborg 1980). The criteria used for diagnosing dysplasia are provided in the form of a table in the WHO classification of tumours for the head and neck. Within the frame-work of the grading system of dysplasia, the more prominent or more numerous these factors are, the more severe the grade. These factors are limited to the lower third of the epithelium in mild dysplasia and extend to lower two-thirds of the epithelium in moderate dysplasia upward to the outer layer (Gale et al 2005). The use of the terms full thickness or almost full thickness architectural abnormalities is also recommended for the diagnosis of CIS.

This grading system of one-third, two-thirds and full thickness was described for the first time in the latest version of WHO's classification of head and neck tumours although it had been clearly referred to in the classification of uterine cervix since 1975 (Poulsen et al 1975). The large number of factors in this grading system would appear to be the basis of the many problems associated with the subjectivity of diagnosis (Pindborg 1980; Karabulut et al 1995; Holmstrup et al 2006). Accordingly, examination of the universality (inter-observer variability) and reproducibility (intra-observer variability) of this grading system for diagnosis has been carried out in recent years (Warnakulasuriya et al 2007; Kujan et al 2006; Kujan et al 2007; Ficher et al 2004; Tabor et al 2003; Abbey et al 1995; Brothwell et al 2003; Speight et al 1996) to sharply discriminate "indolent" low grade lesions, potentially reversible, from throughly preneoplastic high grade lesions. To this end, a novel binary grading system (low risk and high risk) designed to simplify the WHO classification and to raise the reproducibility of diagnosis has been advocated (Jares et al 1994; Califano et al 1996).

The histopathological criteria of dysplasia in the WHO classification are widely accepted among pathologists, and the concept of epithelial dysplasia outlined in the classification is considered to be correct in many cases (Crissman et al 1993; Putney&O'Keefe 1953; Ricci et al 2003; Franchi et al 2001). The notion that atypical cells progress from the basal layer to the surface is widely accepted in terms of the universality and reproducibility of diagnosis. However, it has become clear that there is a fatal flaw in this grading system as it does not, in practice, accurately reflect the clinical behaviour (Crissman&Zarbo 1989; Voravud et al 1993; Nadal&Cardesa 2003; Sanz-Ortega et al 2003; Chatrath et al 2003). The grades do not offer clear therapeutic guidelines to clinicians for appropriate management. For CIS at least, the WHO grading system diagnoses CIS showing maturation and differentiation as lower risk lesions, and these lesions account for a large proportion of cases in the oral mucosa (Hellquist et al 1982; Gillis et al 1983; Yoo et al 2004; Kleist&Poetsh 2004; Jeannon et al 2004; Chi et al 2004).

6. SIN/dysplasia classification

In response to the concept of CIN in the uterine cervix, a similar view developed for the oral mucosa. In 2002, Kuffer&Lombardi stated that malignant transformation is a multistep process that should be approached from the histological — not merely clinical — standpoint. Intraepithelial neoplasia, a concept created in relation to the uterine cervix and already extended to other mucosae, should also be adapted for the case of the oral mucosa and used as diagnostic term: the use of the term oral intraepithelial neoplasia represents not only a change in terminology, but also progress in unifying the concept of the precursors of squamous cell carcinoma, while at the same time suppressing the futile debate about severe dysplasia and CIS. Furthermore, grading lesions as low- and high-grade OIN increases diagnostic consistency.

SIN/dysplasia in the oral cavity has been found to take two distinct morphological forms, at opposite ends of the SIN/ dysplasia spectrum: hyperplastic keratinizing SIN/dysplasia and atrophic SIN/dysplasia, which are clinically compatible with leukoplakia and erythroplakia, respectively. The former is keratinizing dysplasia and the latter is the classic (WHO type) form of dysplasia. As a complication, the features of these extremes overlap. Caution must be exercised as the admixture type of these two ends of the spectrum is commonly underdiagnosed and may not be recognized as high-grade SIN/dysplasia.

The SIN/dysplasia classification, a modification of the WHO grading system, proposes a category of keratinizing dysplasia to designate lesions showing superficial keratinization in association with high-grade cytological atypia in the lower epithelium (Crissman et al 1988; Blackwell et al 1995; Kambic˘ et al 1992; Kambic˘ 1997; Hellquist et al 1999; Crissman 1982). The authors suggesting this modification reported that these lesions have a high incidence of local relapse and a high progression rate to invasive SCC and, as such, they are included in the high-grade group as high-grade keratinizing SIN (Crissman&Sakr 2001; Sakr et al 2009). The authors further stressed that abnormal differentiation is present in these lesions in the form of aberrant keratinization (dyskeratosis), manifesting as single-cell keratinization and keratin pearls, occurring in the midst of the epithelium. The histopathological features used for grading OIN according to WHO are listed below:

1. Loss of polarity of the basal cells
2. Proliferation of the basal cells
3. Increased nucleus-to-cytoplasm ratio
4. Epithelial hyperplasia with drop-shaped submucosal rete extension
5. Irregular epithelial stratification and cellular pleomorphism
6. Premature keratinization of single cells (dyskeratosis) or keratin pearls in the rete pegs
7. Increased mitotic figures and abnormally superficial mitoses
8. Presence of abnormal mitotic figures
9. Variation in nucleus size, shape, and hyperchromatism; increased nucleus size

10. Increased number and size of nucleoli
11. Abnormal variation in cell shape and size.

The transition from normal epithelium to atypical epithelium and SCC is related to the progressive accumulation of genetic changes leading to a clonal population of transformed epithelial cells. Despite extensive research into these genetic changes in oral carcinogenesis, reliable genetic markers with diagnostic and prognostic value are still lacking (Gale et al 2009).

7. Ljubljana classification and SIL classification

The Ljubljana classification was devised by laryngeal pathologists Kambic and Lenart in 1971 (Hellquist et al 1999; Gale et al 2009; Kambic&Lenart 1971, Gale et al 2000). Based on clinical and histological observations, these authors adapted the classification to the specific demands of the oral cavity. The Ljubljana system nominally recognizes four grades: simple hyperplasia and basal/parabasal cell hyperplasia include mainly benign categories with a minimum risk of malignant alteration; atypical hyperplasia is potentially a malignant lesion; CIS is actually a malignant lesion (Kambic 1997; Gale et al 2000; Michaels 1997; Eversole 2009; Fleskens&Slootweg 2009; Koren et al, 2002). The main features that differentiate the Ljubljana grading system from other classifications are the distinction between mainly benign (squamous hyperplasia and basal-parabasal hyperplasia) and potentially malignant (atypical hyperplasia) lesions, and the positive separation of CIS from atypical hyperplasia. These two entities differ in morphology and progression to invasive carcinoma. In this classification, all histopathological change is included until it results in SCC (Crissman et al 1988; Sllamniku et al 1988; Crissman 1982). Although many studies have focused on the usefulness of this classification in relation to the larynx (Blackwell et al, 1995; Michaels 1997; Gale et al 2000; Kambic 1997; Frangez I et al, 1997), there is currently almost no verification of this in the oral mucosa (Mahajan&Hazarey 2004; Zerdoner 2003) so its usefulness cannot be discussed as yet.

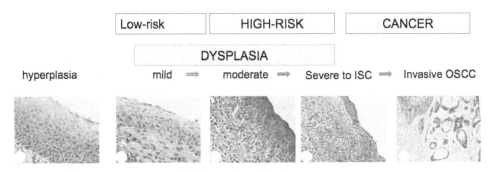

a) Oral mucosa with epithelial hyperplasia without dysplasia. This is not to be considered a preneoplastic lesion.
b) Oral mucosa with mild dysplastic changes. Rare mitoses are appreciable at the basal third of the epithelium.
c) Mild dysplasia of the epithelium of oral mucosa.
d) Severe dysplasia of oral mucosal epithelium (top-right), flanking an area of in situ- and microinfiltrating carcinoma.
e) Deeply infiltrating OSCC
(a-d: hematoxylin and eosin stain; e: immunohistochemical staining for CD44v6)

Fig. 1. Progression of Oral Cancer

The binary system which unites the SIN classification and the Ljubljana classification is advocated mainly by laryngeal pathologists. This system encompasses the Ljubljana classification into the SIN classification, with the concept of SILs being fundamentally the same (Gale et al 2009). The whole spectrum of histological changes, both reversible and irreversible, has recently been cumulatively designated as SIL, ranging from squamous hyperplasia to CIS. In terms of their evolution, some cases of SIL are self-limiting and reversible, some persist, and some progress to SCC despite careful follow-up and treatment. Although it would appear that both classifications can be unified, verification in the case of the oral mucosa remains to be determined.

8. Mechanisms of developing OIN

The genetic changes and the sequence of genetic events underlying the progression of normal mucosa to oral neoplastic tissue are still not entirely recognized. Between six and ten independent genetic events within a single cell have been estimated to be necessary for SCC development in the head and neck region. They are believed to be morphologically expressed as different grades of epithelial abnormality. The latency period between carcinogen exposure and appearance of malignancy may last up to 25 years.

The process of tumorigenesis of solid tumours, including oral neoplasia, involves both activation of proto-oncogene products that stimulate growth, and inactivation of tumour-suppressor genes (TSGs), the products of which normally inhibit cell proliferation (Califano et al 1996; Field 1996; Gallo et al 1997; Califano et al 2000). The identification and characterization of the comprehensive spectrum of genetic aberrations in SCC development may not only elucidate the process of carcinogenesis, but also provide promising diagnostic tools for early detection, prevention and assessment of cancer risk from precursor lesions.

9. Genetic progression model

Califano and co-workers have made two studies of cytogenetic alterations in head and neck carcinogenesis, which showed an increasing number of chromosomal alterations with the progression of Oral Intraepithelial Neoplasia (OILs), ranging from hyperplasia to CIS and invasive SCC. The areas of allelic loss, and less frequently allelic gain, are decisive elements in the progression model involving HNSCC. The results of Califano's studies have revealed that the spectrum of chromosomal loss progressively increases at each histopathological step of squamous intraepithelial lesions from benign hyperplasia to CIS and invasive SCC. The earliest alterations appear on chromosomes 9p21, where the p16 gene resides, at 3p with at least three putative tumour-supressor loci, and at 17p13 where the p53 gene is located. Loss of chromosome region 9p21-22 appeared to be the most common of all genetic changes in HNSCC, with a frequency of 70% (van der Riet P et al 1994). Additional studies of microsatellite DNA allelic imbalance in oral carcinogenesis have confirmed that dysplasia correlates with loss of heterozygosity (LOH) at 3p21, 5q21, 9p21 and 17q13 (Sanz-Ortega J, 2003). Yoo et al. have suggested that 9p21 is the earliest event, already appearing in squamous metaplasia, as well as in invasive and metastatic SCC. LOH at 17p13, 3p35 and 3p14 was observed as an intermediate event, occurring from dysplasia to metastatic SCC (Yoo et al 2004). Micro-satellite instability (MSI), a novel marker of genetic instability, was also applied in a study to assess the risk of malignant progression in laryngeal preinvasive lesions. The authors concluded that MSI is more common in preneoplastic oral lesions that have progressed to invasive SCC. They suggest that MSI assessment may be useful in determining the risk of malignant alteration in patients for whom chemopreventive and multiple endoscopic protocols can be attempted (Sardi et al 2006).

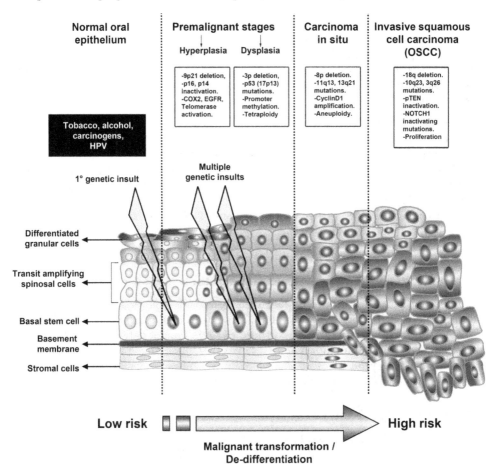

Fig. 2. Genetic Progression Model

10. Key tumour-suppressor genes in oral carcinogenesis

Gene p16 can be inactivated by a variety of mechanisms, such as mutation, homozygous deletion and promoter hypermethylation (Kamb et al 1994; Merlo et al 1995). The p16 gene functions as an inhibitor of cyclin-dependent kinase 4 and 6, with subsequent abrogation of retinoblastoma (Rb) phosphorylation and G1 cell cycle arrest (Serrano et al 1993; Kim et al 2004). Loss of chromosome region 9p21-22 occurs prior to the development of histological atypia, already at the level of hyperplastic mucosa, and is regarded as an early event in the development of HNSCC (Hasina&Lingen 2004).

Another important region identified by allelic loss is chromosome 17p, the site of the p53 gene. It is involved in several key cell functions such as gene transcription, DNA synthesis and repair, cell-cycle coordination and apoptosis. Mutation/inactivation of the p53 gene has been detected in approximately 50%, but may be present in as high as 80% of HNSCCs (Balz et al 2003). It remains unclear whether p53 gene inactivation is an early or late event of oral carcinogenesis. According to Boyle and co-workers, it occurs in the transition from the preinvasive to invasive form (Boyle et al 1993). Some others argue the opposite, presenting alterations of p53 among early steps of neoplastic transformation. Furthermore, gene p53 mutation has been hypothesized to be the earliest event in the development of a genetically altered field in oral mucosa, identifying an area of clonally related cells with malignant potential (Braakhuis et al 2003).

Although LOH frequently appears in head and neck carcinogenesis at chromosome 3p, the genes at this region have not been well defined (Ha&Califano 2002). The fragile histidine triad (FHIT) gene has been identified on chromosome 3p14 as one candidate for TSG, altered by deletions in human tumours. The expression of FHIT protein has recently been studied in HNSCC and premalignant lesions (Yuge et al 2005). Loss of FHIT protein was observed in 42% of SCCs and 23% of premalignant lesions. There was no significant difference among the three grades of dysplasia and FHIT expression. The results of this study indicate that FHIT alterations may play an important role in early events of carcinogenesis.

11. Key oncogenes in malignant alteration of SILs

Chromosome region 11q13 has been identified as the site of several putative oncogenes, such as Bcl-1, int-2, hst-1, EMS-1 and cyclin D1/PRAD1 (Kim&Califano 2004). Amplification of 11q13 is detected in approximately one-third of HNSCCs, but only cyclin D1 has shown consistent overexpression/amplification (Jares et al 1994, Callender et al 1994). The function of proto-oncogene cyclin D1 is to activate Rb via phosphorylation, thus facilitating progression from the G1 phase to the S phase (Kim&Califano 2004).

12. HPV-linked OSCC and OIN

Besides the evident epidemiological meaning, HPV infection linked to OSCC development shows clinical implications as these patients have about half the risk of death with respect to HPV-negative OSCC ones (Fakhry et al 2006). Moreover, the incidence of the subsets of OSCC more frequently found associated with HPV infection, i.e. tongue and tonsillar cancers, has been rising in youngs with men and women 18 to 44 years old (67% of increase) for the past three decades, and the trend is actually most evident for young white women (Patel et al, 2011), whereas the OSCC incidence is declining for nonwhite men, for all age groups. These findings have been justified with the decrease of alcohol and smoke abuse, and the relative prevalence of infection with high-risk HPV strains, particularly in youngs.

This identifies then distinct risk factor profiles for HPV-positive and HPV-negative OIN patients, and justifies the designation of clinical trials to assess the optimal treatment for these groups. From a histopathological point-of-view, the HPV-linked OIN are mostly undifferentiated (Carpenter et al 2011).

This is particularly intriguing, considerating that traditionally undifferentiated cancers have a very worse clinical outcome, being radio- and chemo-resistant, whereas HPV-linked undifferentiated OIN seems to have a better overall prognosis and a good response to post-surgical therapy.

For this reason, it seems fundamental to easily detect this subgroup of cancers and precancerous lesions, to preserve patients from overtreatment of their lesions.

Since high-risk HPVs lead to the intracellular accumulation of p16INK4a protein, due to the E7 block of pRb, it has been proposed to utilize the immunohistochemical evaluation of p16 for the screening of lesional tissue obtained from diagnostic biopsies. This has been shown to reliably predict the high-risk HPV infection in oral biopsies (Hoffmann et al 2010).

The screening with IHC for p16 INK4a protein, then, may be regarded as a precious tool for the proper evaluation of the outcome and responsiveness to therapy of oral cancer and precancerous lesions.

13. Key protein-based alterations in oral carcinogenesis

Protein overexpression can appear as a consequence of gene amplification, increased DNA transcription and translation. Several gene products can influence cancer progression in this manner (Ha&Califano 2002).

Epidermal growth factor receptor (EGFR), located on chromosome 7p12, codes for transmembrane growth-regulating receptor glycoprotein, which influences cell division, migration, adhesion, differentiation and apoptosis through a tyrosine kinase pathway (Pomerantz&Grandis 2004). The EGFR gene was found to be amplified in 25%, and its mRNA was overexpressed in 43% of oral SCCs. Half of the expressed cases occurred in the absence of detectable gene amplification. Both alterations appeared in advanced HNSCC (Irish&Bernstein 1993). Furthermore, overexpression of EGFR protein is an early event in carcinogenesis, rising with increasing degree of epithelial abnormalities, mainly in the progression of oral intraepithelial lesions to SCC (Shin et al 1994; Gale et al 1997).

Eucaryotic initiation factor 4E (eIF4E) is a 24-kDa protein, which binds to mRNA as the initial rate-limiting step in protein synthesis. Amplification and overexpression of the eIF4E gene, located at chromosome 4q21, has been associated with malignant transformation in breast cancer and HNSCC. The proto-oncogene eIF4E was found to be elevated in 100% of HNSCCs and is of prognostic value in predicting recurrence (Sorrells et al 1999).

14. Field cancerization

In early 1953, Slaughter et al. proposed the clinical concept of field cancerization to explain the development of multiple cancers and precursor lesions in the head and neck area, particularly in the oral cavity. Their concept is based on long-term carcinogenic exposure, which causes the independent transformation of multiple epithelial cells at separate sites. Polyclonal tumours may independently arise from these spots. The so called histologically-based field cancerization model has been gradually succeeded by a new one established on the basis of molecular changes of the affected mucosa. This hypothesis advocates a micrometastatic spread or a monoclonal theory, suggesting that a precancerous field of mucosa may derive from an early genetic event that has undergone clonal expansion and lateral migration or expansion (Ha&Califano 2002, Califano et al 1996; 2000; Bedi et al 1996). Subsequent genetic alterations produce genetic divergence and various phenotypic alterations, resulting in a variety of histopathologically diverse regions in the local anatomical area and in the selection of various subclones. The theory, therefore, proposes a clonal origin of premalignant cells with successive lateral migration, and possible multiple primary tumours would not be monoclonal, but clonally related (Almadori et al 2004).

15. Telomerase reactivation in malignant alteration of OIN

The telomerase enzyme is a specialized multisubunit complex, with telomerase catalytic subunit (hTERT) functioning as a reverse transcriptase that can synthesize the telomeric ends at each cell division. Telomerase has been found to be re-activated in 90% of malignant neoplasms, including oral SCC (Meyerson et al 1996; Shay &Wright 1996, Luzar et al 2001). Recent studies have confirmed a close relationship between hTERT mRNA expression and telomerase activity, suggesting that quantification of hTERT gene expression can be used as an alternative to measurement of telomerase activity (De Kok et al 2000). These results suggest that telomerase re-activation is an early event in oral carcinogenesis, already detectable at the stage of precancerous oral epithelial changes.

16. Additional markers of malignant alterations of oral intraepithelial neoplasia

Several studies of OIN generally agree that the severity of epithelial abnormality reflects the degree of risk of SCC development (Jeannon et al 2004). No marker or group of markers has so far been identified as a reliable predictor of malignant progression of SILs. It is therefore under-standable that numerous studies have been devoted to the progression of OIN to invasive SCC. The role of cell-cycle proteins such as p16, p21, p27, p53, cyclin D1 and E have been extensively studied over the last two decades (Shin et al 1994; Fraczek et al 2007; Gorgoulis et al 1994; Gale et al 1997; Dolcetti et al 1992; Barbatis et al 1995; Nadal et al 1995; Poljak et al 1996; Uhlman et al 1996; Hirai et al 2003; Ioachim et al 2004; Wayne&Robinson 2006). However, none of these markers has been found to have reliable predictive value. In addition, detection of proliferative activity, mainly as immunohistochemical labelling for pCNA and Ki67 antigens, can be used only as adjuncts to light microscopy for more objective and reliable histological grading of OIN (Leopardi et al 2001; Peschos et al 2005). Recent study of the transforming growth factor-beta (TGF-bRII) has indicated that its down-regulation is an early event in oral carcinogenesis, which may occur in the loss of TGF- b-mediated inhibition, thereby facilitating progression of precancerous lesions to SCC (Franchi A, 2001). Promising biomarkers for improving cancer detection include minichromosome maintenance proteins (Mcm-2-7), which assemble in the prereplication complex and are essential for DNA replication in eucaryotic cells. All six proteins are abundant throughout the cell cycle, being broken down rapidly on differentiation and more slowly in quiescence. In 2003 Chatrath et al found that Mcm-2 is expressed within the most superficial surface layer in cases of oral CIS and SCC and with minimal expression in basal-parabasal (abnormal) and atypical hyperplasia. The authors suggest that Mcm-2 would be a good biomarker for distinguishing premalignant from malignant lesions. Quantification of cellular DNA by image or flow cytometry has achieved acceptance as an objective and reproducible component in diagnostic pathology. Several studies of oral intraepithelial lesions have shown that a proportion of these lesions show abnormal DNA content and that the incidence of this finding correlates with the degree of oral intraepithelial leions (Brac˘ko 1997; Munck-Wikland et al 1991; Crissman&Zarbo 1991). Brac˘ko has additionally noted that lack of abnormal DNA does not exclude malignant alteration, since malignant tumours exhibit minimal chromosomal abnormalities resulting in DNA changes, which are below the threshold of sensitivity of measurement with the use of image analysis or flow cytometry. In 2004, Kim and co-workers performed a study of quantitative PCR for genes specific to mitochondrial abundance in a spectrum of dysplastic head and neck lesions (Kim et al 2004b). Their study shows that mitochondrial DNA is directly proportional to histo-pathological grade.

17. Next-generation sequencing reveals NOTCH1 as an important tumor suppressor gene in head and neck cancer

Recently (Brakenhoff 2011; License Number 2756960749906), two papers came out on Science (Agrawal 2011; Stransky 2011). Their aim has been to provide new insight into the genetic changes of Head&Neck-SCC that may suggest the development of alternative treatment strategies. By using a high-throughput technique called massively parallel sequencing or next-generation sequencing to analyze the genomes of head and neck cancers in great detail. Both groups sequenced the exons of all known human genes in tumor DNA and compared the sequence to that of the corresponding normal DNA of the same patient. In total, the genomic landscapes of 32 (Agrawal 2011) and 74 (Stransky 2011) tumors were examined, including tumors that were positive or negative for the human papillomavirus. Agrawal et al. also provided genetic profiling data on chromosomal changes, verified the mutations by classical Sanger sequencing, and validated some mutations in an additional panel of tumor and normal tissues. Mutations were found in many of the genes already known to play a role in HNSCC, such as TP53, CDKN2A, PIK3CA, PTEN, and HRAS, but at least one new cancer gene previously not known to be involved in HNSCC, NOTCH1, was identified. In both studies, inactivating mutations of NOTCH1 were found in 10 to 15% of the head and neck tumors, and it was the second most frequently mutated gene after TP53 (which is mutated in 50 to 80% of the tumors). In several tumors, both alleles harbored mutations in NOTCH1.

Why was NOTCH1 not found before in this type of cancer or even in other malignancies (Klinakis et al 2011,) as an important tumor suppressor? Functional studies had identified a role for NOTCH1 in squamous cell carcinogenesis, at least in the skin (Dotto 2008), but robust mutational data in clinical samples were missing. NOTCH1 is a very large gene consisting of 34 coding exons, which hampers classical (Sanger) DNA sequencing, thus demonstrating the major improvement afforded by next- generation sequencing platforms.

The finding of numerous inactivating mutations in NOTCH1 in HNSCCs and the observation that mice with a disrupted NOTCH1 gene in the skin show a skin cancer phenotype (Nicolas et al 2003; Proweller et al 2006) provide strong evidence that NOTCH1 is an important tumor suppressor gene in HNSCC. NOTCH1 encodes a transmembrane receptor that functions in cell-to-cell communication (Ranganathan et al 2011) and is in the skin typically located in the cilia of the squamous cells, the dermal keratinocytes (Okuyama et al 2008; Ezratty et al 2011).

After ligand binding, the cytoplasmic tail of NOTCH is cleaved by a secretase enzyme, translocates to the nucleus, and functions as a transcription factor, driving the expression of numerous genes. All four NOTCH receptors encoded in the human genome are important for cell differentiation. Stransky et al. also found mutations in other cell differentiation-related genes, such as NOTCH2, NOTCH3, and TP63, suggesting that deregulation of the terminal differentiation program of mucosal keratinocytes is critical for squamous cancer development. This is not unexpected because terminal differentiation of normal keratinocytes in skin and mucosal epithelia is characterized by loss of cell organelles and even the nucleus during cornification — events that support the barrier function of squamous epithelia but which would inhibit malignant transformation.

However, some questions remain. A high-throughput sequencing approach can reveal many mutations in a large number of genes, but this does not necessarily imply that these are all "driver mutations" causally related to the malignant transformation process. Tumors are genetically unstable and acquire many mutations including so-called "passenger mutations" (Sjoblom et al 2006; Wood et al 2007) that are a result of malignant transformation and not the cause. Functional studies in animal models are required to elucidate the exact role of the NOTCH receptors and the other genes that are mutated in HNSCC. As an example, Agrawal et al. indicated that they also found mutations in FBXW7 in tumors that lack inactivating NOTCH1 mutations. The FBXW7 protein is a component of a ubiquitin ligase complex that targets NOTCH receptors for degradation by the proteasome, the protein degradation system of the cell, and this could be considered an inhibitory regulatory system of NOTCH1. Surprisingly, these FBXW7 mutations were also inactivating. One would have expected activating mutations in this inhibitory down- stream pathway, assuming that NOTCH1 is the target. Hence, this requires more detailed investigation. Relating mutations to phenotypic consequences is a challenge for all potential cancer genes identified by these high-throughput methods. Even non-synonymous mutations in established cancer genes may not always be driving, unless supported by functional testing in relevant models.

An issue even more relevant to clinical application is that identification of a cancer gene does not mean that it is druggable. As Agrawal et al. note, proteins encoded by oncogenes (genes that, when activated, cause a normal cell to become cancerous) are most suited for treatments because inhibitory drugs will result in a reduction of cellular proliferation. However, in the case of inactivated or lost tumor suppressor proteins, inhibitors are of no use, and reactivation is complex or impossible. Instead, one has to make use of the principal of synthetic lethality — finding another pathway that compensates the effect of, for example, NOTCH pathway inactivation (Iglehart&Silver 2009). Cancer-associated signaling pathways are often so critical for cellular homeostasis that there are mechanisms of redundancy to compensate inactivation, and these take over in tumor cells. Blocking this compensating pathway is then lethal for tumor cells, whereas in normal cells this has less effect as both pathways are active. This principle of synthetic lethality is a highly successful strategy, as shown by the application of poly(ADP-ribose) polymerase inhibitors in BRCA1- and BRCA2-deficient breast cancers (Fong et al 2009). However, the presence of such compensating pathways and their synthetic lethal character need to be identified. Hence, there is more work to be done, but the studies by Agrawal et al and Stransky et alindicate that there are more candidate cancer genes to be identified and we should keep searching for them.

18. Cancer stem cells in oral cancer

The cancer stem cell hypothesis suggests that neoplastic clones are maintained exclusively by a rare fraction of cells with stem cell properties. Stem cells are defined as cells which are able to both extensively self-renew and differentiate into progenitors. Furthermore, stem cells are also attractive candidates as origin of cancers, as in their long lifespan they can acquire mutations and epigenetic changes that could favour the evolution toward malignancy. We discuss the evidences reported in literature on existence of cancer stem cells in oral cancer and mechanisms of the extrinsic and intrinsic circuitry controlling stem cell fate as well as their possible connections to cancer.

Oral cancer is a culmination of continued hyperplasia or uncontrolled proliferation of basal epithelial stem cells. In a well differentiated tumor tissue, the suprabasal cells exhibits basal stem-like phenotype and differ from the terminal highly keratinized cells. Many experiments have compared the expression patterns of epidermal and oral epithelial stem cells (Kaur&Li 2000; Evander et al 1997). Up to now, no true stem cell population could be identified from both normal and tumor tissue of oral epithelium purely based on sorting for stem cell specific surface markers reported from epidermal tissue (Prince&Ailles 2008). The stratified squamous epithelia of the oesophagus and epidermis have different functions and embryological origins. The pursuit for specific oral epithelial stem cell surface markers lead to the identification of CD markers such as CD44 (Prince et al 2007; Naor et al 2008), CD147 (Kose et al 2007; Toole et al 2008), integrins (Evander et al 1997), cytokeratins (Lindberg et al 1989, Presland et al 2002), EpCAM/ESA (Trzpis et al 2007; Munz et al 2004), E-cadherin (Kudo et al 2004), along with transcription factors Oct-4, Nanog (Chiou et al 2008) and Bmi-1 (Prince et al 2007). p75NGFR, a potential oral keratinocyte stem cell marker also co-localizes with BrdU incorporated stem cells and functions to mediate intercellular signaling in cell survival and apoptosis (Hatakeyama et al 2007; Nakamura et al 2007). An ideal cancer stem cell marker should impart all the acquired hallmarks of self-sufficiency in growth signals, anchorage-independent growth, apoptotic/drug resistance, invasiveness, metastatic potential in addition to primacy of high self- renewal conferred by cell of its origin, the normal stem cell. We discuss below several such stem cell markers representing the putative CSCs in oral squamous cell carcinoma and functional attributes bestowed by the expression pattern.

Methods for the identification of CSCs in solid malignancies mirror those strategies employed to differentiate normal stem cells from their differentiated progeny. These include the efflux of vital dyes by multidrug transporters, the enzymatic activity of aldehyde dehydrogenase, colony and sphere-forming assays utilizing specific culture conditions and the most widely used method—the expression of specific cell surface antigens known to enrich for stem cells. Once the subpopulation of tumor cells has been identified and isolated, functional characterization through quantitative xenotransplantation assays, the gold-standard for identification of CSCs, are used to assess the tumorigenicity and self- renewing potential of the putative CSC population in vivo

18.1 Surface antigens

By far the most common method of identifying CSCs has relied on the expression of specific cell-surface antigens that enrich for cells with CSC properties. Many of these antigens were initially targeted because of their known expression on endogenous stem cells. While a multitude of studies have identified CSC markers across a variety of solid malignancies, relatively few of these markers have been studied in HNSCC. **CD133**. A pentaspan transmembrane glycoprotein localized on cell membrane protrusions (Costea et al 2006; Prince&Alley 2008), is a putative CSC marker for a number of epithelial malignancies including brain, prostate, colorectal, and lung (Chiou et al 2008; Kelly et al 2007). In HNSCC cell lines, CD133hi cells display increased clonogenicity, tumor sphere formation and tumorigenicity in xenograft models when compared to their CD133 low counterparts (Ramalho-Santos&Willenbring 2007; Singh et al 2004; Ricci-Vitiani et al 2007). **CD44**. A large cell surface glycoprotein involved in cell adhesion and migration. It is a known receptor for hyaluronic acid and interacts with other ligands such as matrix metalloproteases (Tan&Coussens 2007; Mimeault M, 2007). Initially identified as a solid malignancy CSC marker in breast cancer (Tabor et al 2002), Prince et al. demonstrated that CD44 expression could also be used to isolate a tumor subpopulation with increased tumorigenicity in HNSCC (Pillai&Nair 2000). Although CD44 expression enriches for cells with CSC properties, the relatively high number of cells required for tumor formation as compared with known CSC populations from other epithelial malignancies raises questions about whether CD44 expression alone is sufficient for isolation of a pure CSC population. Using primary human tumor samples as well as utilizing a more natural host microenvironment through an orthotopic xenograft model (Phesse&Clarke 2009) might reduce the number of cells needed to generate tumors.

18.2 Aldehyde dehydrogenase activity

Aldehyde dehydrogenase (ALDH) is an intracellular enzyme normally present in the liver. Its known functions include the conversion of retinol to retinoic acids and the oxidation of toxic aldehyde metabolites, like those formed during alcohol metabolism and with certain chemotherapeutics such as cyclophosphamide and cisplatin (Bosron et al 1988; Thomasson et al 1991; Visus et al 2007). ALDH activity is known to enrich hematopoetic stem/progenitor cells (Chute et al 2006) and more recently has been shown to enrich cells with increased stem-like properties in solid malignancies (Carpentino et al 2009; Croker et al 2009, Deng et al 2010; Ma et al 2008). Chen et al. showed that ALDH activity correlated with disease staging in HNSCC and that higher enzymatic activity correlated with expression of epithelial-to-mesenchymal transition (EMT) genes as well as enriching cells with CSC properties (Chen et al 2009). **Side Population.** Hoechst 33342 is a fluorescent DNA- binding dye that preferentially binds to A-T rich regions. It is actively pumped out of cells by members of the ATP-binding cassette (ABC) transporter superfamily. Once stained with Hoechst dye, cells can be sorted by fluorescent-activated cell sorting (FACS) based upon the activity level of these multidrug transporters. Originally noted to enrich bone marrow for long-term hematopoetic stem cells (Clay et al 2010), this method has also been used to identify cells within solid tumors with increased tumorigenicity (Ho et al 2007; Szotek et al 2006; Wang et al 2007). Side population (SP) cells from oral squamous cell carcinoma have been shown to have increased clonogenicity and tumorigenicity in xenotransplantation assays (Loebinger et al 2008). Furthermore, HNSCC SP cells displayed higher expression of known stem cell related genes — Oct4, CK19, BMI-1 and CD44 — and lower expression of involucrin and CK13, genes associated with a differentiated status (Zhang et al 2009).

18.3 Tumor sphere formation

Under serum-free culture conditions CSCs can be maintained in an undifferentiated state, and when driven toward proliferation by the addition of growth factors, form clonally derived aggregates of cells termed tumor spheres (Singh et al 2003). The ability of CSCs — but not the remaining tumor bulk—to form tumor spheres has been used extensively in neural tumors to identify populations enriched for CSCs. In HNSCC, these spheres have been shown to be enriched for stem markers, including CD44hi (Okamoto et al 2009), Oct-4, Nanog, Nestin, and CD133hi (Zhang et al 2009), as well as exhibiting increased tumorigenicity in orthotopic xenografts (Chiou et al 2008).

19. Cancer stem cells and disease progression

While there exists significant data defining the presence of CSCs within a variety of tumor types and many aspects of the cell and molecular biology of CSC have been elucidated, the manner in which this unique cell population influences clinical disease progression remains unclear. Given that metastases can be formed from implantation of a single tumor cell (Fidler&Talmadge 1986), it seems likely that CSCs, as the progenitor of all tumor cell types, would be responsible for metastatic spread. Central to the CSC hypothesis is the presence of a unique stem cell "niche" or environment necessary to support the growth of stem cells (Li&Xie 2005). It has been shown that a premetastatic niche is established by the attraction of bone marrow derived cells to the future site of metastases by the secretion of factors from cancer cells and that blocking the creation of this premetastatic niche prevents metastases (Kaplan et al 2005). What these secreted factors are and whether they are secreted by CSCs or one of their progeny remains an open question; however, creation of this niche, possibly for the arrival of CSCs to form a metastasis, appears to be a crucial step in metastatic spread.

Another stem cell marker, CD44, has also been implicated in metastatic spread and disease progression in HNSCC (Celetti et al 2005), although the CD44 story is more complex. Recently, three different isoforms, CD44 v3, v6, and v10, have been shown to be associated with progression and metastasis of HNSCC (Wang et al 2009). Increased CD44 v3 expression in primary tumors was associated with lymph node metastasis, while CD44 v10 expression was associated with distant metastasis and CD44 v6 expression was associated with perineural spread. In cell culture, blockade of these CD44 isoforms with isoform-specific antibodies inhibited cellular proliferation, with the greatest inhibition seen with blockade of CD44 v6. Finally, increased expression of CD44 v6 and v10 was associated with shortened disease-free survival (Staibano et al 2007). These studies suggest that alteration in CSC phenotype through variation in CD44 isoform expression may alter the interaction of CSCs with the surrounding microenvironment. This may allow CSCs to more readily invade surrounding tissues or metastasize, thereby promoting disease progression.

20. Treatment and evolution to malignancy

To date, many researchers have reported that the risk of developing cancer from oral leukoplakia could not be significantly reduced by surgical intervention (Holmstrup et al 2006; Vedtofte et al 1987; Schoelch et al 1999). Moreover, some review papers have stated that it is actually unclear whether removal of the lesion decreases malignant transformation of oral leukoplakia because there is a lack of randomized controlled trials comparing the different treatment modalities (Lodi et al 2006, Lodi&Porter 2008).

Nevertheless, the research by Amagasa et al. (2006; 2011a; 2011b) showed that the malignant transformation rate of leukoplakia treated by surgery was significantly lower than that without any treatment or that without surgery, so we believe that surgical excision with an adequate safety margin, coupled with well-timed evaluation of oral leukoplakia on follow-up, is effective in preventing the malignant transformation of these lesions.

21. References

Abbey L, Kaugars GE, Gunsolley JC et al (1995) Intraexaminer and interexaminer reliability in the diagnosis of oral epithelial dysplasia. Oral Surg Oral Med Oral Pathol 80:188–191

Agrawal N, Frederick MJ, Pickering CR, Bettegowda C, Chang K, Li RJ, et al (2011) Exome Sequencing of Head and Neck Squamous Cell Carcinoma Reveals Inactivating Mutations in NOTCH1, Science 333, 1154.

Almadori G, Bussu F, Cadoni G et al (2004) Multistep laryngeal carcinogenesis helps our understanding of the field cancerisation phenomenon: a review. Eur. J. Cancer 40; 2383– 2388.

Amagasa T, Yamashiro M, Ishikawa H (2006) Oral leukoplakia related to malignant transformation. Oral Sci Int 3:45–55

Amagasa T (2011a) Oral premalignant lesions Int J Clin Oncol 16:1–4

Amagasa T, Yamashiro M, Uzawa N (2011b) "Oral premalignant lesions: from a clinical perspective" Int J Clin Oncol 16:5–14

Axell T, Pindborg JJ, Smith CJ et al (1996) Oral white lesions with special reference to precancerous and tobacco-related lesions. J Oral Pathol Med 25:49–54

Balz V, Scheckenbach K, Gotte K, Bockmuhl U, Petersen I, Bier H (2003) Is the p53 inactivation frequency in squamous cell carcinomas of the head and neck underestimated? Analysis of p53 exons 2-11 and human papillomavirus 16/18 E6 tran- scripts in 123 unselected tumor specimens. Cancer Res 63; 1188–1191.

Barbatis C, Loukas L, Grigoriou M et al (1995) p53 overexpression in laryngeal squamous cell carcinoma and dysplasia. Clin. Mol. Pathol 48; M194–M197.

Bedi GC, Westra WH, Gabrielson E, Koch W, Sidransky D. (1996) Multiple head and neck tumors: evidence for a common clonal origin. Cancer Res. 56; 2484–2487.

Blackwell KB, Calcaterra TC, Fu YS (1995) Laryngeal dysplasia: epidemiology and treatment outcome. Ann Otol Rhinol Laryngol 104:596–602

Blackwell KE, Fu YS, Calcaterra TC (1995) Laryngeal dysplasia. A clinicopathologic study. Cancer 75; 457–463.

Bonneretal JA, N.Engl.J.Med.354,567(2006).

Bosman FT (2001) Dysplasia classification: pathology in disgrace? J. Pathol. 194; 143–144.

Bosron WF, Lumeng L, and Li TK (1988) "Genetic polymorphism of enzymes of alcohol metabolism and susceptibility to alcoholic liver disease," Molecular Aspects of Medicine, vol. 10, no. 2, pp. 147–158

Bouquot JE, Gnepp DR (1991) Laryngeal precancer: a review of the literature, commentary, and comparison with oral leukoplakia. Head Neck 13; 488–497.

Bouquot JE, Speight PM, Farthing PM (2006) Epithelial dysplasia of the oral mucosa-diagnostic problems and prognostic features. Curr Diagn Pathol 12:11–21

Boyle JO, Hakim J, Koch W et al (1993) The incidence of p53 mutations increases with progression of head and neck cancer. Cancer Res 53; 4477–4480.

Braakhuis BJ, Tabor MP, Kummer JA, Leemans CR, Brakenhoff RH (2003) A genetic explanation of Slaughter's concept of field cancerization: evidence and clinical implications. Cancer Res 63; 1727–1730.

Brac˘ko M. Evaluation of DNA content in epithelial hyperplastic lesions of the larynx (1997) Acta Otolaryngol 527(Suppl.); 62–65.

Brakenhoff RH (2011) Another NOTCH for Cancer, Science 333, 1102-1103

Broders AC (1932) Carcinoma in situ contrasted with benign penetrating epithelium. J Am Med Assoc 99:1670–1674

Brothwell DJ, Lewis DW, Bradley G et al (2003) Observer agreement in the grading of oral epithelial dysplasia. Commun Dent Oral Epidemiol 31:300–305

Callender T, el-Naggar AK, Lee MS, Frankenthaler R, Luna MA, Batsakis JG (1994) PRAD-1 (CCND1) / cyclin D1 oncogene amplification in primary head and neck squamous cell carcinoma. Cancer 74; 152–158.

Califano J, van der Riet P, Westra W et al (1996) Genetic progression model for head and neck cancer: implications for field cancerization. Cancer Res 56; 2488–2492.

Califano J, Westra WH, Meininger G, Corio R, Koch WM, Sidransky D (2000) Genetic progression and clonal relationship of recurrent premalignant head and neck lesions. Clin. Cancer Res 6; 347–352.

Carpenter DH, El-Mofty SK, Lewis JS Jr (2011) Undifferentiated carcinoma of the oropharynx: a human papillomavirus-associated tumor with a favorable prognosis. Mod Pathol 24(10):1306-12.

Carpentino JE, Hynes MJ, Appelman HD et al (2009) Aldehyde dehydrogenase-expressing colon stem cells contribute to tumorigenesis in the transition from colitis to cancer Cancer Research, vol. 69, no. 20, pp. 8208–8215

Celetti A, Testa D, Staibano S, Merolla F, Guarino V, Castellone MD, et al (2005) Overexpression of the cytokine osteopontin identifies aggressive laryngeal squamous cell carcinomas and enhances carcinoma cell proliferation and invasiveness. Clin Cancer Res 15;11(22):8019-27.

Chatrath P, Scott IS, Morris LS et al (2003) Aberrant expression of minichromosome maintenance protein-2 and Ki67 in laryngeal squamous epithelial lesions. Br. J. Cancer 89; 1048– 1054.

Chen Y-C, Chen Y-W, Hsu H-S et al (2009) Aldehyde dehydrogenase 1 is a putative marker for cancer stem cells in head and neck squamous cancer Biochemical and Biophysical Research Communications, vol. 385, no. 3, pp. 307–313.

Chi FL, Yuan YS, Wang SY, Wang ZM (2004) Study on ceramide expression and DNA content in patients with healthy mucosa, leukoplakia, and carcinoma of the larynx. Arch. Otolaryngol. Head Neck Surg 130; 307–310.

Chiou S-H, Yu C-C, Huang C-Y, Lin S-C, Liu C-J, Tsai T-H et al (2008) Positive correlations of Oct-4 and Nanog in oral cancer stem-like cells and high-grade oral squamous cell carcinoma, Clinical Cancer Research 14 4085–4095.

Chute JP, Muramoto GG, Whitesides J et al (2006) Inhibition of aldehyde dehydrogenase and retinoid signaling induces the expansion of human hematopoietic stem cells," Proceedings of the National Academy of Sciences of the United States of America, vol. 103, no. 31, pp. 11707-11712.

Clay MR, Tabor M, Owen JH et al (2010) Single-marker identification of head and neck squamous cell carcinoma cancer stem cells with aldehyde dehydrogenase Head Neck, vol. 32, no. 9, pp. 1195-1201.

Costea D, Tsinkalovsky O, Vintermyr O, Johannessen A, Mackenzie I (2006) Cancer stem cells new and potentially important targets for the therapy of oral squamous cell carcinoma, Oral Diseases 12 443-454.

Crissman JD (1979) Laryngeal keratosis and subsequent carcinoma. Head Neck Surg 1; 386-391.

Crissman JD (1982) Laryngeal keratosis preceding laryngeal carcinoma. A report of four cases. Arch Otolaryngol 108:445-448

Crissman JD, Fu YS (1986) Intraepithelial neoplasia of the larynx. Arch. Otolaryngol. Head Neck Surg. 112; 522-528.

Crissman JD, Zarbo RJ, Drozdowicz S et al (1988) Carcinoma in situ and microinvasive squamous carcinoma of the laryngeal glottis. Arch Otolaryngol Head Neck Surg 114:299-307

Crissman JD, Zarbo RJ (1989) Dysplasia, in situ carcinoma, and progression to invasive squamous cell carcinoma of the upper aerodigestive tract. Am. J. Surg. Pathol. 13(Suppl. I); 5- 16.

Crissman JD, Zarbo RJ (1991) Quantitation of DNA ploidy in squamous intraepithelial neoplasia of the laryngeal glottis. Arch. Otolaryngol. Head Neck Surg. 117; 182-188.

Crissman JD, Visscher DW, Sakr W (1993) Premalignant lesions of the upper aerodigestive tract: pathologic classification. J. Cell. Biochem. Suppl. 17F; 49-56.

Crissman JD, Sakr WA (2001) Squamous intraepithelial neoplasia of upper aerodigestive tract. In: Gnepp DR (ed) Diagnostic surgical pathology of the head and neck. Saunders, Philadelphia, pp 1-17

Croker AK, Goodale D, Chu J et al (2009) High aldehyde dehydrogenase and expression of cancer stem cell markers selects for breast cancer cells with enhanced malignant and metastatic ability," Journal of Cellular and Molecular Medicine, vol. 13, no. 8 B, pp. 2236-2252 .

De Kok JB, Schalken JA, Aalders TW, Reurs TJM, Willems HL, Swinkels DW (2000) Quantitative measurement of telomerase reverse trascriptase (hTERT) mRNA in urothelial cell carcinomas. Int. J. Cancer 87; 217-220.

Deng S, Yang X, Lassus H et al (2010) Distinct expression levels and patterns of stem cell marker, aldehyde dehydrogenase isoform 1 (ALDH1), in human epithelial cancers," PLoS ONE, vol. 5, no. 4, Article ID e10277, pp. 1-11.

Dolcetti R, Doglioni C, Maestro R et al (1992) p53 over-expression is an early event in the development of human squamous-cell carcinoma of the larynx: genetic and prognostic implications. Int. J. Cancer 52; 178-182.

Dotto GP (2008) Notch tumor suppressor function Oncogene 27,5115.

Ezratty EJ, Stokes N, Chai S, Shah AS, Williams SE, Fuchs E (2011) A role for the primary cilium in Notch signaling and epidermal differentiation during skin development Cell 145, 1129-41.

Evander M, Frazer I, Payne E, Qi Y, Hengst K, McMillan N (1997) Identification of the alpha6 integrin as a candidate receptor for papillomaviruses Journal of Virology 71 2449-2456.

Eversole LR (2009) Dysplasia of the upper aerodigestive tract squamous epithelium. Head Neck Pathol 3:63-68

Fakhry C, D'souza G, Sugar E, Weber K, Goshu E, Minkoff H et al (2006) Relationship between prevalent oral and cervical human papillomavirus infections in human immunodeficiency virus-positive and -negative women. J Clin Microbiol 44(12):4479-85

Field JK (1996) Genomic instability in squamous cell carcinoma of the head and neck. Anticancer Res 16; 2421-2431.

Ficher DJ, Epstein JB, Morton TH et al (2004) Interobserver reliability in the histopathologic diagnosis of oral pre-malignant and malignant lesions. J Oral Pathol Med 33:65-70

Fidler IJ and Talmadge JE (1986) Evidence that intravenously derived murine pulmonary melanoma metastases can originate from the expansion of a single tumor cell Cancer Research, vol. 46, no. 10, pp. 5167-5171

Fong PC, Boss DS, Yap TA, Tutt A, Wu P, Mergui-Roelvink M et al (2009) N.Engl.J.Med. 361, 123-34.

Fleskens S, Slootweg P (2009) Grading systems in head and neck dysplasia: their prognostic value, weakness and utility. Head Neck Oncol 1:1-8

Fraczek M, Wozniak Z, Ramsey D, Krecicki T (2007) Epression patterns of cyclin E, cyclin A and CDC25 phosphatases in laryngeal carcinogenesis. Eur. Arch. Otorhinolaryngol 264; 923-928.

Franchi A, Gallo O, Sardi I, Santucci M (2001) Downregulation of transforming growth factor beta type II receptor in laryngeal carcinogenesis. J. Clin. Pathol 54; 201-204.

Frangez I, Gale N, Luzar B (1997) The interpretation of leukoplakia in laryngeal pathology. Acta Otolaryngol 527:142-144

Friedmann I, Ferlito A (1993) Precursors of squamous cell carcinoma. In Ferlito A ed. Neoplasms of the laryn. Edinburgh: Churchill Livingstone 97-111.

Gale N, Zidar N, Kambic V, Poljak M et al (1997) Epidermal growth factor receptor, c-erbB-2 and p53 overexpressions in epithelial hyperplastic lesions of the larynx. Acta Otolaryngol. 527(Suppl.); 105-110.

Gale N, Kambicˇ V, Michaels L et al (2000) The Ljubljana classification: a practical strategy for the diagnosis of laryngeal precancerous lesions. Adv. Anat. Pathol 7; 240-251.

Gale N, Kambic V, Poljak M, Cor A, Velkavrh D, Mlacak B (2000) Chromosomes 7,17 polysomies and overexpression of epidermal growth factor receptor and p53 protein in epithelial hyperplastic laryngeal lesions. Oncology 58; 117-125.

Gale N, Pilch BZ, Sidransky D, Westra WH, Califano J (2005) Epithelial precursor lesions. In Barnes L, Eveson JW, Reichart P, Sidransky D eds. World Health Organization classification of tumour. Pathology and genetics of head and neck tumours. Lyon: IARC, 140-143.

Gale N, Michaels L, Luzar B et al (2009) Current review on squamous intraepithelial lesions of the larynx. Histopathology 54:639–656

Gallo O, Santucci M, Franchi A (1997) Cumulative prognostic value of p16/CDKN2 and p53 oncoprotein expression in premalignant laryngeal lesions. J. Natl Cancer Inst. 89; 1161–1163.

Gallo A, de Vincentiis M, Della Rocca C et al (2001) Evolution of precancerous laryngeal lesions: a clinicopathologic study with long-term follow-up on 259 patients. Head Neck 23; 42– 47.

Gillis TM, Incze J, Strong MS, Vaughan CW, Simpson GT (1983) Natural history and management of keratosis, atypia, car- cinoma in situ, and microinvasive cancer of the larynx. Am. J. Surg 146; 512–516.

Goodman ML. Keratosis (leukoplakia) of the larynx Otolaryngol (1984) Clin. North Am. 17; 179–183.

Gorgoulis V, Rassidakis G, Karameris A, Giatromanolaki A, Barbatis C, Kittas C (1994) Expression of p53 protein in laryngeal squamous cell carcinoma and dysplasia: possible correlation with human papillomavirus infection and clinicopathological findings. Virchows Arch 425; 481–489.

Grundmann E (1983) Classification and clinical consequences of precancerous lesions in the digestive and respiratory tracts. Acta Pathol. Jpn. 33; 195–217.

Gupta PC, Mehta FS, Daftary DR et al (1980) Incidence rates of oral cancer and natural history of oral precancerous lesions in a 10 year follow-up study of Indian villagers. Community Dent Oral Epidemiol 8:287–333

Ha PK, Califano J (2002) The molecular biology of laryngeal cancer. Otolaryngol. Clin. North Am 3; 993–1012.

Hasina R, Lingen MW (2004) Head and neck cancer: the pursuit of molecular diagnostic markers. Semin. Oncol 31; 718–725.

Hatakeyama S, Yaegashi T, Takeda Y, Kunimatsu K (2007) Localization of bromo-deoxyuridine-incorporating, p63- and p75NGFR- expressing cells in the human gingival epithelium, Journal of Oral Science 49, 287–291.

Hellquist H, Lundgren J, Olofsson J (1982) Hyperplasia, keratosis, dysplasia and carcinoma in situ of the vocal cords–a follow-up study. Clin. Otolaryngol 7; 11–27.

Hellquist H, Cardesa A, Gale N, Kambicˇ V, Michaels L (1999) Criteria for grading in the Ljubljana classification of epithelial hyperplastic laryngeal lesions. A study by members of the Working group on Epithelial Hyperplastic Laryngeal Lesions of the European Society of Pathology. Histopathology 34; 226–233.

Henry RC (1979) The transformation of laryngeal leucoplakia to cancer. J. Laryngol. Otol 93; 447–459

Hirai T, Hayashi K, Takumida M, Ueda T, Hirakawa K, Yajin K (2003) Reduced expression of p27 is correlated with progression in precancerous lesions of the larynx. Auris Nasus Larynx 30; 163–168.

Ho M M, Ng AV, Lam S, and Hung JY (2007) Side population in human lung cancer cell lines and tumors is enriched with stem-like cancer cells Cancer Research, vol. 67, no. 10, pp. 4827–4833.

Hoffmann M, Ihloff AS, Görögh T, Weise JB, Fazel A, Krams M et al (2010) p16(INK4a) overexpression predicts translational active human papillomavirus infection in tonsillar cancer. Int J Cancer Oct 1;127(7):1595-602.

Holmstrup P, Vedtofte P, Reibel J et al (2006) Long-term treat- ment outcome of oral premalignant lesions. Oral Oncol 42:461–474

Iglehart JD, Silver DP (2009) Synthetic lethality--a new direction in cancer-drug development. N.Engl.J.Med. 361, 189.

Ioachim E, Peschos D, Goussia A et al (2004) Expression patterns of cyclins D1, E in laryngeal epithelial lesions: correlation with other cell cycle regulators (p53, pRb, Ki-67 and PCNA) and clinicopathological features. J. Exp. Clin. Cancer Res 23; 277–283.

Irish JC, Bernstein A (1993) Oncogenes in head and neck cancer Laryngoscope 103; 42–52.

Jackson C (1923) Cancer of the larynx: is it preceded by a recognizable precancerous condition? Ann. Surg 77; 1–14.

Jares P, Fernandez PL, Campo E et al (1994) PRAD-1 / cyclin D1 gene amplification correlates with messenger RNA overexpression and tumor progression in human laryngeal carcinomas. Cancer Res 54; 4813–4817.

Jeannon JP, Soames JV, Aston V, Stafford FW, Wilson JA (2004) Molecular markers in dysplasia of the larynx: expression of cyclin-dependent kinase inhibitors p21, p27 and p53 tumour suppressor gene in predicting cancer risk. Clin. Otolaryngol. Allied Sci 29; 698–704.

Kamb A, Shattuck-Eidens D, Eeles R et al (1994) Analysis of the p16 gene (CDKN2) as a candidate for the chromosome 9p melanoma susceptibility locus. Nat. Genet 8; 23–26.

Kambic V, Lenart I (1971) Notre classification des hyperplasies de l'epithelium du larynx au point de vue prognostic. JFORL 20:1145–1150

Kambic̆ V (1978) Difficulties in management of vocal cord precancer- ous lesions. J. Laryngol. Otol 92; 305

Kambic̆V, Gale N, Ferluga D (1992) Laryngeal hyperplastic lesions, follow-up study and application of lectins and anticytokeratins for their evaluation. Path. Res. Pract 188; 1067–1077.

Kambic̆V, Gale N (1995) Epithelial hyperplastic lesions of the larynx. Amsterdam: Elsevier 1–265.

Kambic̆ V (1997) Epithelial hyperplastic lesions – a challenging topicin laryngology. Acta Otolaryngol. (Stockh) 527(Suppl.); 7–11.

Kaplan RN, Riba RD, Zacharoulis S et al (2005) VEGFR1- positive haematopoietic bone marrow progenitors initiate the pre-metastatic niche Nature, vol. 438, no. 7069, pp. 820–827.

Karabulut A, Reibel J, Therkildsen MH et al (1995) Observer variability in the histologic assessment of oral premalignant lesions. J Oral Pathol Med 24:198–200

Kaur P, Li A (2000) Adhesive properties of human basal epidermal cells: an analysis of keratinocyte stem cells, Transit amplifying cells and postmitotic differentiating cells 114 413–420.

Kelly PN, Dakic A, Adams JM, Nutt SL, Strasser A (2007) Tumor growth need not be driven by rare cancer stem cells, Science 317, 337-.

Kim MM, Califano JA (2004) Molecular pathology of head-and-neck cancer. Int. J. Cancer 112; 545-553.

Kim MM, Clinger JD, Masayesva BG et al (2004) Mitochondrial DNA quantity increases with histopathologic grade in premalignant and malignant head and neck lesions. Clin. Cancer Res 10; 8512-8515.

Klinakis A Lobry C, Abdel-Wahab O, Oh P, Haeno H, Buonamici S, van De Walle I et al (2011) A novel tumour-suppressor function for the Notch pathway in myeloid leukaemia. Nature 473, 230-3.

Kleist B, Poetsch M (2004) Divergent patterns of allelic alterations in premalignant laryngeal lesions indicate differences in the impact of morphological grading characteristics. Oncology 67; 420-427.

Koren R, Kristt D, Shvero J et al (2002) The spectrum of laryngeal neoplasia: the pathologist's view. Pathol Res Pract 198:709-715

Kose O, Lalli A, Kutulola AO, Odell EW, Waseem A (2007) Changes in the expression of stem cell markers in oral lichen planus and hyperkeratotic lesions, Journal of Oral Science 49 133-139.

Kramer IRH, Lucas RB, Pindborg JJ et al (1978) Definition of leukoplakia and related lesions: an aid to studies on oral pre- cancer. Oral Surg Oral Med Oral Pathol 46:518-539

Kudo Y, Kitajima S, Ogawa I, Hiraoka M, Sargolzaei S. Keikhaee MR et al (2004) Invasion and metastasis of oral cancer cells require methylation of E-cadherin and/or degradation of membranous a-catenin, Clinical Cancer Research 10 5455-5463.

Kuffer R, Lombardi T (2002) Premalignant lesions of the oral mucosa. A discussion about the place of oral intraepithelial neoplasia (OIN). Oral Oncol 38:125-130

Kujan O, Oliver RJ, Khattab A et al (2007) Why oral histopa- thology suffers inter-observer variability on grading oral epi- thelial dysplasia: an attempt to understand the sources of variation. Oral Oncol 43:224-231

Kujan O, Oliver RJ, Khattab A, Roberts SA, Thakker N, Sloan P (2006) Evaluation of binary system of grading oral epithelial dysplasia for prediction of malignant transformation. Oral Oncol 42; 987-993.

Leopardi G, Serafini G, Simoncelli C, Ludovini V, Pistola L, Altissimi G (2001) Ki67 and p53 in laryngeal epithelial lesions: correlations with risk factors. Acta Otorhinolaryngol. Ital 21; 243-247.

Li L and Xie T (2005) Stem cell niche: structure and function Annual Review of Cell and Developmental Biology, vol. 21, pp. 605-631.

Lindberg K, Rheinwald J (1989) Suprabasal 40 kd keratin (K19) expression as an immunohistologic marker of premalignancy in oral epithelium, The American Journal of Pathology 134, 89-98.

Loebinger MR, Giangreco A, Groot KR et al (2008) Squamous cell cancers contain a side population of stem-like cells that are made chemosensitive by ABC transporter blockade," British Journal of Cancer, vol. 98, no. 2, pp. 380-387.

Lodi G, Sardella A, Bez C et al (2006) Interventions for treating oral leukoplakia. Cochrane Database Syst Rev; CD001829

Lodi G, Porter S (2008) Management of potentially malignant disorders: evidence and critique. J Oral Pathol Med 37(2):63–69

Luzar B, Poljak M, Marin IJ, Fischinger J, Gale N (2001) Quantitative measurement of telomerase catalytic subunit (hTERT) mRNA in laryngeal squamous cell carcinomas. Anticancer Res. 21; 4011–4015.

Ma S, Kwok W C, Lee T KW et al (2008) Aldehyde dehy- drogenase discriminates the CD133 liver cancer stem cell populations Molecular Cancer Research, vol. 6, no. 7, pp. 1146–1153.

Mahajan M, Hazarey VK (2004) An assessment of oral epithelial dysplasia using criteria of 'Smith & Pindborg grading system' & 'Ljubljana grading system' in oral precancerous lesions'. J Oral Maxillofac Pathol 8:73–81

Merlo A, Herman JG, Mao L et al (1995) CpG island methylation is associated with transcriptional silencing of the tumour sup- pressor p16 / CDKN2 / MTS1 in human cancers. Nat. Med 1; 686–692.

Meyerson M, Counter CM, Eaton EN et al (1996) Telomerase activity in human cancer. Curr. Opin. Oncol 8; 66–71.

Michaels L (1997) The Kambic–Gale method of assessment of epithelial hyperplastic lesions of the larynx in comparison with the dysplasia grade method. Acta Otolaryngol 527:17–20

Michaels L, Hellquist HB (2001) Ear, nose and throat histopathology, 2nd edn. London: Springer, 378–387.

Mimeault M, Hauke R, Mehta PP, Batra SK (2007) Recent advances in cancer stem/progenitor cell research: therapeutic implications for overcoming resistance to the most aggressive cancers, Journal of Cellular and Molecular Medicine 11, 981–1011.

Munck-Wikland E, Kuylenstierna R, Lindholm J, Auer G (1991) Image cytometry DNA analysis of dysplastic squamous epithelial lesions in the larynx. Anticancer Res 11; 597–600.

Munz M, Kieu C, Mack B, Schmitt B, Zeidler R, Gires O (2004) The carcinoma- associated antigen EpCAM upregulates c-myc and induces cell proliferation, Oncogene 23, 5748–5758.

Nadal A, Campo E, Pinto J et al (1995) p53 expression in normal, dysplastic, and neoplastic laryngeal epithelium. Absence of a correlation with prognostic factors. J. Pathol 175; 181–

Nadal A, Cardesa A. Molecular biology of laryngeal squamous cell carcinoma. Virchows Arch. 2003; 442; 1–7.

Nakamura T, Endo K.i, Kinoshita S (2007) Identification of human oral keratinocyte stem/progenitor cells by neurotrophin receptor p75 and the role of neurotrophin/p75 signaling, Stem Cells 25, 628–638.

Naor D, Wallach-Dayan SB, Zahalka MA, Sionov RV (2008) Involvement of CD44, a molecule with a thousand faces, in cancer dissemination, Seminars in Cancer Biology 18, 260–267.

Napier SS, Speight PM (2008) Natural history of potentially malignant oral lesions and conditions: an overview of the lit- erature. J Oral Pathol Med 37:1–10

Nicolas M, Wolfer A, Raj K, Kummer JA, Mill P, van Noort M, et al (2003) Nat. Genet. 33, 416-21.

Olde Kalter P, Lubsen H, Delemarre JFM, Snow GB (1987) Squamous cell hyperplasia of the larynx. A clinical follow-up study. J. Laryngol. Otol 101; 579–588.

Okamoto A, Chikamatsu K, Sakakura K, Hatsushika K, Takahashi G, and Masuyama K (2009) Expansion and characterization of cancer stem-like cells in squamous cell carcinoma of the head and neck Oral Oncology, vol. 45, no. 7, pp. 633–639.

Patel SC, Carpenter WR, Tyree S, Couch ME, Weissler M, Hackman T et al (2011) Increasing incidence of oral tongue squamous cell carcinoma in young white women, age 18 to 44 years. J Clin Oncol 29(11):1488-94.

Peschos D, Stefanou D, Vougiouklakis T, Assimakopoulos DA, Agnantis NJ (2005) Cell cycle proteins in laryngeal cancer: role in proliferation and prognosis. J. Exp. Clin. Cancer Res. 24; 431–437.

Phesse TJ, Clarke AR (2009) Normal stem cells in cancer prone epithelial tissues, British Journal of Cancer 100 221-227.

Pillai MR, Nair MK (2000) Development of a condemned mucosa syndrome and pathogenesis of human papillomavirus-associated upper aerodigestive tract and uterine cervical tumors, Experimental and Molecular Pathology 69, 233–241.

Pindborg JJ, Renstrup G, Poulsen HE et al (1963) Studies in oral leukoplakias. V. Clinical and histologic signs of malignancy. Acta Odont Scand 21:407–414.

Pindborg JJ (1980) Oral cancer and precancer. John Wright & Sons, Bristol

Pindborg JJ, Reichart PA, Smith CJ et al (1997) World Health Organization: Histological typing of cancer and precancer of the oral mucosa, 2nd edn. Springer, Berlin

Pomerantz RG, Grandis JR (2004) The epidermal growth factor receptor signaling network in head and neck carcinogenesis and implications for targeted therapy. Semin. Oncol 31; 734-443.

Poljak M, Gale N, Kambic V, Ferluga D, Fischinger J (1996) Over- expression of p53 protein in benign and malignant laryngeal epithelial lesions. Anticancer Res. 16; 1947–1951.

Poulsen HE, Taylor CW, Sobin LH (1975) Histological typing of female genital tract tumours, International histological classification of tumours, No. 13. World Health Organization, Geneva

Proweller A, Tu L, Lepore JJ, Cheng L, Lu MM, Seykora J et al (2006) CancerRes. 66, 7438-44.

Presland R, Jurevic R (2002) Making sense of the epithelial barrier: what molecular biology and genetics tell us about the functions of oral mucosal and epidermal tissues, Journal of Dental Education 66, 564–574.

Prince ME, Sivanandan R, Kaczorowski A, Wolf GT, Kaplan MJ, Dalerba P et al (2007) Identification of a subpopulation of cells with cancer stem cell properties in head and neck squamous cell carcinoma, Proceedings of the National Academy of Science 104, 973–978.

Prince ME, Ailles LE (2008) Cancer stem cells in head and neck squamous cell cancer, Journal of Clinical Oncology 26 2871-2875.

Putney FJ, O'Keefe JJ (1953) The clinical significance of keratosis of the larynx as a premalignant lesion. Ann. Otol. Rhinol. Laryngol 62; 348–357.

Ramalho-Santos M, Willenbring H (2007) On the origin of the term "stem cell", Cell Stem Cell 1 35–38.

Renan MJ (1993) How many mutations are required for tumorigen- esis? Implications from human cancer data. Mol. Carcinog 7; 139–146.

Ranganathan P, Weaver KL, Capobianco AJ (2011), Nat.Rev. Cancer 11, 338-51.

Reagan JW, Hamonic MJ (1956) Dysplasia of the uterine cervix. Ann N Y Acad Sci 63:1236–1244

Resta L, Colucci GA, Troia M, Russo S, Vacca E, Pesce Delfino V (1992) Laryngeal intraepithelial neoplasia (LIN). An analytical mor- phometric approach. Path. Res. Pract.; 188; 517–523.

Ricci G, Molini E, Faralli M, Simoncelli C. (2003) Retrospective study on precancerous laryngeal lesions: long-term follow-up. Acta Otorhinolaryngol. Ital 23; 362–367.

Ricci-Vitiani L, Lombardi DG, Pilozzi E, Biffoni M, Todaro M, Peschle C (2007) Identification and expansion of human colon-cancer-initiating cells, Nature 445 111–115.

Richart RM (1966) Influence of diagnostic and therapeutic procedures on the distribution of cervical intraepithelial neo- plasia. Cancer 19:1635–1638

Richart RM (1973) Cervical intraepithelial neoplasia. Pathol Ann 8:301–328

Robbins SL, Cortran RS (1979) Pathologic basis of disease, 2nd edn. Saunders, Philadelphia

Saglam O, Shah V, Worsham MJ (2007) Molecular differentiation of early and late stage laryngeal squamous cell carcinoma: an exploratory analysis. Diagn. Mol. Pathol 16; 218–221.

Sakr WA, Gale N, Gnepp DR et al (2009) Squamous intraepi- thelial neoplasia of the upper aerodigestive tract. In: Gnepp DR (ed) Diagnostic surgical pathology of the head and neck, 2nd edn. Saunders, Philadelphia, pp 1–44

Sanz-Ortega J, Valor C, Saez MC et al (2003) 3p21, 5q21, 9p21 and 17p13 allelic deletions accumulate in the dysplastic spectrum of laryngeal carcinogenesis and precede malignant transforma- tion. Histol. Histopathol 18; 1053–1057.

Sardi I, Franchi A, De Campora L, Passali GC, Gallo O (2006) Microsatellite instability as an indicator of malignant progres sion in laryngeal premalignancy. Head Neck 28; 730–739.

Shafer WG (1975) Oral carcinoma in situ. Oral Surg 39:227–238

Shear M, Pindborg JJ (1980) Verrucous hyperplasia of the oral mucosa. Cancer (Phila) 46:1855–1962

Schoelch ML, Sekandari N, Regezi JA et al (1999) Laser management of oral leukoplakias: a follow-up study of 70 patients. Laryngoscope 109:949–953

Serrano M, Hannon GJ, Beach D (1993) A new regulatory motif in cell-cycle control causing specific inhibition of cyclin D/CDK4. Nature 366; 704–707.

Shay JW, Wright WE (1996) Telomerase activity in human cancer. Curr. Opin. Oncol 8; 66–71.

Shin DM, Kim J, Ro JY et al (1994) Activation of p53 gene expression in premalignant lesions during head and neck tumorigenesis. Cancer Res 54; 321–326.

Silverman S Jr, Bhargava K, Mani NJ et al (1976) Malignant transformation and natural history of oral leukoplakia in 57518 industrial workers of Gujarat, India. Cancer (Phila) 38:1790–1795

Singh SK, Clarke ID, Terasaki M et al (2003) Identification of a cancer stem cell in human brain tumors," Cancer Research, vol. 63, no. 18, pp. 5821–5828.

Singh SK, Hawkins C, Clarke ID, Squire JA, Bayani J, Hide T et al (2004) Identification of human brain tumour initiating cells, Nature 432 396–401.

Sjoblom T, Jones S, Wood LD, Parsons DW, Lin J, Barber TD, et al , Science 314, 268-74 (2006).

Sllamniku B, Bauer W, Painter C, Sessions D (1989) The transformation of laryngeal keratosis into invasive carcinoma. Am. J. Otolaryngol 10; 42–54.

Somers KD, Merrick MA, Lopez ME, Incognito LS, Schechter GL, Casey G (1992) Frequent p53 mutations in head and neck cancer Cancer Res 52; 5997–6000.

Sorrells DL Jr, Ghali GE, De Benedetti A, Nathan CA, Li BD (1999) Progressive amplification and overexpression of the eukaryotic initiation factor 4E gene in different zones of head and neck cancers J. Oral Maxillofac. Surg 57; 294–299.

Speight PM, Farthing PM, Bouquot JE (1996) The pathology of oral cancer and precancer. Curr Diagn Pathol 3:165–176

Staibano S, Merolla F, Testa D, Iovine R, Mascolo M, Guarino V et al (2007) OPN/CD44v6 overexpression in laryngeal dysplasia and correlation with clinical outcome. Br J Cancer 97(11):1545-51

Stransky N, Egloff AM, Tward AD, Kostic AD, Cibulskis K, Sivachenko A, et al (2011) The Mutational Landscape of Head and Neck Squamous Cell Carcinoma, Science 333, 1157 10.1126/ science.1208130.

Szotek PP, Pieretti-Vanmarcke R, Masiakos PT et al (2006) Ovarian cancer side population defines cells with stem cell-like characteristics and Mullerian inhibiting substance respon- siveness Proceedings of the National Academy of Sciences of the United States of America, vol. 103, no. 30, pp. 11154–11159,.

Tabor MP, Brakenhoff RH, Ruijter-Schippers HJ, Van Der Wal JE, Snow GB, Leemans CR, et al (2002) Multiple head and neck tumours frequently originate from a single preneoplastic lesion. Am J Pathol 161: 1051–1060

Tabor MP, Braakhuis BJM, Van der Waal JE et al (2003) Comparative molecular and histological grading of epithelial dysplasia of the oral cavity and the oropharynx. J Pathol 199:354–360

Tan T-T, Coussens LM (2007) Humoral immunity, inflammation and cancer, Current Opinion in Immunology 19 209–216.

Thomasson HR, Edenberg HJ, Crabb DW et al (1991) Alcohol and aldehyde dehydrogenase genotypes and alcoholism in Chinese men American Journal of Human Genetics, vol. 48, no. 4, pp. 677–681.

Toole BP, Wight TN, Tammi MI (2002) Hyaluronan-cell interactions in cancer and vascular disease, The Journal of Biological Chemistry 277 4593–4596.

Trzpis M, McLaughlin PMJ, de Leij LMFH, Harmsen MC (2007) Epithelial cell adhesion molecule: more than a carcinoma marker and adhesion molecule, The American Journal of Pathology 171, 386–395.

Uhlman DL, Adams G, Knapp D, Aeppli DM, Niehans G (1996) Immunohistochemical staining for markers of future neoplastic progression in the larynx. Cancer Res 56; 2199–2205.

van der Riet P, Nawroz H, Hruban RH et al (1994) Frequent loss of chromosome 9p21-22 early in head and neck cancer progres- sion. Cancer Res 54; 1156–1158.

van der Waal I, Schepman KP, van der Meiji EH et al (1997) Oral leukoplakia: a clinicopathological review. Oral Oncol 33:291–301

Vedtofte P, Holmstrup P, Hjorting-Hansen E et al (1987) Surgical treatment of premalignant lesions of the oral mucosa. Int J Oral Maxillofac Surg 16:656–664

Velasco JRR, Nieto CS, de Bustos CP, Marcos CA (1987) Premalignant lesions of the larynx: pathological prognostic factors. J. Otolaryngol 16; 367–370.

Vinitha Richard, M. Radhakrishna Pillai (2010) The stem cell code in oral epithelial tumorigenesis: "The cancer stem cell shift hypothesis" Biochimica et Biophysica Acta 1806 146–162

Visus C, Ito D, Amoscato A et al (2007) Identification of human aldehyde dehydrogenase 1 family member a1 as a novel CD8+ T-cell-defined tumor antigen in squamous cell carcinoma of the head and neck Cancer Research, vol. 67, no. 21, pp. 10538–10545.

Voravud N, Shin DM, Ro JY, Lee JS, Hong WK, Hittelman WN (1993) Increased polysomies of chromosomes 7 and 17 during head and neck multistage tumorigenesis. Cancer Res 53; 2874–2883.

Wahi PN, Cohen B, Luthra UK et al (1971) Histological typing of oral and oropharyngeal tumours. International histological classification of tumours no. 4. World Health Organization, Geneva

Wang SJ, Wong G, De Heer AM, Xia W, and Bourguignon LYW (2009) CD44 Variant isoforms in head and neck squamous cell carcinoma progression Laryngoscope, vol. 119, no. 8, pp. 1518–1530.

Wang J, Guo LP, Chen LZ, Zeng YX, and Shih HL (2007) Identification of cancer stem cell-like side population cells in human nasopharyngeal carcinoma cell line Cancer Research, vol. 67, no. 8, pp. 3716–3724.

Warnakulasuriya S (2001) Histological grading of oral epithelial dysplasia: revisited. J Pathol 194:294–297

Warnakulasuriya S, Johnson NW, van der Wall I (2007) Nomenclature and classification of potentially malignant disorders of the oral mucosa. J Oral Pathol Med 36:575–580

Warnakulasuriya S, Reibel J, Bouquot J et al (2008) Oral epithelial dysplasia classification system: predictive value, utility, weakness and scope for improvement. J Oral Pathol Med 37:127–133

Wayne S, Robinson RA (2006) Upper aerodigestive tract squamous dysplasia: correlation with p16, p53, pRb, and Ki-67 expres- sion. Arch. Pathol. Lab. Med 130; 1309–1314.

Wood LD, Parsons DW, Jones S, Lin J, Sjöblom T, Leary RJ, et al (2007) Science 318, 1108-13.

Yoo WJ, Cho SH, Lee YS et al (2004) Loss of heterozygosity on chromosomes 3p,8p,9p and 17p in the progression of squamous cell carcinoma of the larynx. J. Korean Med. Sci 19; 345–351.

Yuge T, Nibu K, Kondo K, Shibahara J, Tayama N, Sugasawa M (2005) Loss of FHIT expression in squamous cell carcinoma and premalignant lesions of the larynx. Ann. Otol. Rhinol. Laryngol 114; 127–131.

Zerdoner D (2003) The Ljubljana classification: its application to grading oral epithelial hyperplasia. J Craniomaxillofacial Surg 31:75–79

Zhang P, Zhang Y, Mao L, Zhang Z, and Chen W (2009) Side population in oral squamous
    cell carcinoma possesses tumor stem cell phenotypes Cancer Letters, vol. 277, no. 2,
    pp. 227– 234.
Zhang Q, Shi S, Yen Y, Brown J, Ta JQ, and Le AD (2009) A subpopulation of CD133+ cancer
    stem-like cells characterized in human oral squamous cell carcinoma confer
    resistance to chemotherapy Cancer Letters, vol. 289, no. 2, pp. 151–160.

Part 2

Intraepithelial Neoplasia of Eye and Ocular Adnexa

2

Ocular Surface Squamous Neoplasia

Napaporn Tananuvat[1] and Nirush Lertprasertsuke[2]
Departments of [1]Ophthalmology and [2]Pathology,
Faculty of Medicine, Chiang Mai University,
Thailand

1. Introduction

The ocular surface is composed of the conjunctiva and the cornea. The conjunctiva is a mucous membrane which covers the globe and inner part of the eyelids. The morphology of conjunctival epithelial cells are nonkeratinized stratified epithelia which vary from cuboidal over the tarsus, to columnar in the fornices, to squamous epithelia on the globe. Goblet cells account for up to 10% of the basal cells of the conjunctival epithelia. The substantia propia of the conjunctiva consists of loose connective tissue. The cornea is a transparent, avascular tissue which acts as both the anterior eye wall and an optical media for light to enter the eye. The corneal epithelium layer is composed of stratified squamous epithelial cells and makeup about 5 % (0.05 mm) of the total corneal thickness. The corneal epithelial stem cells located at the basal layer of the limbal epithelia proliferate continuously and give rise to the superficial layer that subsequently differentiate into superficial cells. Regulation of cell growth and metabolism are critical to maintain an intact ocular surface and transparent cornea.

Primary tumors of the conjunctiva and cornea can be grouped into two major categories: congenital and acquired. The acquired lesions are composed of a variety of neoplasms which originate from squamous epithelia, melanocytes, and lymphocyte cells. Tumors of squamous epithelium occupy a large spectrum of lesions, ranging from benign lesions like squamous papilloma, to precancerous lesions which are confined to the surface epithelium (intraepithelial neoplasia or dysplasia, previously known as Bowen's disease). There are even more invasive squamous cell carcinomas that break through the basement membrane to the underlying substantria propia of the conjunctiva or corneal stroma.

The term ocular surface squamous neoplasia (OSSN) was first described in 1995 by Lee and Hirst to denote a spectrum of neoplasm originate from squamous epithelium ranging from simple dysplasia to invasive squamous cell carcinoma(SCC), involving the conjunctiva, the limbus, and the cornea.(Lee & Hirst 1995) Similar to cancer of cervix, it has a relative high recurrence after treatment and may metastasize. This tumor is considered as a low grade malignancy but invasive lesion can spread to the globe or orbit. This chapter highlights the epidemiology, etiologies and related factors, clinical manifestations, diagnostic tools, and standard care of management of these tumors. Squamous papilloma is also included as some conjunctival papilloma may have dysplastic potential.

2. Epidemiology and pathogenesis

OSSN is considered an uncommon disease with geographic incidences which vary from 0.2 to 3.5 per 100,000, with greater frequency near the equator. (Lee & Hirst 1995) It is the most common ocular surface tumor in many series.(Lee & Hirst 1995; Shields et al. 2004; Shields & Shields 2004) Prior to HIV pandemic, OSSN was noted to occur predominantly in the elderly for whom it was the third most common oculo-orbital tumor after malignant melanoma and lymphoma. (Lee & Hirst 1995) This tumor is rare in the United States, with an incidence rate of 0.03 per 100,000 persons, although the rate was approximately 5-fold higher in males and Caucasians (Sun et al. 1997).

Pathogenesis of OSSN has yet to be attributed to specific etiologic factors, the main associated factors being exposure to ultraviolet (UV) radiation, human papilloma virus infection, and human immunodeficiency virus (HIV) seropositivity.

2.1 Ultraviolet-B

Chronic exposure to UV–B radiation (290-320 nm) is an established cause of many eye diseases such as pingecular, pterygium, cataract, and age-related macular degeneration. (Taylor et al. 1992) Evidence from epidemiologic studies and worldwide cancer registries have confirmed that the incidence rate of OSSN increased with proximity to the equator, presumably from increased solar UV radiation. (Lee et al. 1994; Newton et al. 1996) One population-based cancer study found that the incidence of squamous cell carcinoma(SCC) of the eye declined by 49% for each 10 degree increase in latitude, falling from more than 12 cases per million per year in Uganda, to less than 0.2 cases per million per year in the UK. The incidence of SCC decreased by 29% per unit reduction in UV exposure. (Newton et al. 1996) There is considerable evidence linking cutaneous malignancy and UV exposure. (English et al. 1997) These lesions occur predominantly in sun-exposured areas of the skin. Lesions of OSSN are often found at the corneal limbus in the interpalpebral area, where sun-exposure is greater. The corneal limbus is a transitional area, from the conjunctival to corneal epithelial, analogous to the squamocollumnar junction of the uterine cervix which is prone to dysplastic change. The role of limbal stem cells in development of OSSN is controversial. These cells are long-lived and have great potential to clonagenic division. OSSN may arise from dysfunction limbal stem cells and from mutagenic agents such as UV radiation leading to mutations in the P53 tumor suppressor gene, also known as *TP53* gene. One pilot case-control study found that the *TP53* mutation was detected in 56% of cancer cases (SCC) and 14% of control. 50% of mutations were CC-TT transition which was a molecular signature of mutagenesis by solar UV rays. This prevalence was high compared to any cancer type (not exceed 6%), but matched that of skin cancer in subjects with xeroderma pigmentosum.(Ateenyi-Agaba et al. 2004) Solar elastosis was also found more frequently in pathological specimens from the conjunctival squamous cell neoplasia (53.3% of cases and 3.3% of controls). (Tulvatana et al. 2003) One immunohistochemical study showed that UV radiation may play a role as a stimulating agent in the expression of some proteolytic enzymes, such as matrix metalloproteinases (MMPs) and their tissue inhibitors (TIMPs), which are relevant to neoplasia. (Ng et al. 2008)

2.2 Human papilloma virus

Human papilloma viruses (HPV) are oncogenic viruses and their role in human cervical carcinoma is well-established, however, their role in OSSN is unclear. Nakamura demonstrated that 50% of squamous tumors of the ocular surface and lacrimal sac were associated with HPV. (Nakamura et al. 1997) Biopsy specimens together with analyses of archrival embedded tissue revealed that the low risk HPV type 6 and 11 were the most common types of viruses associated with conjunctival papilloma. (Sjo et al. 2007; Verma et al. 2008) The high risk HPV type 16 and 18 have also been demonstrated in conjunctival papilloma, however, both are commonly found in high grade dysplasia, or invasive squamous cell carcinoma of the conjunctiva. (Sjo et al. 2007; Verma et al. 2008) One study identified the DNA of HPV 16, 18, and mRNA from the *E6* region, which represented active transcribed viruses from all specimens of conjunctival intraepithelial neoplasia by using the PCR technique (n = 10). (Scott et al. 2002)

In contrast, several studies have failed to demonstrate HPV in malignant conjunctival epithelial tumors and suggested that HPV was not associated with malignant conjunctival lesions and posed other mechanism, such as UVB being more important to the etiology of these lesions. (Eng et al. 2002; Tulvatana et al. 2003; Sen et al. 2007; Manderwad et al. 2009) Thus, the association between HPV and OSSN is variable in different geographic areas, and perhaps depends on the method of detection used. (Eng et al. 2002; Sen et al. 2007; Guthoff et al. 2009; Manderwad et al. 2009)

2.3 Human immunodeficiency virus

OSSN is now recognized as an AIDS-related cancer and its incidence has increased with the HIV pandemic in Africa. (Porges & Groisman 2003) One study revealed that HIV was strongly associated with conjunctival squamous neoplasia in Africa with an odds ratio of 13 (HIV was positive in 71% of cases and 16% of controls). (Waddell et al. 1996) A case-control study of conjunctival SCC in Uganda demonstrated a 10 fold increased risk of conjunctival SCC in HIV-infected patients. (Newton et al. 2002) These tumors occurred at an earlier age in HIV-infected individuals and was often more aggressive than immunocompetent patients. OSSN may be the primary or only apparent manifestation of HIV infection in sub-Saharan Africa. (Spitzer et al. 2008) SCC can also involve other non-ocular sites such as the oropharynx, cervix, and anorectum.(Jeng et al. 2007) One study from the US found that there was an increased prevalence of HIV among patients with CIN who were younger than 50 years of age. (Karp et al. 1996) A HIV/AIDS Cancer Match Registry Study in the USA, however, demonstrated that the risk of conjunctival SCC was elevated regardless of HIV category, CD4 lymphocyte count, and time relative to AID-onset. The risk was highest with age≥ 50, Hispanic ethnicity, and residence in regions with high solar-UV radiation. (Guech-Ongey et al. 2008) Tissue analysis from OSSN specimens in HIV-1 patients identified multiple oncogenic viruses including HPV, EBV, and KSHV, suggested that these infectious agents may contribute to the development of this malignancy in HIV patients. (Simbiri et al. 2010)

2.4 Immunosuppression

Of note, OSSN shares some striking similarities to skin neoplasm. It is believed that localized immune suppression of the skin from sun damage may lead to increased susceptibility to HPV infection, causing neoplasia. Additional risks have also been reported in immunosuppressed cancer patients and organ transplant patients. (Shelil et al. 2003; Shome et al. 2006) As well, there have been reports of OSSN after corneal grafts, which may partly be related to local immunosuppression, HPV, or possibly that neoplastic cells had been in the donor corneal epithelia at the time of transplantation. (Ramasubramanian et al. 2010)

2.5 Others

Other factors associated with this condition include old age, the male sex (Lee & Hirst 1995; Sun et al. 1997), and fair skin pigmentation (Lee et al. 1994; Sun et al. 1997), as well as heavy cigarette smoking (Napora et al. 1990), exposure to petroleum products (Napora et al. 1990), and some genetic conditions like xeroderma pigmentosum. The latter is an uncommon genetic disorder, where excessive reactivity to UV light-induced damage results in a more malignant course. It is common in early childhood with severe photosensitivity and photophobia. (Kraemer et al. 1987; Chidzonga et al. 2009) Long standing use of ocular prosthesis (Jain et al. 2010)and contact lens wear (Guex-Crosier & Herbort 1993) have also been implicated in the pathogenesis of OSSN, although evidence is scant.

3. Clinical manifestations

The clinical spectrum of OSSN varies from benign lesions like squamous papilloma, precancerous lesions like conjunctiva-corneal intraepithelial dysplasia (CCIN), carcinoma in situ, and invasive squamous cell carcinoma (SCC).

3.1 Conjunctival papilloma

Squamous papillomas are among the most common benign acquired lesions of the conjunctiva. There are two forms of conjunctival papilloma: pendunculated and sessile. Both have different etiology and clinical courses. A pedunculated conjunctival papilloma is a fleshy, exophytic mass with a fibrovascular core which gives rise to a stalk. (Fig.1) It often arises in the inferior fornix, but can be present on the tarsus or bulbar conjunctiva. This lesion is associated with HPV subtype 6 or 11 (Sjo et al. 2007), and often occurs in children. It can regress spontaneously, or may recur after surgical excision.

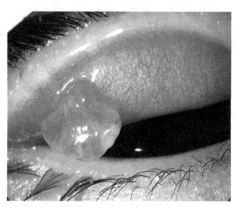

Fig. 1. Penduculate conjunctival papilloma arising from upper palpebral conjunctiva.

A sessile papilloma is more typically found at the limbus and has a broad base. The glistening surface and numerous red dots resemble a strawberry. (Fig.2) In contrast, a sessile lesion usually occurs in adults and more prone to dysplastic change. This lesion is related to HPV subtype 16 or 18. The latter oncogenic virus strains are strongly associated with human cervical carcinoma.

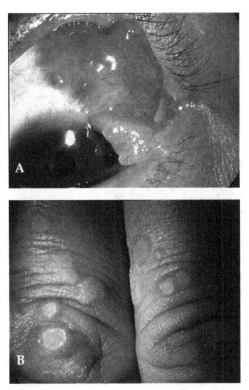

Fig. 2. A. Sessile mass arising from bulbar conjunctiva. B. Multiple papillomas involve skin of two fingers in the same patient.

3.2 Conjunctival-corneal intraepithelial neoplasia

The clinical symptoms are generally nonspecific, vary from asymptomatic to chronic irritation, redness, and varying degrees of visual involvement determine by the extension of lesions to the visual axis. Clinical patterns may be in a papilliform, as well as velvety, gelatinous, leukoplakic, nodular or even diffuse fashion. (Fig. 3-5) The lesions most commonly arise in the interpalpebral area of perilimbal conjunctiva, but are less common in the forniceal or palpebral conjunctiva. A white plaque (leukoplakia) may occur on the surface of the lesion, representing secondary hyperkeratosis, which results from squamous cell dysfunction. The conjunctival lesion is mobile with feeder vessels supplying the mass. These tumors may appear as slowly growing localized lesions that mimic benign conjunctival degenerations, and sometimes coexist with pterygia or pingecula. (Hirst et al. 2009) Sometimes, the lesions can have pigmentation and masquerade as malignant melanoma. (Shields et al. 2008) (Fig.6) OSSN can be diffused or have bilateral involvement. (Fig.7) Corneal OSSN is usually an extension of conjunctival squamous neoplasia. Rarely, isolated corneal involvement has been reported with the potentially aggressive form. (Fig.8) Bowman's layer usually is a barrier to invasive lesions. (Cha et al. 1993)

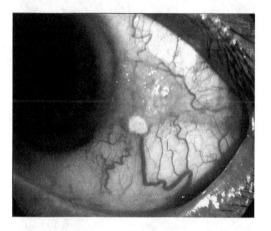

Fig. 3. Conjunctival intraepithelial neoplasia is present as a nodular mass with foci of leukoplakia on the surface of the lesion.

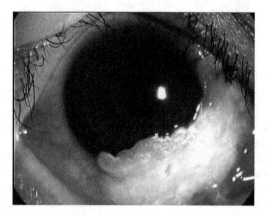

Fig. 4. Conjunctival-corneal intraepithelial neoplasia: a flat gelatinous mass with surface leukoplakia involves 2 quadrants of limbus.

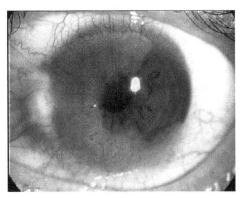

Fig. 5. Corneal intraepithelial neoplasia involving 270 degrees of the limbus (note vascular tuffs present on the mass)

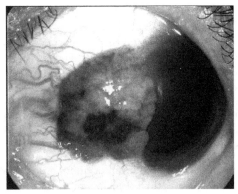

Fig. 6. Conjunctival-corneal intraepithelial neoplasia presents as a nodular mass with papillomatous pattern and hyperpigmentation (note feeder vessels are present).

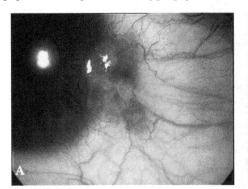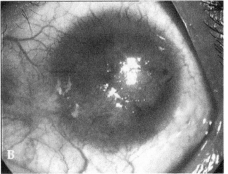

Fig. 7. Bilateral conjunctival-corneal intraepithelial neoplasia in an HIV-infected patient. A. Pigmented lesion with fibrovascular fond arising at the limbus. B. Diffused, flat lesion involving 360 degrees of the limbus (note central corneal epithelia defect is present in the photograph).

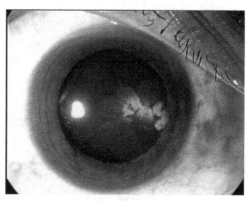

Fig. 8. Corneal intraepithelial neoplasia is present as a flat grayish mass with a fimbriated border and surface keratinization.

3.3 Squamous cell carcinoma

Squamous cell carcinoma is the final stage of this tumor where dysplastic epithelial invade beyond the basement membrane to the conjunctival substantia propia or corneal stroma. Clinically, invasive squamous cell carcinoma is generally larger and more elevated than CIN. (Fig.9) In practice, it may not be possible to distinguish invasive squamous cell carcinoma from intraepithelial lesion or carcinoma in-situ by using clinical features alone. However, an advanced lesion or mass that is immobile and fix to the globe should be suspected as an invasive lesion. A long term neglected mass or incomplete excised mass can invade through the globe or orbit. (Fig.10)

Local invasion is the most prevalent mechanism of tumor spread. Intraocular invasion may be associated with iritis, glaucoma, retinal detachment, or rupture of the globe. Metastases are rare, and the first site of extraocular involvement is regional lymph nodes.

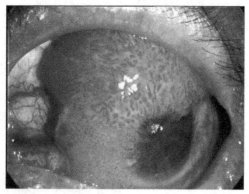

Fig. 9. Invasive squamous cell carcinoma involves two quadrants of conjunctiva and cornea (note papillary vascular pattern present on the mass with feeder vessels).

A rare variant of conjunctival squamous cell carcinoma is the mucoepidermoid carcinoma. Clinically, this tumor occurs in older patients and has a yellow globular cystic component due to the presence of abundant mucous-secreting cells within cysts. It tends to be more aggressive than the standard squamous cell carcinoma, thus deserves wider excision and closer follow-up. The spindle cell variant of squamous cell carcinoma is likewise aggressive. (Shields et al. 2007)

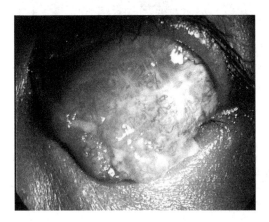

Fig. 10. An advanced squamous cell carcinoma involves the entire cornea and conjunctival surface with protrusions of the mass onto the lower eyelid.

4. Diagnosis and investigations

There are several points to cover before reaching diagnostic and management planning for OSSN, including clinical and pathologic findings, as well as the extension and complications of the tumors.

- Clinical feature of the lesion: morphology, size, site, surface, feeder vessels, and exact anatomical location whether it is conjunctival (move with conjunctiva when applying topical anesthesia with cotton tip applicator) or scleral involvement (fixed to the globe).
- Assessment of extension of the lesions
  - Intraocular invasion: perform gonioscopy to assess the invasion angle of the tumor. (Fig.11) Dilated fundus examination should be done to assess the intraocular invasion. In cases with media opacity, B-scan ultrasonography is helpful to assess sclera and intraocular spread.
  - Orbital invasion: by using CT scans or MRI scans, the accuracy and extension of the mass can accurately assess for the orbital or anterior eye involvement.
  - Regional lymph node spread: it is critical to assess the regional lymph nodes (preauricular, submandibular and cervical lymph nodes) as part of the clinical examination.

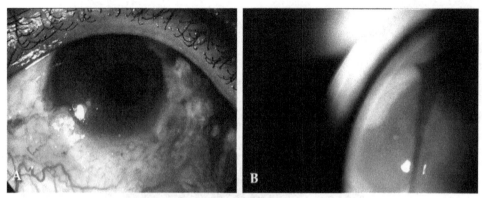

Fig. 11. Squamous cell carcinoma A. Diffused mass involves more than two quadrants of the limbus. B. Gonioscopic findings in the same eye show angle invasion by the mass.

- Pathologic diagnosis

  Since clinical appearance alone may not differentiate intraepithelial from invasive lesions, the gold standard for definite diagnosis is tissue histology, which can be performed by incisional or excisional biopsy. For relatively small tumors (≤ 4 clock hours of limbal involvement or ≤ 15 mm basal diameter), excisional biopsy is generally preferred to incisional biopsy. Larger lesions can be approached by wedge or punch biopsy. Incisional biopsy is also appropriate for conditions that are ideally treated with topical chemotherapy, or other treatments, such as radiation.

4.1 Histology

Histologic features of OSSN can be classified according to the presence of dysplastic cells originating in the basal cell layers which extend toward the surface. There are various patterns of dysplastic changes, ranging from the small squamous cells with increased nuclear-to-cytoplasmic (N/C) ratio, large squamous cells with hyperchromatic nuclei, and spindle cells bearing oval-shaped nuclei. The dysplastic cells contain abnormal nuclei either with nuclear pleomorphism or anisonucleosis. In addition, mitotic figures are increased and gradually pushed upward to the surface along with the degree of dysplasia. Many mitotic figures are abnormal. The histologic terms used to describe the OSSN include(Font et al. 2006):

- *Dysplasia*: dysplastic epithelial lesions of the conjunctiva and cornea divides into three grades based on the thickness of intraepithelial involvement. Koilocytes are rarely identified but suggestive for HPV infection when encountered. The thickness of involvement can be estimated using Periodic acid-Schiff (PAS) stain to demonstrate the presence of glycogen in non-neoplastic superficial squamous cells. Moreover, proliferating cell nuclear antigen (PCNA), Ki-67 and p53 immunostaining as well as argyrophillic nucleolar organizer region (AgNOR) staining may be useful for grading the dysplastic lesions as well as for correlation with clinical morphologic findings. (Aoki et al. 1998) Grading of dysplasia is described as:
  - Mild – less than a third thickness of the epithelium is occupied by atypical cells.(Fig.12A)

- Moderate - within three quarters thickness of the epithelium is occupied by atypical cells.
- Severe – nearly full thickness of the epithelium is occupied by atypical cells.(Fig.12B)
- *Carcinoma in situ*: full-thickness epithelial neoplasia with loss of the normal surface layer. (Fig.12C) Arborizarion of the proliferating blood vessels and extension of connective tissue along the neoplastic area may mimic the sessile papilloma.(Pizzarello & Jakobiec 1978)
- *Invasive squamous cell carcinoma*: the entire thickness of the epithelium has been replaced by the dysplastic cells and the basement membrane of the basal epithelial layer has been breached due to invasion of dysplastic cells into the substantia propia. Formation of cancer cell nests and single cancer cells with bizarre nuclei in the stroma is definitive of invasive carcinoma.(Tunc et al. 1999) (Fig.12D)

4.2 Cytology

Ocular surface cytology can be performed by two major techniques:first is exfoliative cytology by using spatula scrapings or a cytobrush to collect the sample, and second is impression cytology by using the collecting devices to collect the sample by contact with the surface of the lesions. The cytologic features of OSSN have been reviewed by several authors.(Lee & Hirst 1995)

- *Dysplasia*: Squamous cells with enlarged nuclei bearing fine to coarse granulation of the nuclear chromatins, irregular nuclear borders, scanty cytoplasm. The background is clean.
- *Carcinoma in situ*: Variable numbers of dysplastic cells with an admixture of intact and well preserved malignant cells. They are variable in size with scanty cytoplasm, usually < 1 nuclear diameter in width. The enlarged nuclei displays neoplastic features of hyperchromatism, irregular nuclear membrane thickening, or crusting of nuclear membranes. The other nuclear features include abnormal clearing or condensation of nuclear chromatins and large acidophilic nucleoli. However, background of the smear is clean.
- *Invasive squamous cell carcinoma*: Cytologic features of the SCC have been graded into two groups.
    - Grade 1-2: Marked cytologic aberration with bizarre malignant cell features including tadpole cells with cytolplasmic tails, fiber or spindle cells, hyperkeratinized cells with opaque refractile red or orange cytoplasm, and malignant nuclei.
    - Grade 3-4: Large or small cancer cells with scanty cytoplasm. Nonkeratinized cells maybe partially destructed cells, or complete loss of cytoplasm bearing large to huge pleomorphic nuclei. With deeper invasion and ulceration, tumor "diathesis" background- necrotic tumor cells, debris, blood, and leukocyte exudates are more prominent.

The advantages of cytology are a simple technique in diagnosis and follow-up after treatment in OSSN, particularly for detection of recurrences. However, some problems have been reported in exfoliative cytology techniques which may include a degree of uncomfort for the patient, problems with drying artifacts, problems with cellular overlap (difficult to interpret the specimens reliably) and non-localized lesions.

Impression cytology (IC) is a technique for collecting the superficial layers of the ocular surface by applying collecting devices. Commonly used are cellulose acetate filter paper with a pore size ranging from 0.025 – 0.45 micron or other materials (nitrocellulose filters, Biopore membranes, or polyether sulfone filters)(Calonge et al. 2004) so that cells adhere to the surface of the device and can be removed and processed further for analysis by a diversity of methods. IC represents a simple and non-invasive technique for both diagnosis and follow-up after treatment of several disorders of the ocular surface. The main advantages are that it allows relatively easy collection of epithelial samples with minimal

Fig. 12. Histologic features. A. Mild dysplasia; the basal cells are disordered with increased nuclear sizes and coarse nuclear chromatin. B. Severe dysplasia; the epithelial cells are varied in shapes and sizes with large pleomorphic nuclei. The surface cells are flattened with pyknotic nuclei. C. Carcinoma in situ: the entire thickness of the epithelium is composed of dysplastic cells bearing pleomorphic nuclei. Note the inflammatory reaction in the stroma. D. Invasive squamous cell carcinoma; the invasive nest in the stroma is composed of bizarre cells similar to those in the epithelium. The nuclei are plemorphic with thick nuclear membranes and prominent nucleoli (Hematoxylin and Eosin stain. Original magnification X40)

discomfort to the patient, can be performed on an outpatient basis, and allows more precise localization of the area being studied. In addition, a cell to cell relationship can be assessed, which allows one to see cells the way they exist in vivo.

Successful results of IC in diagnosis of OSSN in histologic-confirmed cases have been reported, with positive results of 77% - 80% of the cases.(Nolan et al. 1994; Tole et al. 2001) One study of ocular surface tumors found that IC had a positive and negative predictive value of 97.4% and 53.9%, respectively, when compared to histology.(Tananuvat et al. 2008) The limitations of IC are that, first, IC may be less sensitive for cases with keratotic lesions, because keratotic lesions are common in OSSN (68%) compared with a much lower incidence in cervical cancer. Second, IC may not distinguish carcinoma in situ from minimally invasive disease, because only the superficial cells are collected in the IC method. Therefore, a tissue biopsy remains necessary in cases with negative cytology.

At present, no cytologic criteria have been identified that reliably differentiate invasive carcinoma from in situ in IC samples. Squamous cell abnormalities may be classified into 4 groups, using a modification of the Bethesda system in cervical cytology(Solomon et al. 2002): (1) atypical squamous cells (ASC) (Fig.13 B); (2) low grade squamous intraepithelial lesions (LSIL), which encompass squamous papilloma and mild dysplasia(Fig.13 C); (3) high grade squamous intraepithelial lesions (HSIL), which encompass moderate to severe dysplasia and carcinoma in situ (CIS) (Fig.13 D-E); and (4) squamous cell carcinoma (SCC).(Fig.13 F) One series of OSSN found that SCC from cytology had a highest rate of correlation(91.7%) with histology followed by HSILs (45.5%), ASCs(42.9%),normal epithelia (33%), and LSILs (21.4%), respectively.(Tananuvat et al. 2008) Barros and coworkers used a scoring index modified from the Bethesda system which revealed a predictive index score of ≥4.5 represented the best cut-off point for diagnosis of SCC by using IC with a sensitivity of 95%, specificity of 93%, positive predictive value of 95%, and a negative predictive value of 93%.(Barros et al. 2009) However, the skill and the experience of cytologist are necessary for interpretation of the IC specimens.

4.3 Immunohistochemical analysis: Ki-67 proliferative index

Ki-67 nuclear antigen is expressed in all phases of the cell cycle, except the G0 phase. Ki-67 immunohistochemical analysis has been applied in the histopathologic diagnosis of malignant tumors. In normal cervical squamous mucosa, Ki-67 positive cells are found mainly in the parabasal layer. In cervical squamous intraepithelial lesions (SILs), the number of Ki-67 positive cells increased as the cell grading went from normal to low grade SIL(LSIL) to high grade SIL(HSIL). Similar findings have been reported in case of conjunctival SCC and intraepithelial neoplasia. One study compared tissue specimens obtained from SCC, CIN, and non-CIN (pterygium) lesions, revealed that Ki-67 proliferative index (Ki-67 PI) was significantly higher in SCC and CIN than in pterygium.(Ohara et al. 2004) In another study, the Ki-67 PI of CINs accounted for 20-48% which was significantly higher than non-CIN lesions (8-12%) and normal conjunctivae (8-12%). This study also showed that there was no statistical significance of P53-positive cells in CIN lesion compared to non-CIN lesions and normal conjunctiva due to the wide standard deviations. (Kuo et al. 2006) Therefore, Ki-67 PI may serve as a meaningful diagnostic marker for OSSN.

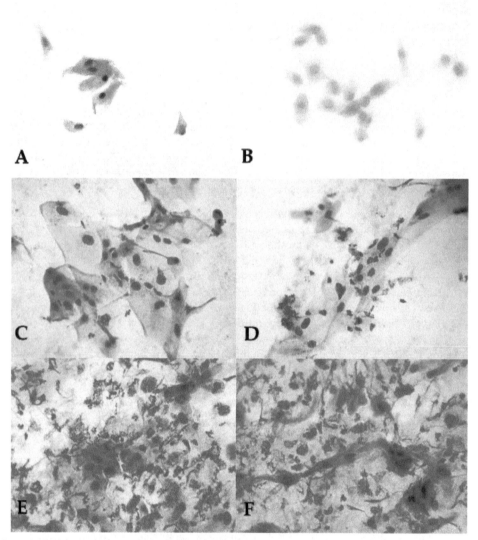

Fig. 13. Cytologic features from impression cytology specimens. A. Normal squamous cells with small nuclei and fine keratohyaline granules. B. Atypical cells with increased nuclear-to-cytoplasmic (N/C)ratio and glassy cytoplasm. C. Low grade corneal intraepithelial lesion; the dysplastic cells are varied in sizes with increased N/C ratio. They are similar to basal cells. Large polygonal squamous cells with small nuclei are also included. D. High grade corneal intraepithelial lesion; the nuclei are pleomorphic with coarse nuclear chromatins. E. High grade corneal intraepithelial lesion with inflammatory exudates on the background. The dysplastic cells cluster together with pleomorphic nuclei. F. Squamous cell carcinoma. The small and spindle cancer cells are aggregated together with the inflammatory background. Nuclear details are hardly noted as the cells overlap one another. (Papanicolaou stain. Original magnification X40)

4.4 Other investigation tools

Recently, in vivo confocal microscopy has proved useful as a noninvasive technique to investigate various ocular surface lesions including OSSN. Two studies found that confocal microscopic findings highly correlated with histologic features in CIN, thus provided real-time monitoring of the condition during treatment. (Alomar et al. 2011; Parrozzani et al. 2011)When compared to histology, however, there were some limitations. First, confocal microscopy provides en face images of cells compared to cross-sectional images from tissue histology. Second, fixation process required for histology results in shrinkage of tissue, therefore, morphometric comparison between living and fixed tissue have to be viewed in this context. Third, it is difficult to obtain in vivo confocal microscopic images and histologic images from exactly the same site of the tissue being examined.

The ultra high resolution (UHR) optical coherence tomography (OCT), a novel diagnostic technique for assessment of anterior eye segment lesions, was used for diagnosis and follow up after treatment of conjunctival-corneal intraepithelial neoplasia (CCIN) in a prospective case series. The UHR OCT images correlated well with the histologic specimens obtained from incisional biopsy before treatment. The UHR OCT was able to detect residual disease that was clinically invisible. The limitation of this machine was its capability to detect microinvasive lesions because the resolution of the current UHR OCT is approximately 2 micron, thus could not detect intracellular features. (Shousha et al. 2011)

Differential diagnosis

Because of the noninvasive nature of OSSN, the diagnosis is often missed or delayed. The patients' symptoms are sometimes treated as chronic conjunctivitis. Other conditions that are commonly mistaken include pterygium, pingecular, corneal pannus, viral keratoconjunctivitis, and corneal dystrophy.

5. Management

5.1 Conjunctival papilloma

Many conjunctival papilloma regress spontaneously. A pedunculated papilloma that is small, cosmetically acceptable and asymptomatic may be observed, although it may take months to years for spontaneous resolution. Larger and more peduculated lesions are generally symptomatic and of poor cosmetic acceptance, thus surgery adjunct with cryotherapy is recommended. A sessile papilloma must be observed closely. If there is any evidence of dysplastic change, excision with cryotherapy should be preformed.

Complete excision without manipulation of the tumor (no touch technique) is a crucial part of the surgical excision to minimize the risk of the virus spreading to uninvolved healthy conjunctiva. Double freeze-thaw cryotherapy is applied to the remaining conjunctiva to prevent tumor recurrence. An incomplete excision can stimulate growth and lead to a recurrence of the lesion and a worse cosmetic outcome. (Fig.14) Topical interferon- alpha 2b (Schechter et al. 2002; Kothari et al. 2009) and mitomycin C(Hawkins et al. 1999; Yuen et al. 2002) have been employed in the treatment of conjunctival papilloma. Immunomodulating agents such as oral cimetidine have led to regression of viral related papilloma. (Chang & Huang 2006)

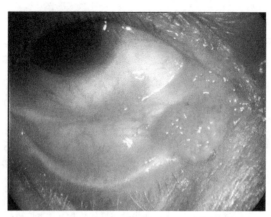

Fig. 14. Multifocal recurrence conjunctival papillomas involving lower palpebral conjunctiva, fornix, canruncle, and lower punctum after two previous excisions.

5.2 Preinvasive and invasive squamous neoplasia

5.2.1 Surgery

The management of OSSN varies with the extent of the lesion. The most accepted method of OSSN remains complete surgical excision. However, residual tumor cells left at the bordering tissue can induce tumor recurrence. Adjuvant therapies such as cryotherapy, alcohol abrasion, or topical agents are used in order to absolutely eradicate tumor cells from the ocular surface. Thus, the main treatment strategy is complete excision of the tumor with a wide surgical free margin followed by double freeze-thaw cryotherapy at the conjunctival margin and alcohol epitheliectomy for the corneal component. In case the tumor is adherent to the globe, a thin lamella of underlying sclera should be removed.

In order to decrease the chance of tumor recurrence, the standard surgical technique should be emphasized in all cases. The "no touch" technique purposed by Shield et al (Shields et al. 1997) is a widely accepted surgical approach as the conjunctival components, along with Tenon's fascia, should be excised with minimal manipulation of the tumor because cells from these friable tumors can seed into adjacent tissue. In addition, the surgery should be performed using microscopic techniques and the operative field should be left dry until after the tumor is completely removed to minimize spreading of tumor cells. Cryotherapy is thought to act through its direct destructive effects on cells, as well as the obliteration of microcirculation in the areas treated, resulting in ischemic infraction of the abnormal tissue. This is performed by freezing the surrounding bulbar conjunctiva as it is lifted away from the sclera using the cryoprobe. When the ice ball reaches the size of 4-5 mm, it is allowed to thaw and the cycle repeated. The complications that may occur from misuse of this technique or when the globe is accidentally frozen include cataract, uveitis, sclera and corneal thinning, and phthisis bulbi.

In cases of advanced tumors, the large conjunctival defect created by excision , particularly those over 4 clock hours, often require tissue replacement from a transpositional conjunctival flap, a free conjunctival autograft from the opposite eye, buccal mucosa graft, or amniotic membrane transplantation.

However, OSSN can be diffused or multifocal, with borders that are difficult to detect clinically, and there is also a chance for skipped areas from histopathologic examination. Reported recurrence rate after surgical treatment is significant (range between 15%-52%). (Lee & Hirst 1995; Tabin et al. 1997; Sudesh et al. 2000; McKelvie et al. 2002) Incomplete excision with positive surgical margins has been identified as a major risk factor for recurrence. (McKelvie et al. 2002) The more severe grades of OSSN appear to recur at higher rates. With adjunctive cryotherapy, the recurrent rate appears to be reduced (from 28.5% and 50% after simple excision, to 7.7% and 16.6% after excision with cryotherapy in primary and recurrence OSSN, respectively). (Sudesh et al. 2000)

The drawbacks of surgical treatment are complications resulted from the healing process, particularly in advanced lesions, including tissue granulation, symblepharon, pseudopterygium, diplopia from tissue shortening, blepharoptosis, limbal stem cell deficiency, and other complications. These surgical problems instigate further investigation into safer, alternative treatments.

5.2.2 Chemotherapy

Due to the relatively high rate of recurrence after surgical excision, various topical treatments have been advocated as a sole therapy for OSSN. Topical therapy offers a nonsurgical method for treating the entire ocular surface with less dependence on defining the tumor margin, potentially eliminating subclinical lesions. Topical treatment can offer a high drug concentration, avoiding systemic side effects. Furthermore, the increased cost, stress, pain, and trauma associated with surgical procedures are avoided. Topical medications have been used effectively for treating this condition comprised of mitomycin C (MMC), 5-fluorouracil (5-FU), and interferon, with MMC used most commonly by a group of external disease specialists. (Stone et al. 2005) These agents have been used as a sole therapy or a surgical adjuvant (preoperatively, intraoperatively, and postoperatively) for treatment of OSSN.

Mitomycin C

Mitomycin C (MMC) is an ankylating antibiotic that binds to DNA during all phases of the cell cycle leading to irreversible cross-linking and inhibition of nucleotide synthesis. When applied to conjunctival surfaces as a surgical adjunct, MMC has been shown to inhibit fibroblast cell migration, decrease extracellular matrix production, and to induce apoptosis in Tenon's capsule fibroblast. It is well known that chronic tissue effects from topical MMC administration can persist for many years after cessation of the treatment, thereby mimicking the effect of ionizing radiation. (McKelvie & Daniell 2001)

MMC has been widely used in glaucoma and pterygium surgery for its anti-fibrotic effect on subconjunctival fibroblast. The use of MMC for treatment of OSSN was first described in 1994.(Frucht-Pery & Rozenman 1994) Since then several case series using different concentrations and durations have been published. Common protocol ranges from topical MMC 0.02%-0.04% given four times a day to the affected eye for 7 to 28 days.(Fig.15) One case series demonstrated that even a smaller concentration of 0.002% of MMC was effective in treatment of primary and recurrent OSSN. (Prabhasawat et al. 2005) Several studies (similar to those used in fractionation of radiation in treatment of systemic cancers) preferred a cycle of 7 days in alternate weeks (1 week on and 1 week off) to allow cells of the ocular surface to recover/repair. (McKelvie & Daniell 2001; Shields & Shields 2004) One randomized control trial found that MMC 0.04% eye drops used 4 times a day for 3 weeks was effective and caused early resolution of noninvasive OSSN. A relative resolution rate in MMC versus placebo was 40.87 and the mean time for tumor resolution in this study was 121 days, and there was no serious complication in midterm follow-up. (Hirst 2007) MMC has also been used as a surgical adjunct for OSSN: preoperative, to decrease the size of the extensive lesions before surgical excision (chemoreduction), intraoperative, and postoperative to decrease recurrences.(Kemp et al. 2002; Chen et al. 2004; Gupta & Muecke 2010)

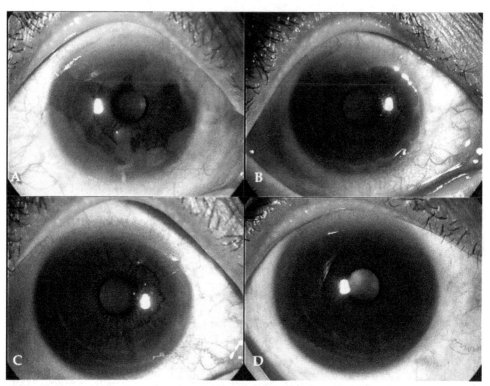

Fig. 15. Severe corneal intraepithelial neoplasia treated with mitomycin C 0.02% four times daily, alternating weeks: A. Appearance before treatment; B. Lesion partially resolved two months after treatment ; C. Completely resolved mass three months after treatment; D. Cornea is clear without recurrence eight years later.

Reported complications of MMC in treatment of OSSN included conjunctival hyperemia, punctuated epithelial erosion, and keratoconjunctivitis. A large retrospective series (n= 100 eyes) of ocular surface tumors treated with topical MMC 0.04% revealed that allergic reaction and punctual stenosis were two common complications. (Khong & Muecke 2006) Some of these side effects can be managed by stopping the medication and adding topical steroid three to four times daily. No significant changes were found on corneal endothelial cells after treatment with topical MMC 0.04% in a cyclic manner. (Panda et al. 2008)

However, MMC was found to have deleterious effects on endothelium cells after pterygium surgery, thus its judicious use and long term follow-up are mandatory.(Bahar et al. 2009) Even though common side effects related to topical MMC are self-limited, limbal stem cell deficiency appeared to be a significant long-term complication. (Dudney & Malecha 2004; Russell et al. 2011) Mckelvie and coworker reported the effects of MMC in treatments of OSSN on impression cytology; MMC appeared to produce cell death by apoptosis and necrosis. Cellular changes related to MMC mimic those caused by radiation-cytolmegaly, nucleomegaly, and vacuolation. These changes may persist at least 8 months after cessation of MMC therapy. (McKelvie & Daniell 2001) MMC-induced long term cytologic changes on the ocular surface have been demonstrated in another study. (Dogru et al. 2003) Serious complications of MMC such as scleromalacia, corneal perforation, cataract, glaucoma, and anterior uveitis have been reported in pterygium treatment and should be of concern if this agent is used in an open conjunctival wound or used excessively.(Rubinfeld et al. 1992)(Fig.16)

When MMC is prescribed as a treatment for OSSN, certain precaution should be taken. Patients and their families are advised to carefully handle the medication. Pregnant women and young children should avoid direct contact with the medication. Patients should be instructed to close their eyes for at least 5 minutes after instillation of MMC or punctal plugs are placed in both superior and inferior puncta to avoid nasolacrimal and systemic absorption of the drug. Since MMC is a chemotherapeutic agent, all residual bottles should be returned to the pharmacy for proper disposal.

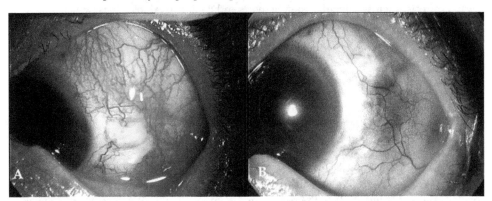

Fig. 16. A. Scleritis in eye with conjunctival intraepithelial neoplasia after excisional biopsy and postoperative mitomycin C. B. Scleral thinning in the same eye one year later after scleritis resolved.

5-Fluorouracil

Similar to MMC, topical 5-fluorouracil (5-FU) has been used to inhibit subconjunctival fibroblasts in glaucoma surgery. 5-FU is an antimetabolite used to treat many epithelial cancers because of its rapid action on rapidly proliferating cells. It acts by the inhibition of thymidylate synthetase during the S phase of the cell cycle, preventing DNA and RNA synthesis in rapidly dividing cells because of a lack of thymidine. Pulse 1% topical 5-FU in cycle of 4 days "on" followed by 30 days "off" until resolution of the lesion was a well-tolerated and effective method in treatment of OSSN, alone or as an adjunct to excision or debulking therapy. (Yeatts et al. 2000; Al-Barrag et al.; Parrozzani et al.; Rudkin & Muecke) Local side effects associated with topical 5-FU, such as lid toxicity, superficial keratitis, epiphora, and corneal epithelial defect have been reported. (Rudkin & Muecke 2011) By using confocal microscopy, there was no long-term corneal toxicity associated with 1% topical 5-FU compared to the controlled eye. (Parrozzani et al. 2011)The advantages of this agent are its few side effects, plus the medication is inexpensive, easy to handle by both medical personnel, as well as the patients.

Interferon

Interferons (IFN) are a group of proteins that bind to surface receptors of target cells, triggering a cascade of intracellular antiviral and antitumor activities. Systemic interefon-alpha has been used in treatment of hairy cell leukemia, condyloma acuminate, Karposi's sarcoma in AIDS, and hepatitis (both B and C). Recombinant topical IFNα-2b (1 million IU/ml) 4 times a day has been used effectively in treatment of primary OSSN. (Sturges et al. 2008) The antiviral effects of IFNα-2b may explain why it may be less effective as a primary treatment for lesions not linked to HPV infections. Topical IFNα-2b has been used effectively in management of recurrent or recalcitrant lesions where surgical excision or MMC have failed. (Holcombe & Lee 2006) This agent is well tolerated and does not markedly damage the limbal stem cells. Subconjunctival/perilesional IFN-α-2b (1-3 million IU/ml) has also been used effectively for treatment of both primary and recurrent OSSN. (Nemet et al. 2006; Karp et al. 2010) Topical instillation of IFN appears to be associated with few side effects, such as follicular conjunctivitis and conjunctival injections, which appeared to completely resolve after cessation of the medication. (Schechter et al. 2008) There was a report of corneal epithelial microcyst after topical administration interferon identical to that which had been reported with systemic interferon therapy. (Aldave & Nguyen 2007) Subconjunctival IFNα-2b has been associated with transient fever and myalgias , similar to systemic applications.

Topical chemotherapeutic agents have demonstrated acceptable efficacy in treatment of OSSN. Comparison of these three drugs for treatment of noninvasive OSSN reveals that MMC is the most effective (88%), followed by 5-FU(87%), and IFNα-2b (80%). MMC has the highest rate of side effects, perhaps because MMC is the most frequently used topical agent. IFNα-2b is the least toxic, however, it is the costliest of the three agents. (Sepulveda et al. 2010) The relative indications of using topical treatments in OSSN are: 1) >2 quadrants conjunctival involvement, 2) > 180 degree limbal involvement, 3) extension into the clear cornea involving the papillary axis, 4) positive margin after excision, and 5) patient unable to undergo surgery. (Sepulveda et al. 2010) However, some clinicians prefer surgical excision as an initial treatment of invasive lesions if the extension is less than 6 clock hours of involvement, because this provides confirmation of the diagnosis with little cosmetic disfigurement if properly performed.(Shields et al. 2002) When topical agents are considered as a treatment regimen of OSSN, they should be used with caution as long-term effects on the ocular surface of the eye, as well as the adjacent eyelids and nasolacrimal drainage system, have not yet been completely defined.

Other treatment modalities in management of OSSN include plaque brachytherapy with Iodine-125 (Walsh-Conway & Conway 2009), beta-radiation therapy, gamma radiation, and immunotherapy with dinitrochlorobenzene (DNCB). (Lee & Hirst 1995) Aggressive treatments such as enucleation or exenteration are considered in cases with ocular or orbital invasion. (Shields & Shields 2004)

6. Clinical course

OSSN is a slow growing tumor; however in neglected cases it can invade the globe and orbit and may lead to death. It has a potential for recurrence after treatment. In a series of OSSN, both intraepithelial and invasive lesions, it was found that sclera involvement occurred in 37%, orbital invasion 11%, and no metastasis or death was related to the tumors. (Tunc et al. 1999) In a series of 26 conjunctival SCC, intraocular invasion occurred in 11% of the patients, corneal or sclera involvement 30%, and orbital invasion 15%. Exenteration was required in 23% of cases, and 8% died of metastatic diseases. (McKelvie et al. 2002) Predicting factors related to significantly increased tumor recurrence include old age, large diameter lesions, high proliferation index (Ki-67 score), and positive surgical margin. (McKelvie et al. 2002)

A long-term study of CCIN also found that the recurrence rate after surgery was higher in cases with positive surgical margins than those with free margins (56% versus 33%). Timing for recurrence ranged from 33 days to 11.5 years after primary treatment, and those with incomplete excision recurred earlier than those with free margins. (Tabin et al. 1997) The slow growth of recurrent tumors and evidence of late recurrence 10 years after surgery warranted the need to have annual patient follow-ups for the remainder of their lives.

OSSN in immunosuppressed individuals seem to have an aggressive course in contrast to a relatively benign clinical course in classic OSSN.(Masanganise & Magava 2001; Gichuhi & Irlam 2007) The tumors often grow rapidly and have a tendency to invade the globe or orbit. This problem is exacerbated by poor health care facilities, and patient compliance, which are often present in HIV endemic areas. Management with standard approaches with these patients is often associated with higher rates of recurrence and intraocular or orbital invasion. Thus, treatment regimens may need a wide excision with a histological analysis of the margin, as well as other adjuncts such as cryotherapy, topical chemotherapeutic agents to prevent local recurrence, intraocular or orbital invasion, and metastasis. In addition, it is crucial for every HIV patient to have a detailed eye examination at presentation and maintain a close follow-up to detect recurrent disease early in its course.

7. Conclusion

OSSN is a spectrum of diseases ranging from simple dysplasia to invasive carcinoma. This lesion is considered a low grade malignancy, but its invasive counterpart can spread to the globe or orbit. It is the most common ocular surface tumor and its incidence varies in different geographic locations. The main risk factor is UV-B exposure as its incidence increases in areas close to the equator. Other important risk factors are the human papilloma virus and human immunodeficiency virus. However, it is unclear whether host factors (e.g. genetic factors and HIV-related immune impairment) or characteristics of the ocular surface epithelia may also be part of the etiopathogenesis of OSSN. Symptoms range from none at all to severe pain or visual loss. Clinically, these tumors most commonly arise in the interpalpebral area, particularly at the limbal region. Early diagnosis and management decrease the risk of locally aggressive and can improve the patients' prognosis for local control and preservation of vision. In clinical practice, OSSN is generally evaluated by tissue histology. The developments of pre-operative diagnostic techniques such as impression cytology are of value in diagnosis and follow-up after treatment. Surgical excision adjunct with cryotherapy combined with alcohol abrasion in cases of corneal involvement are the main treatment strategy. Recurrence rates are higher for more severe grades of OSSN and have been related to the adequate of surgical margins at the initial excision. The standard management care of OSSN appears to shift toward topical chemotherapy such as MMC, 5 FU, and interferon as a sole therapy, or a surgical adjunct, particularly in diffused or unoperable cases. These alternative treatments continue to evolve despite a paucity of long term results in published literature. Invasive disease may cause intraocular or orbital involvement with eye loss, and occasionally may lead to death. Recurrence after initial treatment is variable and warrants life-long follow-up in all case of OSSN.

8. References

Al-Barrag, A.; Al-Shaer,M.; Al-Matary,N. & Al-Hamdani, M. (2010). 5-Fluorouracil for the treatment of intraepithelial neoplasia and squamous cell carcinoma of the conjunctiva, and cornea. *Clin Ophthalmol*, vol. 4, (July,2010), pp 801-8, ISSN 1177-5483 (Electronic)

Aldave, AJ. & Nguyen, A. (2007). Ocular surface toxicity associated with topical interferon alpha-2b. *Br J Ophthalmol*, vol. 91, No.8, (Aug,2007), pp 1087-8, ISSN 0007-1161

Alomar, TS.; Nubile, M. ; Lowe, J. & Dua, HS. (2011). Corneal intraepithelial neoplasia: in vivo confocal microscopic study with histopathologic correlation. *Am J Ophthalmol*, vol. 151, No.2, (Feb,2011), pp 238-47, ISSN 1879-1891 (Electronic)

Aoki, S.; Kubo, E.; Nakamura, S.; Tsuzuki, A.; Tsuzuki, S.; Takahashi, Y. & Akagi, Y. (1998). Possible prognostic markers in conjunctival dysplasia and squamous cell carcinoma. *Jpn J Ophthalmol*, vol. 42, No.4, (Jul-Aug,1998), pp 256-61, ISSN 0021-5155

Ateenyi-Agaba, C.; Dai, M.; Le Calvez, F.; Katongole-Mbidde, E.; Smet, A.; Tommasino, M.; Franceschi, S.; Hainaut, P. & Weiderpass, E. (2004). TP53 mutations in squamous-cell carcinomas of the conjunctiva: evidence for UV-induced mutagenesis. *Mutagenesis*, vol. 19, No.5, (Sep,2004), pp 399-401, ISSN 0267-8357

Bahar, I.; Kaiserman, I.; Lange, AP.; Slomovic, A.; Levinger, E.; Sansanayudh, W. & Slomovic, AR. (2009). The effect of mitomycin C on corneal endothelium in pterygium surgery. *Am J Ophthalmol*, vol. 147, No.3, (Mar,2009), pp 447-452 e1, ISSN 1879-1891 (Electronic)

Barros, JN.; Lowen, MS.; Ballalai,PL.; Mascaro, VL.; Gomes, JA. & Martins, MC. (2009). Predictive index to differentiate invasive squamous cell carcinoma from preinvasive ocular surface lesions by impression cytology. *Br J Ophthalmol*, vol. 93, No.2, (Feb,2009), pp 209-14, ISSN 1468-2079 (Electronic)

Calonge, M.; Diebold, Y.; Saez, V.; Enriquez de Salamanca, A.; Garcia-Vazquez, C.; Corrales, RM. & Herreras, JM. (2004). Impression cytology of the ocular surface: a review. *Exp Eye Res*, vol. 78, No.3, (Mar,2004), pp 457-72, ISSN 0014-4835

Cha, SB.; Shields, CL.; Shields, JA.; Eagel, Jr., RC.; De Potter, P. & Talansky, M. (1993). Massive precorneal extension of squamous cell carcinoma of the conjunctiva. *Cornea*, vol. 12, No.6, (Nov,1993), pp 537-40, ISSN 0277-3740

Chang, SW. & Huang, ZL. (2006). Oral cimetidine adjuvant therapy for recalcitrant, diffuse conjunctival papillomatosis. *Cornea*, vol. 25, No.6, (Jul,2006), pp 687-90, ISSN 0277-3740

Chen, C.; Louis, D.; Dodd, T. & Muecke, J. (2004). Mitomycin C as an adjunct in the treatment of localised ocular surface squamous neoplasia. *Br J Ophthalmol*, vol. 88, No.1, (Jan, 2004), pp 17-8, ISSN 0007-1161

Chidzonga, MM.; Mahomva,L.; Makunike-Mutasa, R. & Masanganise, R. (2009). Xeroderma pigmentosum: a retrospective case series in Zimbabwe. *J Oral Maxillofac Surg*, vol. 67, No.1, (Jan, 2009), pp 22-31, ISSN 1531-5053 (Electronic)

Dogru, M.; Erturk, H.; Shimazaki,J.; Tsubota, K. & Gul, M. (2003). Tear function and ocular surface changes with topical mitomycin (MMC) treatment for primary corneal intraepithelial neoplasia. *Cornea*, vol. 22, No.7, (Oct,2003), pp 627-39, ISSN 0277-3740

Dudney, BW. & Malecha, MA. (2004). Limbal stem cell deficiency following topical mitomycin C treatment of conjunctival-corneal intraepithelial neoplasia. *Am J Ophthalmol*, vol. 137, No.5, (May,2004), pp 950-1, ISSN 0002-9394

Eng, HL.; Lin, TM.; Chen, SY.; Wu, SM. & Chen, WJ. (2002). Failure to detect human papillomavirus DNA in malignant epithelial neoplasms of conjunctiva by polymerase chain reaction. *Am J Clin Pathol*, vol. 117, No.3, (Mar,2002), pp 429-36, ISSN 0002-9173

English, DR.; Armstrong, BK.; Kricker, A. & Fleming, C. (1997). Sunlight and cancer. *Cancer Causes Control*, vol. 8, No.3, (May,1997), pp 271-83, ISSN 0957-5243

Font, RL.; Croxatto, JO. & Rao, NA. (2006). Tumors of the conjunctiva and caruncle. In: *Tumors of the eye and ocular adnexa*. SG Silverberg, pp. 7-10, American Registry of Pathology,ISBN 1-881041-99-9, Washington DC

Frucht-Pery, J. & Rozenman, Y. (1994). Mitomycin C therapy for corneal intraepithelial neoplasia. *Am J Ophthalmol*, vol. 117, No.2, (Feb,1994), pp 164-8, ISSN 0002-9394

Gichuhi, S. & Irlam, JJ. (2007). Interventions for squamous cell carcinoma of the conjunctiva in HIV-infected individuals. *Cochrane Database Syst Rev*, vol.18, No.2,(April,2007), pp CD005643, ISSN 1469-493X (Electronic)

Guech-Ongey, M.; Engels, EA.; Goedert,JJ.; Biggar, RJ. & Mbulaiteye, SM. (2008). Elevated risk for squamous cell carcinoma of the conjunctiva among adults with AIDS in the United States. *Int J Cancer*, vol. 122, No.11, (Jun ,2008), pp 2590-3, ISSN 1097-0215 (Electronic)

Guex-Crosier, Y. & Herbort, CP. (1993). Presumed corneal intraepithelial neoplasia associated with contact lens wear and intense ultraviolet light exposure. *Br J Ophthalmol*, vol. 77, No.3, (Mar,1993), pp 191-2, ISSN 0007-1161

Gupta, A. & Muecke, J. (2010). Treatment of ocular surface squamous neoplasia with Mitomycin C. *Br J Ophthalmol*, vol. 94, No.5, (May,2010), pp 555-8, ISSN 1468-2079 (Electronic)

Guthoff, R.; Marx, A. & Stroebel, P. (2009). No evidence for a pathogenic role of human papillomavirus infection in ocular surface squamous neoplasia in Germany. *Curr Eye Res*, vol. 34, No.8, (Aug,2009), pp 666-71, ISSN 1460-2202 (Electronic)

Hawkins, AS.; Yu, J.; Hamming, NA. & Rubenstein, JB. (1999). Treatment of recurrent conjunctival papillomatosis with mitomycin C. *Am J Ophthalmol*, vol. 128, No.5, (Nov,1999), pp 638-40, ISSN 0002-9394

Hirst, LW. (2007). Randomized controlled trial of topical mitomycin C for ocular surface squamous neoplasia: early resolution. *Ophthalmology*, vol. 114, No.5, (May,2007), pp 976-82, ISSN 1549-4713 (Electronic)

Hirst, LW.; Axelsen, RA. & Schwab, I. (2009). Pterygium and associated ocular surface squamous neoplasia. *Arch Ophthalmol*, vol. 127, No.1, (Jan,2009), pp 31-2, ISSN 1538-3601 (Electronic)

Holcombe, DJ. & Lee, GA. (2006). Topical interferon alfa-2b for the treatment of recalcitrant ocular surface squamous neoplasia. *Am J Ophthalmol*, vol. 142, No.4, (Oct,2006), pp 568-71, ISSN 0002-9394

Jain, RK.; Mehta, R. & Badve, S. (2010). Conjunctival squamous cell carcinoma due to ocular prostheses: a case report and review of literature. *Pathol Oncol Res*, vol. 16, No.4, (Dec,2010), pp 609-12, ISSN 1532-2807 (Electronic)

Jeng, BH.; Holland, GN.; Lowder, CY.; Deegan, 3rd, WF.; Raizman, MB. & Meisler, DM. (2007). Anterior segment and external ocular disorders associated with human immunodeficiency virus disease. *Surv Ophthalmol*, vol. 52, No.4, (Jul-Aug,2007), pp 329-68, ISSN 0039-6257

Karp, CL.; Galor, A.; Chhabra, S.; Barnes, SD. & Alfonso, EC. (2010). Subconjunctival/perilesional recombinant interferon alpha2b for ocular surface squamous neoplasia: a 10-year review. *Ophthalmology*, vol. 117, No.12, (Dec,2010), pp 2241-6, ISSN 1549-4713 (Electronic)

Karp, CL.; Scott, IU.; Chang, TS. & Pflugfelder, SC. (1996). Conjunctival intraepithelial neoplasia. A possible marker for human immunodeficiency virus infection? *Arch Ophthalmol*, vol. 114, No.3, (Mar,1996), pp 257-61, ISSN 0003-9950

Kemp, EG.; Harnett, AN. & Chatterjee, S. (2002). Preoperative topical and intraoperative local mitomycin C adjuvant therapy in the management of ocular surface neoplasias. *Br J Ophthalmol*, vol. 86, No.1, (Jan,2002), pp 31-4, ISSN 0007-1161

Khong, JJ. & Muecke, J. (2006). Complications of mitomycin C therapy in 100 eyes with ocular surface neoplasia. *Br J Ophthalmol*, vol. 90, No.7, (Jul,2006), pp 819-22, ISSN 0007-1161

Kothari, M.; Mody, K. & Chatterjee, D. (2009). Resolution of recurrent conjunctival papilloma after topical and intralesional interferon alpha2b with partial excision in a child. *J AAPOS*, vol. 13, No.5, (Oct,2009), pp 523-5, ISSN 1528-3933 (Electronic)

Kraemer, KH.; Lee, MM. & Scotto, J. (1987). Xeroderma pigmentosum. Cutaneous, ocular, and neurologic abnormalities in 830 published cases. *Arch Dermatol*, vol. 123, No.2, (Feb,1987), pp 241-50, ISSN 0003-987X

Kuo, KT.; Chang, HC.; Hsiao, CH. & Lin, MC. (2006). Increased Ki-67 proliferative index and absence of P16INK4 in CIN-HPV related pathogenic pathways different from cervical squamous intraepithelial lesion. *Br J Ophthalmol*, vol. 90, No.7, (Jul,2006), pp 894-9, ISSN 0007-1161

Lee, GA. & Hirst, LW. (1995). Ocular surface squamous neoplasia. *Surv Ophthalmol*, vol. 39, No.6, (May-Jun,1995), pp 429-50, ISSN 0039-6257

Lee, GA.; Williams, G.; Hirst, LW. & Green, AC. (1994). Risk factors in the development of ocular surface epithelial dysplasia. *Ophthalmology*, vol. 101, No.2, (Feb,1994), pp 360-4, ISSN 0161-6420

Manderwad, GP.; Kannabiran, C.; Honavar, SG. & Vemuganti, GK. (2009). Lack of association of high-risk human papillomavirus in ocular surface squamous neoplasia in India. *Arch Pathol Lab Med*, vol. 133, No.8, (Aug,2009), pp 1246-50, ISSN 1543-2165 (Electronic)

Masanganise, R. & Magava, A. (2001). Orbital exenterations and squamous cell carcinoma of the conjunctiva at Sekuru Kaguvi Eye Unit, Zimbabwe. *Cent Afr J Med*, vol. 47, No.8, (Aug,2001), pp 196-9, ISSN 0008-9176

McKelvie, PA. & Daniell, M. (2001). Impression cytology following mitomycin C therapy for ocular surface squamous neoplasia. *Br J Ophthalmol*, vol. 85, No.9, (Sep,2001), pp 1115-9, ISSN 0007-1161

McKelvie, PA.; Daniell, M.; McNab, A.; Loughnan, M. & Santamaria, JD. (2002). Squamous cell carcinoma of the conjunctiva: a series of 26 cases. *Br J Ophthalmol*, vol. 86, No.2, (Feb,2002), pp 168-73, ISSN 0007-1161

Nakamura, Y.; Mashima, Y.; Kameyama, K.; Mukai, M. & Oguchi, Y. (1997). Detection of human papillomavirus infection in squamous tumours of the conjunctiva and lacrimal sac by immunohistochemistry, in situ hybridisation, and polymerase chain reaction. *Br J Ophthalmol*, vol. 81, No.4, (Apr,1997), pp 308-13, ISSN 0007-1161

Napora, C.; Cohen, EJ.; Genvert, GI.; Presson, AC.; Arentsen, JJ.; Eagle, RC. & Laibson, PR. (1990). Factors associated with conjunctival intraepithelial neoplasia: a case control study. *Ophthalmic Surg*, vol. 21, No.1, (Jan,1990), pp 27-30, ISSN 0022-023X

Nemet, AY.; Sharma, V. & Benger, R. (2006). Interferon alpha 2b treatment for residual ocular surface squamous neoplasia unresponsive to excision, cryotherapy and mitomycin-C. *Clin Experiment Ophthalmol*, vol. 34, No.4, (May-Jun,2006), pp 375-7, ISSN 1442-6404

Newton, R.; Ferlay, J.; Reeves, G.; Beral , V.& Parkin, DM. (1996). Effect of ambient solar ultraviolet radiation on incidence of squamous-cell carcinoma of the eye. *Lancet*, vol. 347, No.9013, (May ,1996), pp 1450-1, ISSN 0140-6736

Newton, R.; Ziegler, J.; Ateenyi-Agaba, C.; Bousarghin, L.; Casabonne, D.; Beral, V.; Mbidde, E.; Carpenter,L.; Reeves,G.; Parkin, DM.; Wabinga, H.; Mbulaiteye,S.; Jaffe,H.; Bourboulia,D.; Boshoff,C.; Touze, A. & Coursaget, P. (2002). The epidemiology of conjunctival squamous cell carcinoma in Uganda. *Br J Cancer*, vol. 87, No.3, (Jul ,2002), pp 301-8, ISSN 0007-0920

Ng, J.; Coroneo, MT.; Wakefield, D. & Di Girolamo, N. (2008). Ultraviolet radiation and the role of matrix metalloproteinases in the pathogenesis of ocular surface squamous neoplasia. *Invest Ophthalmol Vis Sci*, vol. 49, No.12, (Dec,2008), pp 5295-306, ISSN 1552-5783 (Electronic)

Nolan, GR.; Hirst, LW.; Wright, RG. & Bancroft, BJ. (1994). Application of impression cytology to the diagnosis of conjunctival neoplasms. *Diagn Cytopathol*, vol. 11, (1994), pp 246-249, ISSN 8755-1039

Ohara, M.; Sotozono, C.; Tsuchihashi, Y. & Kinoshita, S. (2004). Ki-67 labeling index as a marker of malignancy in ocular surface neoplasms. *Jpn J Ophthalmol*, vol. 48, No.6, (Nov-Dec,2004), pp 524-9, ISSN 0021-5155

Panda, A.; Pe'er, J.; Aggarwal, A.; Das, H.; Kumar, A. & Mohan, S. (2008). Effect of topical mitomycin C on corneal endothelium. *Am J Ophthalmol*, vol. 145, No.4, (Apr,2008), pp 635-638, ISSN 0002-9394

Parrozzani, R.; Lazzarini, D.; Alemany-Rubio, E.; Urban, F. & Midena, E. (2011). Topical 1% 5-fluorouracil in ocular surface squamous neoplasia: a long-term safety study. *Br J Ophthalmol*, vol. 95, No.3, (Mar,2011), pp 355-9, ISSN 1468-2079 (Electronic)

Parrozzani, R.; Lazzarini, D.; Dario, A. & Midena, E. (2011). In vivo confocal microscopy of ocular surface squamous neoplasia. *Eye (Lond)*, vol. 25, No.4, (Apr,2011), pp 455-60, ISSN 1476-5454 (Electronic)

Pizzarello, L. & Jakobiec, FA. (1978). Bowen's disease of the conjunctiva: a misnomer. In: *Ocular and adnexal tumors.* FA Jakobiec, pp. 553-71, Aesculapius Pub,ISBN 9780912684154, Birmingham

Porges, Y. & Groisman, GM. (2003). Prevalence of HIV with conjunctival squamous cell neoplasia in an African provincial hospital. *Cornea*, vol. 22, No.1, (Jan,2003), pp 1-4, ISSN 0277-3740

Prabhasawat, P.; Tarinvorakup, P.; Tesavibul,N.; Uiprasertkul, M.; Kosrirukvongs, P.; Booranapong, W. & Srivannaboon, S. (2005). Topical 0.002% mitomycin C for the treatment of conjunctival-corneal intraepithelial neoplasia and squamous cell carcinoma. *Cornea*, vol. 24, No.4, (May,2005), pp 443-8, ISSN 0277-3740

Ramasubramanian, A.; Shields, CL.; Sinha, N. & Shields, JA. (2010). Ocular surface squamous neoplasia after corneal graft. *Am J Ophthalmol*, vol. 149, No.1, (Jan,2010), pp 62-5, ISSN 1879-1891 (Electronic)

Rubinfeld, RS.; Pfister,RR.; Stein,RM.; Foster,CS.; Martin, NF.; Stoleru, S.; Talley, AR. & Speaker, MG. (1992). Serious complications of topical mitomycin-C after pterygium surgery. *Ophthalmology*, vol. 99, No.11, (Nov,1992), pp 1647-54, ISSN 0161-6420

Rudkin, AK. & Muecke, JS. (2011). Adjuvant 5-fluorouracil in the treatment of localised ocular surface squamous neoplasia. *Br J Ophthalmol*, vol. 95, No.7, (Jul,2011), pp 947-50, ISSN 1468-2079 (Electronic)

Russell, HC.; Chadha,V.; Lockington, D. & Kemp, EG. (2011). Topical mitomycin C chemotherapy in the management of ocular surface neoplasia: a 10-year review of treatment outcomes and complications. *Br J Ophthalmol*, vol. 94, No.10, (Oct,2011), pp 1316-21, ISSN 1468-2079 (Electronic)

Schechter, BA.; Koreishi, AF.; Karp, CL. & Feuer, W. (2008). Long-term follow-up of conjunctival and corneal intraepithelial neoplasia treated with topical interferon alfa-2b. *Ophthalmology*, vol. 115, No.8, (Aug,2008), pp 1291-6, 1296 e1, ISSN 1549-4713 (Electronic)

Schechter, BA.; Rand, WJ.; Velazquez, GE.; Williams , WD.& Starasoler, L. (2002). Treatment of conjunctival papillomata with topical interferon Alfa-2b. *Am J Ophthalmol*, vol. 134, No.2, (Aug,2002), pp 268-70, ISSN 0002-9394

Scott, IU.; Karp, CL. & Nuovo, GJ. (2002). Human papillomavirus 16 and 18 expression in conjunctival intraepithelial neoplasia. *Ophthalmology*, vol. 109, No.3, (Mar,2002), pp 542-7, ISSN 0161-6420

Sen, S.; Sharma, A. & Panda, A. (2007). Immunohistochemical localization of human papilloma virus in conjunctival neoplasias: a retrospective study. *Indian J Ophthalmol*, vol. 55, No.5 (Sep-Oct,2007), pp 361-3, ISSN 0301-4738

Sepulveda, R.; Pe'er, J.; Midena, E.; Seregard, S.; Dua, HS. & Singh, AD. (2010). Topical chemotherapy for ocular surface squamous neoplasia: current status. *Br J Ophthalmol*, vol. 94, No.5, (May,2010), pp 532-5, ISSN 1468-2079 (Electronic)

Shelil, AE.; Shields,CL.; Shields , JA.& Eagle, Jr., RC. (2003). Aggressive conjunctival squamous cell carcinoma in a patient following liver transplantation. *Arch Ophthalmol*, vol. 121, No.2, (Feb,2003), pp 280-2, ISSN 0003-9950

Shields, CL.; Demirci, H.; Karatza, E. & Shields, JA. (2004). Clinical survey of 1643 melanocytic and nonmelanocytic conjunctival tumors. *Ophthalmology*, vol. 111, No.9, (Sep,2004), pp 1747-54, ISSN 1549-4713 (Electronic)

Shields, CL.; Manchandia, A.; Subbiah, R.; Eagle, Jr., RC. & Shields, JA. (2008). Pigmented squamous cell carcinoma in situ of the conjunctiva in 5 cases. *Ophthalmology*, vol. 115, No.10, (Oct,2008), pp 1673-8, ISSN 1549-4713 (Electronic)

Shields, CL.; Naseripour, M. & Shields, JA. (2002). Topical mitomycin C for extensive, recurrent conjunctival-corneal squamous cell carcinoma. *Am J Ophthalmol*, vol. 133, No.5, (May,2002), pp 601-6, ISSN 0002-9394

Shields, CL. & Shields, JA. (2004). Tumors of the conjunctiva and cornea. *Surv Ophthalmol*, vol. 49, No.1, (Jan-Feb,2004), pp 3-24, ISSN 0039-6257

Shields, JA.; Shields , CL.& De Potter, P. (1997). Surgical management of conjunctival tumors. The 1994 Lynn B. McMahan Lecture. *Arch Ophthalmol*, vol. 115, No.6, (Jun,1997), pp 808-15, ISSN 0003-9950

Shields, JA.; Eagle, RC.; Marr, BP.; Shields, CL.; Grossniklaus, HE. & Stulting, RD. (2007). Invasive spindle cell carcinoma of the conjunctiva managed by full-thickness eye wall resection. *Cornea*, vol. 26, No.8, (Sep,2007), pp 1014-6, ISSN 0277-3740

Shome, D.; Honavar, SG.; Manderwad, GP. & Vemuganti, GK. (2006). Ocular surface squamous neoplasia in a renal transplant recipient on immunosuppressive therapy. *Eye (Lond)*, vol. 20, No.12,(Dec,2006), pp 1413-4, ISSN 0950-222X

Shousha, MA.; Karp,CL.; Perez, VL.; Hoffmann, R.; Ventura,R.; Chang, V.; Dubovy, SR. & Wang, J. (2011). Diagnosis and Management of Conjunctival and Corneal Intraepithelial Neoplasia Using Ultra High-Resolution Optical Coherence Tomography. *Ophthalmology*, vol.118, No. 8 (August,2011), pp 1531-7, ISSN 1549-4713 (Electronic)

Simbiri, KO.; Murakami, M.; Feldman, M.; Steenhoff, AP.; Nkomazana, O.; Bisson, G. & Robertson, ES. (2010). Multiple oncogenic viruses identified in Ocular surface squamous neoplasia in HIV-1 patients. *Infect Agent Cancer*, vol. 5, (Mar,2010), pp 6, ISSN 1750-9378 (Electronic)

Sjo, NC.; von Buchwald, C.; Cassonnet, P.; Norrild,B.; Prause, JU.; Vinding, T. & Heegaard, S. (2007). Human papillomavirus in normal conjunctival tissue and in conjunctival papilloma: types and frequencies in a large series. *Br J Ophthalmol*, vol. 91, No.8, (Aug,2007), pp 1014-5, ISSN 0007-1161

Solomon, D.; Davey, D.; Kurman, R.; Moriarty,A.; O'Connor, D.; Prey,M.; Raab, S.; Sherman, M.; Wilbur, D.; Wright, Jr.,T. & Young, N. (2002). The 2001 Bethesda System: terminology for reporting results of cervical cytology. *JAMA*, vol. 287, No.16, (Apr ,2002), pp 2114-9, ISSN 0098-7484

Spitzer, MS.; Batumba, NH.; Chirambo, T.; Bartz-Schmidt, KU.; Kayange, P.; Kalua, K. & Szurman, P. (2008). Ocular surface squamous neoplasia as the first apparent manifestation of HIV infection in Malawi. *Clin Experiment Ophthalmol*, vol. 36, No.5, (Jul,2008), pp 422-5, ISSN 1442-9071 (Electronic)

Stone, DU.; Butt, AL. & Chodosh, J. (2005). Ocular surface squamous neoplasia: a standard of care survey. *Cornea*, vol. 24, No.3, (Apr,2005), pp 297-300, ISSN 0277-3740

Sturges, A.; Butt, AL.; Lai, JE. & Chodosh, J. (2008). Topical interferon or surgical excision for the management of primary ocular surface squamous neoplasia. *Ophthalmology*, vol. 115, No.8, (Aug,2008), pp 1297-302, 1302 e1, ISSN 1549-4713 (Electronic)

Sudesh, S.; Rapuano, CJ.; Cohen,EJ.; Eagle, Jr.,RC. & Laibson, PR. (2000). Surgical management of ocular surface squamous neoplasms: the experience from a cornea center. *Cornea*, vol. 19, No.3, (May,2000), pp 278-83, ISSN 0277-3740

Sun, EC.; Fears, TR. & Goedert, JJ. (1997). Epidemiology of squamous cell conjunctival cancer. *Cancer Epidemiol Biomarkers Prev*, vol. 6, No.2, (Feb,1997), pp 73-7, ISSN 1055-9965

Tabin, G.; Levin, S.; Snibson, G.; Loughnan, M. & Taylor, H. (1997). Late recurrences and the necessity for long-term follow-up in corneal and conjunctival intraepithelial neoplasia. *Ophthalmology*, vol. 104, No.3, (Mar,1997), pp 485-92, ISSN 0161-6420

Tananuvat, N.; Lertprasertsuk, N.; Mahanupap, P. & Noppanakeepong, P. (2008). Role of impression cytology in diagnosis of ocular surface neoplasia. *Cornea*, vol. 27, No.3, (Apr,2008), pp 269-74, ISSN 0277-3740

Taylor, HR.; West, S.; Munoz,B.; Rosenthal, FS.; Bressler, SB. & Bressler, NM. (1992). The long-term effects of visible light on the eye. *Arch Ophthalmol*, vol. 110, No.1, (Jan,1992), pp 99-104, ISSN 0003-9950

Tole, DM.; McKelvie, PA. & Daniell, M. (2001). Reliability of impression cytology for the diagnosis of ocular surface squamous neoplasia employing the Biopore membrane. *Br J Ophthalmol*, vol. 85, No.2, (Feb,2001), pp 154-8, ISSN 0007-1161

Tulvatana, W.; Bhattarakosol,P.; Sansopha,L.; Sipiyarak,W.; Kowitdamrong, E.; Paisuntornsug, T. & Karnsawai, S. (2003). Risk factors for conjunctival squamous cell neoplasia: a matched case-control study. *Br J Ophthalmol*, vol. 87, No.4, (Apr,2003), pp 396-8, ISSN 0007-1161

Tunc, M.; Char, DH.; Crawford, B & Miller, T. (1999). Intraepithelial and invasive squamous cell carcinoma of the conjunctiva: analysis of 60 cases. *Br J Ophthalmol*, vol. 83, No.1, (Jan,1999), pp 98-103, ISSN 0007-1161

Verma, V.; Shen, D.; Sieving, PC. & Chan, CC. (2008). The role of infectious agents in the etiology of ocular adnexal neoplasia. *Surv Ophthalmol*, vol. 53, No.4 (Jul-Aug,2008), pp 312-31, ISSN 0039-6257

Waddell, KM.; Lewallen, S.; Lucas, SB.; Atenyi-Agaba, C.; Herrington, CS. & Liomba, G. (1996). Carcinoma of the conjunctiva and HIV infection in Uganda and Malawi. *Br J Ophthalmol*, vol. 80, No.6, (Jun,1996), pp 503-8, ISSN 0007-1161

Walsh-Conway, N. & Conway, RM. (2009). Plaque brachytherapy for the management of ocular surface malignancies with corneoscleral invasion. *Clin Experiment Ophthalmol*, vol. 37, No.6, (Aug,2009), pp 577-83, ISSN 1442-9071 (Electronic)

Yeatts, RP.; Engelbrecht, NE.; Curry, CD.; Ford, JG. & Walter, KA. (2000). 5-Fluorouracil for the treatment of intraepithelial neoplasia of the conjunctiva and cornea. *Ophthalmology*, vol. 107, No.12, (Dec,2000), pp 2190-5, ISSN 0161-6420

Yuen, HK.; Yeung, EF.; Chan, NR.; Chi, SC. & Lam, DS. (2002). The use of postoperative topical mitomycin C in the treatment of recurrent conjunctival papilloma. *Cornea*, vol. 21, No.8, (Nov,2002), pp 838-9, ISSN 0277-3740

3

Conjunctival Intraepithelial Neoplasia – Clinical Presentation, Diagnosis and Treatment Possibilities

Valentín Huerva[1] and Francisco J. Ascaso[2]
[1]Department of Ophthalmology,
Universitary Hospital "Arnau de Vilanova" and IRB, Lleida
[2]Department of Ophthalmology,
"Lozano Blesa" University Clinic Hospital, Zaragoza,
Spain

1. Introduction

Conjunctival tumors are one of the most frequent tumors of the eye and adnexa. They comprise a large variety of conditions, from benign lesions such as papilloma to malignant lesions such as epidermoid carcinoma or melanoma which may threaten visual function and patient's life if not diagnosed early. Although conjunctival tumors may arise from any type of the conjunctival cells, epithelial and melanocytic are the most frequent origins. Epithelial tumors account for a third to half of all tumors, with a higher prevalence in countries with larger actinic exposure. Aproximately 40% of the tumors have an epithelial origin and 64.5 % of them were pre-cancerous lesions (Saornil et al, 2009). The clinical differentiation between pre-cancerous benign and malign lesions is difficult, requiring a biopsy for a definitive diagnosis.

Squamous neoplasia of the conjunctiva / cornea is a rare malignancy of conjunctival limbal stem cells, and the management of this malignancy may affect the ultimate outcome. The clinical distinction of squamous conjunctival neoplasia from other amelanocytic conjunctival tumors is based on certain clinical features of the tumor, and its correct management requires an understanding of normal anatomy and histology of the cornea and conjunctiva, as well as knowledge of the principles of tumor management.

Conjunctiva is a thin and flexible mucous membrane that extends from the internal surface of the eyelids to the fornix and anterior ocular surface up to the corneoscleral limbus. Histologically, conjunctiva is similar to other mucous membranes and comprises a non-keratinized stratified epithelium having two or more layers over the stroma formed by fibrovascular connective tissue containing vessels, nervous and lymphatic tissue. Basal layer of epithelium comprises melanocytes which produces melanine and inject it in the surrounding cells. Throughout the length of epithelium we can observe cup-shaped cells in charge of producing the mucoid component of the lacrimal film. These cells are called *goblet cells*.

1.1 Definition of Ocular Surface Squamous Neoplasias (OSSN)

Squamous cell neoplasia may occur as a localized lesion confined to the surface epithelium (conjunctival intraepithelial neoplasia) or as a more invasive squamous cell carcinoma that has broken through the basement membrane and invaded the underlying stroma (Shields & Shields, 2004).

Currently, the accepted term for the localized variety is conjunctival intraepithelial neoplasia (CIN). However, other authors prefer the terms dysplasia (mild, moderate, or severe) and carcinoma-in-situ. Where there are no longer normal surface cells then the process may be termed carcinoma-in-situ. Those cases where the cornea is invaded by the process are usually called conjunctiva-cornea intraepithelial neoplasia (CCIN). Squamous neoplasia constitutes the most frequent primary malignancy of the ocular surface.

1.1.1 Conjuntival Intraepithelial Neoplasia (CIN)

CIN is confined to the epithelium by definition. The term CIN was suggested in 1978, according with the general pathologic classification of intraepithelial tumors developed for cervical intraepithelial neoplasia (Pizzarello & Jakobiec, 1978). CIN includes previous terms referred to this epithelial neoplasia such as: Bowen's disease, Bowenoid epithelioma, intraepithelial epithelioma, intraepithelioma, dysplasia and carcinoma in situ (CIS).

Subjective symptoms referred by the patients include: foreign body sensation, redness, irritation, and a growth on the ocular surface (Giaconi & Karp, 2003).

Clinically, CIN appears as a fleshy, sessile or minimally elevated lesion usually at limbus in the interpalpebral fissure and less commonly in the forniceal or tarsal conjunctiva (Shields & Shields, 2004). The limbal lesion may extend for a variable distance into the epithelium of the adjacent cornea. A white plaque (leukoplakia) may occur on the surface of the lesion due to secondary hyperkeratosis.

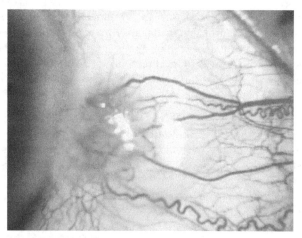

Fig. 1. Conjunctival intraepithelial neoplasia showing corneal invasion.

1.1.2 Squamous Cell Carcinoma (SCC)

Squamous cell carcinoma is characterized by an extension of abnormal epithelial cells through the basement membrane to gain access to the conjunctival stroma (Shields & Shields, 2004). Clinically, invasive squamous cell carcinoma is similar to CIN; however, it may be larger and more elevated than CIN. Even though the cells of invasive squamous cell carcinoma gain access to the blood vessels and lymphatic channels, regional and distant metastases are both rather uncommon. Clinically it is very difficult to distinguish between CIN and SCC (Erie et al, 1986). In many occasions it is necessary to perform a biopsy.

1.2 Incidence

OSSN accounts for only 5% of all ocular malignancies (Lee & Hirts, 1995). CIN is the most common conjunctival malignancy (Grossniklaus et al, 1987). CIN occurred more commonly in pale-skinned groups than in more pigmented people, with an increased incidence in males (75%) vs females (25%), and a mean age of 60 years (Grossniklaus et al, 1987). OSSN associated with human immunodeficiency virus (HIV) is seen at younger ages (average 35 years), usually not in a bulbar location, and is more aggressive from a clinical point of view. Its incidence can vary from 0.13 to 1.9/100.000 inhabitants (Lee & Hirts, 1995), (Giaconi & Karpp, 2003), (Saornil et al, 2009). OSSN incidence varies geographically, increasing with closer distance to the equator. For example, Uganda has 1.2 cases/100.000 persons/year compared to the United Kingdom with less than 0.02 cases/100.000 persons/year. This might suggest a role of ultraviolet light exposure in the etiology of these tumors. US data indicate an incidence of 0.03/100.000 people/year, with a 6-fold increase in association with HIV infection (Sun et al, 1997), (Verma et al, 2008). The lesions are more common in males and elderly, with the majority occurring at the limbus. In Africa the incidence is changing. The tumor is more common, aggressive, more frequent in young persons, especially women (Ateenyi-Agaba, 1995). This is relationed with the coexistence of pandemic AIDS and exposition to the human papillomavirus (HPV) and ultraviolet radiations. Africa has the highest prevalence of HPV infection in the world (with more than 25 % of women from 15 to 74 years affected), followed by South America (14.3%), Asia (8.7%) and Europe (5.2%) (Clifford, 2005). A study in the Kampala Cancer Registry in Uganda showed an increase from 6 cases of OSSN/1.000.000 persons per year between 1970 and 1988 to 35 cases/1.000.000 persons per year in 1992 (Ateenyi-Agaba, 1995). In Australia, a study found that 78.5% of affected people were male with a mean age of 60 years (Lee & Hirst, 1997). Similarly, another study in United Kingdom showed that the 77% were male, being 69% of them older than 60 years (McKelvie et al, 2002). Nevertheless, a study in Zimbabwe found that a 70% of patients were young women with a mean age of 35 years (Pola et al, 2003), while in South Africa mean age was 37 years (Mahomed & Chetty, 2002). A study in Tanzania showed that the 45.8% of 168 conjunctival biopsies were OSSN (Poole, 1999).

2. Etiologic factors for CIN

To date, CIN etiology remains unclear. The most probably explanation may be multifactorial causes. There are many known factors which may contribute to the development of these neoplasias.

The first one is the age, with an average of 60 years (Lee & Hirst, 1995). However, it ranges from 4 to 90 years. The second factor attributed is the UV light exposure (Lee et al, 1994). This justify a higher prevalence of CIN in the ecuatorial areas, as we have previously commented. The exposition to the petroleum products, heavy cigarrete smoking, light hair and ocular pigmentation have also been associated (Napora et al, 1990).

Younger patients affected by Xeroderma pigmentosum (Herle et al, 1991) and HIV may show a higher incidence (Karp et al, 1996). The majority of CIN cases reported in Africa are HIV-positive: 71% in Uganda, 86% in Malawi (Waddell, et al 1996), 70.6% in South Africa (Mahomed & Chetty 2002) and 92.3% in Zimbabwe (Porges & Groisman 2003). On the other hand, the prevalence of CIN in a HIV-positive population in Kenya was 7.8% (Chisi et al, 2006). These findings suggest that CIN is a marker for HIV infection. OSSN in HIV/AIDS patients presents at a younger age (35-40 years old) than in HIV-negative patients (Timm et al, 2004). Additionally, malignancy seems to be more aggressive in HIV/AIDS patients (Kaimbo 1998). It is unclear whether immunosuppression or HIV itself plays a role in this pathogenesis. Although it has been speculated about the role of HIV in conjunctival squamous cell carcinoma in immunosuppression and activation of oncogenic viruses such as HPV in the conjunctiva, thus far only oral and anogenital HPV has been shown to occur more frequently in HIV-positive patients (Verma et al, 2008). Immunosuppression itself may contribute to the carcinogenesis. Several studies have also demonstrated an association between immunosupression secondary to HIV infection and increased risk of cervical intraepithelial neoplasia (Palefsky, 1991).

The role of HPV remains also unclear in the etiology of CIN. It has been proved the causal relationship between HPV type 16 and 18 and uterine cervical carcinoma (Scott el al, 2002), (Giaconi & Karp, 2003), (Verma et al, 2008). However, multiple studies worldwide have failed to document an unequivocal association of HPV with conjunctival squamous cell carcinoma (Tuppurainen, 1992), (Eng et al, 2002). A small study of CIN has demonstrated mRNA from the $E6$ region of HPV, which signals actively transcribed virus (Scott et al 2002). Furthermore, this study in United States demonstrated the lack of such mRNA from normal conjunctivas, whereas African case series have revealed a high prevalence of HPV DNA in clinically normal conjunctivas for HPV 6 and 11, but not HPV 16 and 18 were found (Verma et al, 2008). In controversy HPV types 16 and 18 may be detected in CIN, in non neoplasic lesions and in apparently healthy conjunctiva (Karcioglu & Issa, 1997). Another study in Thailand concluded that solar elastosis is more frequently founded in OSSN cases that in controls, and HPV DNA was not found in any of the specimens (Tulvatana et al, 2003). HPV 5 and 8 were the most common in nearly half of OSSN in a series recently reported in Uganda (Ateenyi-Agaba, et al 2010). The frequency was the same in infected VIH than in non infected VIH. HPV 5 is not reported in caucasian CIN. HPV 16 and 18 may considerer as disease of sexual transmission whereas HPV 5 may appear by other possible vias. It has been shown in cervical cancer that the high risk variants HPV 16 and HPV 18 lead to carcinogenesis by inactivating tumor suppressor gene products $p53$ and Rb in the host with the viral oncoproteins E6 and E7, respectively (Verma et al, 2008). Furthermore, integration of viral sequences into the host genome leads to the constitutive expression of E6/E7 in transformed cervical cells. HPV 5 show highest downregulation of basal interleukin-8 secretion in primary human keratinocytes. This may weaken the response to UV-induced damage and consecutively mutations. Given the conflicting association studies, it appears that UV-B radiation plays a much greater role than HPV in the etiology of CIN (Verma et al, 2008).

3. Clinical presentation of CIN

The clinical presentation of CIN may be variable. There are many different pictures on the ocular surface that constitute a CIN. The subjective symptoms may be also variable in intensivity. Appart from the presence of a growth or mass in their ocular suface, patients may complain no symptoms. Size, color and growth may be variable in each case. The presence of a red eye make, sometimes, that the patient was treated as a conjunctivitis. Foreing body sensation, redness, or irritation may be referred many times. CIN is characterized by a slowly progressive course with low malignant potential. In general, two forms of CIN have been described: nodular (or well localized) and diffuse. The diffuse type is less common and very difficult to diagnose in early stages. This situation may be similar to a chronic conjunctivitis and its surgical treatment may result complicated because clinical borders of the lesion may be indistinguishable (Lee & Hirst, 1995), (Giaconi & Karp, 2003).

The typical location of this slow-growing lesion is the interpalpebral limbus, but it may also arise in the forniceal and palpebral conjunctiva. Limbal lesions may spread onto the cornea. The abnormal corneal epithelium has a frosted appearance with fringed borders and usually demonstrates diffuse punctate staining. Flat or elevated, the lesion may appear relatively translucent, gelatinous, or pearly white. Secondary hyperkeratosis over the surface of the lesion may give rise to a white plaque-like appearance clinically named leukoplakia. Often, there are surrounding corkscrew-like vascular tufts. Pigmentation may be seen and the lesion may be clinically misdiagnosed as melanoma (Shields et al, 2008).

The percentages of CIN that develops into SCC have not been reported in the literature. In cases of SCC the tumor may reach to eye globe, the orbit and cranial extension, with vision loss due to a enucleation or exenteration (Lopez-Garcia et al, 2000). Up to 4% rates of metastasis to cervical lymph nodes have been reported, while metastases to distance are less common (Bhattacharyya et al, 1997). There is not a particular simptomatology for every macroscopic form of CIN. Some of the characteristic forms of presentation of precancerous and malignant lesions are described:

3.1 Precancerous lesions: Actinic keratosis and conjunctival keratotic plaque

Both lesions, impossible to differentitate clinically, consist in a white plaque on the limbal or bulbar conjunctiva, in the exposed interpalpebral conjunctiva. They have a low grade of proliferation and very few possibilities to convert into CIN or SCC. Definitive diagnosis consisted in the histological study (Mauriello el al 1996), (Shields et al, 2004).

3.2 Leucoplakic lesions

Leucoplakia (white plaque) consist in a conjuctival lesion, generally at the limbus, which may be round or irregular. A process of keratosis is involved (Shields & Shields, 2008). These lesion may also extend onto the cornea. Likewise, leukoplakic lesions may appear onto a very diffuse CIN (Huerva et al, 2006). Extensive leukoplakia should raise suspiction of invasive SCC (Shields & Shields, 2008).

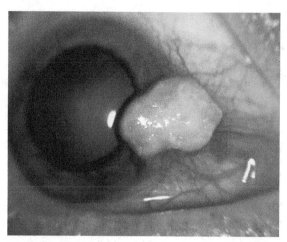

Fig. 2. Leukoplakic CIN, occuping conjunctiva, limbus and cornea at the interpalpebral fissure. Histopatology showed complete dysplasia of the epithelium.

3.3 Papillomatous lesions

CIN may developed simulating a sessile papilloma. The lesion consist in a fleshy red appearance owing numerous fine vascular channels that ramify throughtout the stroma beneath the epithelial surface of the lesion (Shields & Shields, 2008). The presence of displasic epithelial cells helps to the differential diagnosis between papilloma and CIN. In rare occasions papillomas may developed into a CIN.

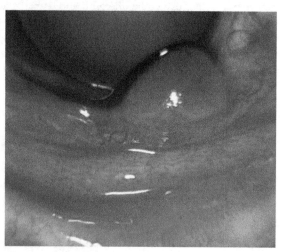

Fig. 3. CIN with papillomatous appearance at the exposed interpalpebral fissure affecting the limbus.

3.4 Fleshly lesions

Clinically, CIN appears as a fleshy, sessile, or minimally elevated lesion usually at the limbus in the interpalpebral fissure and less commonly in the forniceal or palpebral conjunctiva. The size of extension may be variable in each case. The presence of redness may simulate an inflammation. Extensive cases consist in a red gelatinous mass with vascular dilatations that may invade superior and nasal bulbar conjunctiva, including the caruncula, inferior conjuctiva and fornix invading tarsal conjunctiva and even corneal extension. Plaques of leukoplakia may also be present. (Erie et al, 1986), (Shields & Shields, 2004), (Huerva et al, 2006).

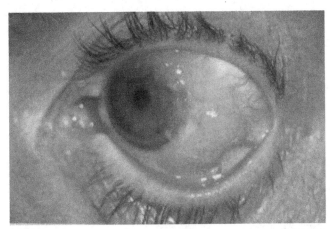

Fig. 4. Fleshy nodular gelatinous mass involving bulbar conjunctiva and limbus.

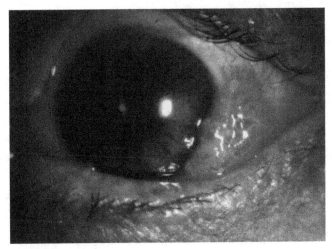

Fig. 5. Fleshy nodular mass at the interpalpebral bulbar conjunctiva.

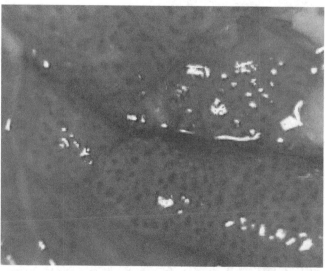

Fig. 6. Fleshy diffuse CIN affecting the inferior fornix resembling a chronic conjunctivitis.

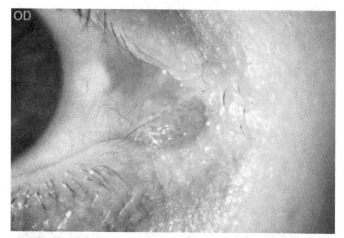

Fig. 7. Nodular fleshy CIN at the caruncle.

3.5 Corneal invasion (Conjunctiva-Cornea Intraepithelial Neoplasia) (CCIN)

This condition is called when the fleshy or papillomatous CIN lesions invading the superficial cornea. The lesions are well documented at the limbus occuping different degrees. Generally, in the extensive cases the cornea may be invaded (Huerva et al, 2006). The form of invasion may be variable: nodular, frothy vascular irregular extension and pedunculated and may simulate other conjunctival lesions as a pterigium or pannus (Shields & Shields, 2008).

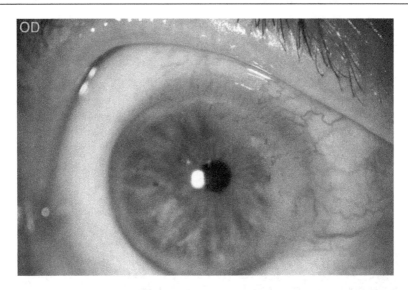

Fig. 8. CCIN: dysplasia in 180 degrees at the corneoscleral limbus resembling a corneal pannus.

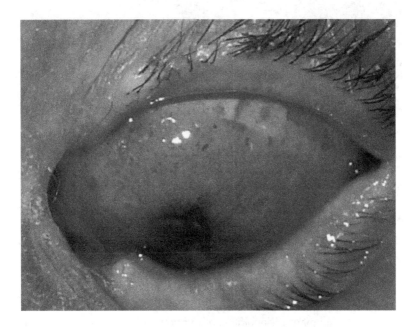

Fig. 9. CCIN. The tumor invade almost 3/4 size of the corneal surface.

3.6 Squamous cell carcinoma

Clinical presentation of invasive SCC is the same that the CIN. As the CIN, it is frequently observed at the interpalpebral region. Definitive diagnosis is only histopathologic (Shileds & Shields, 2008).

There are some different histological types with very aggressive potential effect. Mucoepidermoid or adenoid SSC has an epidermal component and variable quantities of mucin. It often presents with inflammatory signs (Mauriello et al, 1997). Spindle cell SCC is composed by pleomorphic spindle cells. Both are very aggressive with high potential of ocular invasion and distant metastases (Shields & Shields, 2008).

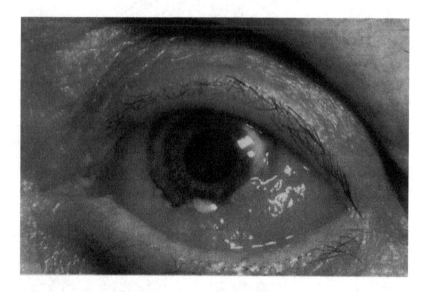

Fig. 10. Diffuse SCC involving the whole bulbar conjunctiva. Leukoplakic plaques are also present. Chronic conjunctivitis may be misdiagnosed. It clinically resembles a diffuse CIN. For definitive diagnosis a incisional biopsy is necessary.

4. Diferential diagnosis

Clinical differentiation between CIN and other limbal lesions is based on characteristic clinical features (Erie et at, 1986). However, it is generally admitted that the grade of dysplasia cannot be consistently determined on clinical inspection alone (Lee & Hirts, 1995). The main differential diagnoses for localized CIN include pinguecula, pterygium and squamous papilloma. The differential diagnosis of conjuctival amelanotic tumors includes CIN, invasive SCC, malignant melanoma and a variety of benign described entities, which include squamous papilloma, solar elastosis and epithelial hyperplasia, keratosis or reactive atypia. Conjunctival pseudoepitheliomatous hyperplasia, keratoacantoma, and conjunctival hereditary benign intraepithelial dyskeratosis may be also considered in CIN diferential diagnosis (Shields & Shields, 2008). In these cases the hyperkeratosis and inflammatory cells are present in the histologic samples. Solar elastosis is a pathognomonic sign in the pathological diagnosis of degenerative diseases of the conjunctiva such as pingueculae and pterygium. In a study, the clinical diagnosis of OSSN may be accurate in 89.5% of the cases (Tulwatana et al, 2003). Solar elastosis was found in 53.3% of OSSN cases compared to 3.3% of matched controls. Solar elastosis has also been identified as a risk factor for OSSN (Tulwatana et al, 2003). On this basis, the clinical diagnosis alone cannot distinguish benign conjuntival limbal tumors from OSSN or reliably exclude, albeit an uncommon diagnosis, amelanotic malignant melanoma. The difficulty in distinguishing clinically between pterygium and CIN was illustrated in a histopathological review of 533 cases of pterygium, in which 9.8% were shown to have evidence of dysplasia (Hirst et al, 2009). The capacity of a clinician to distinguish between grades of CIN, or between CIN and invasive SCC, is also limited (Rudkin et al, 2011). Clinical diagnosis of CIN may be increased by the use of exfoliation or impression cytology. However, histopathology of the excised tumor is the only reliable diagnostic method and it is generally accepted to be the most appropriate approach to lesions presumed to be CIN. The main hazard of clinical misdiagnosis of an excised benign limbal lesion is exposing the patient to unnecessary surgery. For an experienced ocular oncologist, the misdiagnosis of localized limbal OSSN occurs in 10.5% of cases (Rudkin et al, 2011).

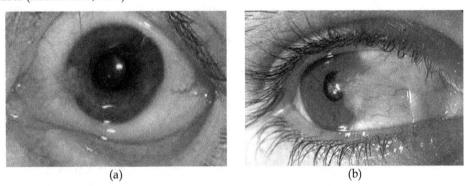

(a)   (b)

Fig. 11. Images of Pterigium (a,b) with corneal invasion may resembling in some a cases a CIN. On the other hand in already of 10 % of the cases may show epithelial conjuntival atypia.

5. Impression cytology in diagnosis of CIN

The management of ocular surface neoplasia depends on the ability to distinguish between benign, preinvasive, and invasive lesions. However, follow-up of suspicious lesions by repeated biopsies may cause complications such as scarring, lid deformity, limbal deficiency, and patient discomfort.

As it has been described, the clinical appearance of the lesion may be suggestive of CIN. However, a tissue biopsy is necessary to confirm the diagnosis. Because many patients with primary or recurrent CIN may be treated with topical chemotherapy without surgical excision of the lesion, impression cytology has been used to confirm the diagnosis without performing histological evaluation of the excisional biopsy. Impression cytology is a simple technique for removing one to three superficial layers of the epithelium by applying collecting devices, either cellulose acetate filter papers or Biopore membrane device. Rates of positivity between 77 and 80% have been reported (Nolan et al, 1994), (Tole et al, 2001). The disadvantage is that the superficial nature of the sample, which sometimes only contains keratinized cells, may be falsely negative. Cytological assessment does not provide enough information regarding the deepest structure of the lesion, in particular, evidence of invasion. Abundance of surface keratin may make sampling inaccurate. Another limitation is that impression cytology may not distinguish in situ from minimally invasive disease, because only superficial cells are collected in the method. However, high-grade dysplasia in OSSN cytology findings have a high correlation with histology findings, and the presence of abundance dysplastic cells in cytology suggest preinvasive or invasive disease in subsequent histology (Tananuvat et al, 2008). Although impression cytology have a high sensitivity for the diagnosis of ocular surface squamous neoplasia, including CIN, there are still cases in which impression cytology yields false negative results. The keratotic surface of the lesion or the presence of dysplastic cells deep within the epithelium are the reason for these false negative results. Repeated consecutive applications of the collecting filter paper to the surface of CIN by approaching the deeper epithelium may result in higher sensitivity of the technique to confirm the diagnosis (Kheirkhah et al, 2011). For the diagnosis of CIN, the second and third applications of impression citology may be significantly more sensitive than the first application. Consecutive repeated applications of the filter paper resulted in a significant higher sensitivity due to access to deeper epithelium. Keratinizing CIN lesions may result in a false negative impression cytology test due to the small number of cells present in the sample. It seems that keratinization leads to more false negative results at first application and repeated sampling in this population of CIN cases is more likely to result in subsequent positive due to the progressive elimination of the keratotic material.

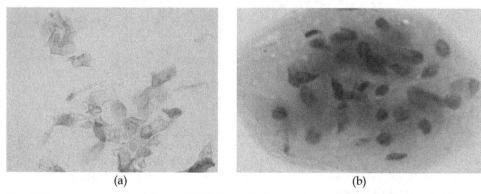

| (a) | (b) |

Fig. 12. Impression cytology from a CIN. Papanicolaou staining. A. False negative for CIN, squamous superficial keratinized material (a). Atypical dysplasic squamous cells from a CIN. Some pleomorphic and hiperchromatic nuclei (b).

In conclusion, repeated consecutive applications of impression citology will lead to a more significant sensitivity for diagnosis in eyes with CIN, thereby obviating the need for excisional biopsy. An additional advantage of impression cytology is the preservation of limbal stem cells, which are located in the basal layer of the limbal epithelium and are responsible for renewal of corneal epithelium throughout life. In most cases of OSSN, the lesions affect predominantly the limbus and have a tendency to recur. Moreover, the technique may be used in the follow-up of patients after treatment to determine the recurrence of the disease, as well as the effects of treatment such as topical chemotherapy.

6. Histopathological findings

The definitive term of CIN or SCC corresponds to the histologic study. When the abnormal conjunctival epithelial cellular proliferation involves only partially the epithelium thickness is classified as mild CIN, a condition also called mild or moderate dysplasia. When it affects full thickness epithelium it is called severe CIN, a condition also called severe dysplasia. In these cases there may be an intact surface layer of cells. Where there are no longer normal surface cells then the process is termed carcinoma-in-situ. Histopathologically, mild CIN (dysplasia) is characterized by a partial thickness replacement of the surface epithelium by abnormal epithelial cells which lack of normal maturation. Severe CIN (severe dysplasia) is characterized by a nearly full-thickness replacement of the epithelium by similar cells. Carcinoma-in-situ represents full thickness replacement by abnormal epithelial cells (Shields & Shields, 2004). Squamous cell carcinoma is an extension of abnormal epithelial cells through the basement membrane to gain access to the conjunctival stroma and have grown in sheets or cords into the stromal tissue. A rare variant of squamous cell carcinoma of the conjunctiva is the mucoepidermoid carcinoma wich presents abundant mucous-secretory cells within cysts. Another rare variety is the spindle cell variant of squamous cell carcinoma that is likewise aggressive. Histopathologically, invasive squamous cell carcinoma is characterized by malignant squamous cell that have violated the basement membrane

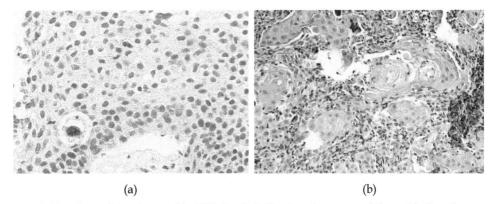

(a)   (b)

Fig. 13. Histological specimens. (a): CIN Grade 3. Total replacement of the epithelium by displasic cells. Hematoxilin-eosin x 40. (b): SCC, Displasic cell islets of squamous cells after invading the basement membrane, presence of keratosic perls. Hematoxilin-eosin x 10.

(Shields & Shields, 2004). According to the definition, CIN may be classified in four stages (Kheirkhah et al, 2011):

- *CIN grade I:* mild dysplasia limited to the basal one third of the thickness of the corneal or conjunctival epithelium.
- *CIN grade II:* moderate dysplasia confined to the basal two thirds of the corneal or conjunctival epithelium.
- *CIN grade III:* or SCC in situ: severe dysplasia that may involve the entire thickness of the corneal or conjunctival epithelium without invading the basement membrane.
- *Invasive SCC:* severe dysplasia with invasion through the basement membrane.

7. Human Papilloma Virus (HPV) detection

As it has been described in the chapter of etiologic factors, the presence of HPV in cases of CIN remains controversial. DNA of HPV may be detectable by in situ hybridization. HPV types 16 and 18, commonly detectable, in uterine cervix may also be detectable in CIN. However, in non neoplasic lesions and in apparently healthy conjunctiva it may also be detectable (Karcioglu & Issa, 1997). In African case series there is a high prevalence of DNA HPV 6 and 11, but not HPV 16 and 18 (Verma et al, 2008). On the other hand, in a series reported recently in Uganda, HPV 5 and 8 were the most common in nearly half of OSSN (Ateenyi-Agaba, et al, 2010). We have detected the presence of DNA HPV type 11. It is possible that different HPV associated to other risk factors may contribute to the development of CIN.

The presence of DNA HPV is not strictly necessary in the diagnosis of CIN. However, when it is possible its determination may clarify the role of these different types of virus in the development of CIN.

8. Staging for conjunctival intraepithelial neoplasia

CIN constitutes a localized malignant situation that, in absence of treatment, may growth progressively with possible transformation into SCC. It develops rarely metastases at distance or produces ocular, orbital or intraccraneal invasion. The clinical TNM classification of the conjunctival carcinoma is as follows (McGowan, 2009) :

Clinical classification (TNM):

Primary tumor (T)

TX Primary tumor cannot be assessed

T0 No evidence of primary tumor

Tis Carcinoma *in situ*

T1 Tumor 5 mm or less in greatest dimension

T2 Tumor more than 5 mm. in greatest dimension, without invasion of adjacent structures

T3 Tumor invades adjacent structures (excluding the orbit)

T4 Tumor invades the orbit with or without further extension

T4a Tumor invades orbital soft tissues, without bone invasion

T4b Tumor invades bone

T4c Tumor invades adjacent paranasal sinuses

T4d Tumor invades brain

Regional lymph nodes (N)

NX Regional lymph nodes cannot be assessed

N0 No regional lymph node metastasis

N1 Regional lymph node metastasis

Distant metastasis (M)

MX Distant metastasis cannot be assessed

M0 No distant metastasis

M1 Distant metastasis

According to this classification with the difference of clinical appearance, the cases of CIN may be: Tis, T1 or T2, N0 and M0.

9. Treatment

The management of CIN or SCC of the conjunctiva varies with the extent or recurrence of the lesion.

9.1 Surgical treatment

While the extent of the lesion determines the management of lesions in the limbal area involves alcohol epitheliectomy for the corneal component and partial lamellar scleroconjunctivectomy, with wide margins (4-5 mm) for the conjunctival component followed by freeze-thaw cryotherapy to the remaining adjacent bulbar conjunctiva (The no touch technique) (Shields & Shields, 2004). In some cases, microscopically controlled excision (Mohs surgery) may be performed at the time of surgery to ensure tumor free margins (Buus et al, 1994). Those tumors in the forniceal region can be managed by wide local resection and cryotherapy. Following surgical excision, large conjunctival defects may be successfully reconstructed with transpositional conjunctival flaps, free conjunctival grafts, oral mucosal grafts, and amniotic membrane grafts (Gündüz et al, 2006). In all cases, the full conjunctival component along with the underlying Tenon's fascia should be excised using the "no touch technique". A thin lamella of underlying sclera should be removed, in the limbal región, when the tumor is adherent to the globe. Intraoperative mitomycine-C (MMC) application has also been combined with excision of ocular surface neoplasia to prevent postoperative recurrences (Siganos et al, 2002). However, studies show a 53%

recurrence rate in pathologic studies which revealed involved margins and a 5% recurrence rate when clear margins are confirmed (Erie et al, 1986). In extensive lesions, surgical excision is difficult, and additional procedures have been employed. Extensive resections in very extensive CIN may produce a limbal stem deficiency (Huerva et al, 2006). Adjuvant radiation has the potential complications of cataracts, scleral necrosis, corneal rupture, scarring of the cornea and conjunctiva, moderate to severe conjunctivitis, and loss of eyelashes (Giaconi & Karp, 2003). For those patients with extensive tumors or those tumors that are recurrent, treatment with topical mitomycin C, 5-fluorouracil, or interferon alfa 2b have been employed.

9.2 Topical chemoteraphy

Topical chemotherapy has a number of advantages over surgical approach. It enables to treat the entire ocular surface and is not dependent upon surgical margins. Primary treatment with a chemotherapeutic agent avoids potential complications of surgery, which can include scarring of the conjunctiva and cornea, limbal stem cell failure and incomplete excision of the lesion. Topical chemotherapics may be preferred over surgery by some patients, and when the patient refuse surgery, topical chemotherapics have been successfully used as primary treatment.

9.2.1 Topical mitomycin C (MMC)

For tumors with extensive involvement, where surgical removal bears significant risks for postoperative problems, topical MMC should have been considered for a long time. Topical MMC 0.02% or 0.04% 4 times daily in 7 to 14-day for two cycles (Shields & Shields, 2004) have been successfully employed for preoperative chemoreduction and to manage recurrent and residual tumors following surgical resection (Shields et al, 2002), (Frucht-Pery et al, 2002), (Shields et al,2005). MMC had been effectively used to treat primary CIN, with reported success rates between 85% (Wilson et al, 1997) and 100% (Frucht-Pery & Rozenmam, 1994), (Ramos-Lopez et al, 2004). Another large study has shown topical MMC to be an efficient treatment of most, but not all cases, of CIN. Tumor regrowth occurred in approximately 17% of cases (Frucht-Pery et al,1997). To avoid possible complications, the lacrimal punctal occlusion is mandatory during topical treatment. Chemoreduction with MMC cycles reduced the tumor size, especially in the surrounding thinner portions, and allowed for a subsequent limited surgical excision in all cases (Shields et al, 2005). Possible complications with topical MMC include superficial punctate epitheliopathy (Shields & Shields, 2004), conjunctival hyperemia, pain, allergy, corneal-scleral, melting disturbance of tear film stability, goblet cell loss, squamous metaplasia and limbal stem cells depletion (Frucht-Pery & Rozenmam, 1994), (Wilson et al, 1997), (Dogru et al, 2003), (Dudney & Malecha, 2004), (Khong & Muecke, 2006). Edema and endothelial apoptosis have been observed in experimental models (Chang, 2004). MMC toxicity seems to be dose dependent, occurring with the repetition of treatment cycles. Chemoreduction with topical MMC, followed by interferon alfa 2b (1 million IU/mL) 4 times daily, is an effective treatment in extensive CIN cases where surgical resection with safety margins is infeasible and corneal extension resection and the repetitive cycles of MMC adjunctive could cause a depletion of limbal stem cells and other commented side effects on the ocular surface (Huerva et al,

2006). In a follow-up of 18 months, topical Cyclosporine A (0,05%) combined with topical low dose of MMC (0,01%) four times a day for 12 weeks after positive margins following surgical excision showed no recurrence of the tumor (Tunc & Erbilen, 2006). In these cases Cyclosporine A has been employed by the antineovascular effect on the ocular surface.

9.2.2 5-Fluoracil (5-FU)

Other treatment options in the management of CIN include 5-fluorouracil (5-FU). However, compared with MCC, the experience with this alternative treatment is limited. Topical 1% 5-FU drops used 4 times daily for 2 to 4 days for each cycle and repeated at 30 to 45 day intervals have been reported. Following initial treatment, 4 patients were disease-free with a mean follow-up of 18.5 months. Of the 3 patients with tumor recurrence, 2 remained tumor-free following additional topical 5-FU treatment and 1 patient had a persistent tumor despite additional treatment with 5-FU and became tumor-free following treatment with topical MMC (Yeatts et al, 2000). No adverse reactions to pulsed treatment were reported. Another study using topical 1% 5-FU drops 4 times daily for 4 weeks in 8 eyes with recurrent, incompletely excised, and untreated conjunctival OSSN showed complete clinical regression at 3 months in all cases. OSSN recurred in 1 patient at 6 months but this was successfully treated with another course of 5-FU (Midena et al, 2000). Transient toxic keratoconjunctivitis that was noticeable with this treatment. Short-term complications include lid toxicity in 52% of patients, keratopathy in 11% and epiphora in 5% (Rudkin et al, 2010).

9.2.3 Interferon (INF) alpha 2b

Topical MMC and 5-fluorouracil have been used to reduce recurrence rates when used as an adjunct to surgical escisión and as a primary treatment; however, their use can be associated with marked ocular surface toxicity. Topical (1.000.000 IU/ ml/ four times a day) or subconjuctival INF alfa 2b (3 million IU/ml/ weekly) have been employed to treat CIN. In general, topical INF alpha-2b is well tolerated. Subconjunctival administration presents more side effects as flu-like symptoms (fatigue, fever, myalgias, malaise) and mild liver disturbances (Huerva & Mangues, 2008). Local conjunctival injection and follicular conjunctivitis are the most frequently reported side effects (Schechter et al, 2002) after topical administration. Redness and increase of CIN volume without ocular discomfort have been reported in a case (Huerva el al, 2007). Fine, diffuse, clear epithelial microcysts in the cornea after instillation of topical interferon a-2b have recently documented in other case (Aldave & Nguyen, 2007).

Topical INF alpha 2-b, sometimes combined with subconjunctival INF alpha 2-b, seems to be effective as primary treatment for CIN, in recurrent cases, and also in retreatment after recurrence when INF has been used previously for a short period of time (Huerva & Mangues, 2008). Approximately, 9% of CIN treated with subconjunctival and/or topical INF alpha 2b showed recurrences, and 33 % of them were successfully retreated with topical IFN alpha 2b (Huerva & Magues, 2008). Another one (16,6%) achieved complete remission after intraperioperative MMC (Hawkins et al, 1999). For INF alpha 2b topical treatment, the average time to complete tumor response is 11 weeks (range 2-59). For INF alpha 2b subconjunctival and topical treatment, the average time to complete tumor response is 5.5 weeks (range 2-12), (Huerva & Mangues, 2008). Previous studies found the same observation (Karp et al, 2001). The time to clinical resolution using topical INF alpha 2-b was longer (11.6 weeks) that the combined intralesional and topical interferon (4.5 weeks), but that INF alpha 2b treatment involved fewer side effects. In general, it seems that the disadvantage with topical treatment is the long duration. We must emphasize the importance of long term follow-up for CIN patients because recurrences can occur anywhere from 33 days to 11.5 years (Tabin et al, 1997), although most recurrent CIN occurs within 2 years of initial excision (Schechter et al, 2005).

Many surgeons add adjunctive topical therapy to their surgical regimens for larger lesions (Stone el al, 2005). However, all sizes of lesions could be treated with topical INF alpha as the primary treatment because it is an effective, non-invasive treatment alternative to surgery that increases quality of life with low costs (Huerva et al, 2006), (Huerva et al, 2007), (Huerva et al, 2009). Actually, no clear consensus on the best way to manage the disorder has been established, because long-term, well designed studies are still needed. However, two recent studies have addressed the above questions and confirmed the effectiveness of this topical therapy for CIN. The first study (Schechter, et al, 2008) demonstrated total resolution of the tumor in 96.4% of cases treated with INF alfa 2b with a mean follow-up of 42.4 months. The second study (Sturges et al, 2008) demonstrated that topical treatment with INF and surgical excision have the same effectiveness as primary treatment for CIN for a mean follow-up of 35.6 months. The authors concluded that topical IFN alfa-2b and aggressive surgical excision can be considered equally effective as first choice for treating CIN. Topical INF alfa-2b has some advantages over conventional excision, including the reduction of risk to loose limbar stem cells secondary to surgical trauma and, thus, compromising the integrity of the ocular surface. This therapeutic mode can be recommended particularly for patients who reject any type of surgery, or mentally retarded patients in whom surgery is complicated as well as extended cases where an aggressive excision could cause the loss of limbar stem cells (Huerva, 2008).

Topical INF or subconjunctival INF remains a controversial issue. A recent report (Karp et al, 2010) concluded that subconjunctival 0.5 ml injection of 3 million IU IFN alfa 2b is a viable medical alternative for the treatment of ocular surface squamous neoplasia (OSSN) with a mean duration of follow-up of 55 months. The authors state that the advantages of perilesional INF alfa 2b injection include more rapid tumor resolution, ensured compliance, and perhaps more direct delivery to the tumor site when compared with topical INF drops. However, some patients may be apprehensive about receiving injections around the eye and may prefer eyedrops. A single weekly injection of INF may have better compliance than 4 eye-drops per day dosing for a mean of three months in many patients. Direct delivery to the tumor site may occur in well-localized lesions, while annular lesions or multifocal disease requires injection over the entire involved area, increasing the risk of conjunctival hemorrhage. By contrast, topical therapy is delivered to the entire ocular surface and has very good success rates. Topical therapy could be recommended for patients who reject any surgical procedure or those who are apprehensive about injections.

Weekly subconjunctival Pegilated INF alpha 2b might be an alternative in resistant cases of CIN or recurrent conjunctival papillomatosis avoiding a mutilating surgery (Tseng, 2009), (Karp et al, 2010).

9.2.4 Other treatment possibilities

Other treatment options in the management of conjuctival OSSN include topical retinoids, cidofovir and photodynamic therapy (PDT). Topical unguent of trans-reinoic acid (0,01%) showed complete resolution of CIN in 20% of cases, whereas 40% showed only partial response (Espana el al, 2003). This treatment may be then only adjuvant to surgery

Regression of diffuse conjunctival CIN was demonstrated following a 6 week course of topical cidofovir eyedrops (2.5 mg/ml) with later residual lesion after surgical excision (Sherman et al, 2002).

Following PDT, using verteporfin, a complete clinical CIN regression, supported with angiographic evidence, has been reported at 1 month, without any recurrence for a mean follow-up of 8.6 months (Barbazetto et al, 2004). Likewise, histopathological evidence showing tumor regression following treatment with PDT in a patient with in situ CIN has been reported (Sears et al, 2008).

10. References

Aldave AJ, Nguyen A. Ocular surface toxicity associated with topical interferon a-2b. Br J Ophthalmol 2007;91: 1087-88.

Ateenyi-Agaba C. Conjunctival squamous cell carcinoma associated with HIV infection in Kampala, Uganda. Lancet 1995; 345: 695-96.

Ateenyi-Agaba C, Franceschi S, wabwire-Mangen F et al. Human papillomavirus infection and squmaus cell carcinoma of the conjunctiva. Br J Cancer 2010; 102: 262-67.

Bahattacharyya N, Wenokur RK, Rubin PA. Metastasis of squamous cell carcinoma of the conjunctiva. Case report and review of the literature. Am J Otolaryngol 1997; 18: 217-19.

Barbazetto IA, Lee TC, Abramson DH. Treatment of conjunctival squamous cell carcinoma with photodynamic therapy. *Am J Ophthalmol* 2004; 138: 183-89.

Buus DR, Tse DT, Folberg R, Buuns DR. Microscopically controlled excision of conjunctival squamous cell carcinoma. *Am J Ophthalmol* 1994; 117: 97-102.

Chang SW. Early corneal edema following topical application of mitomycin-C. J Cataract Refract Surg 2004; 30: 1742-50.

Chisi SK, Kollmann MKH, Karimurio J. Conjunctival squamous cell carcinoma in patients with Human Immunodeficiency Virus infection seen at two hospitals in Kenya. East African Med J 2006; 83: 267-70.

Clifford GM, Gallus S, Herrero R, Munoz N, Snijders PJF, Vaccarella S, et al. Worldwide distribution of human papillomavirus types in cytologically normal women in the International Agency for Research on Cancer HPV prevalence surveys: a pooled analysis. Lancet 2005; 366: 991-98.

Dogru M, Erturk H, Shimazaki J, Tsubota K, Gul M. Tear function and ocular surface changes with topical mitomycin (MMC) treatment for primary corneal intraephitelial neoplasia. Cornea 2003; 22: 627-39.

Dudney BW, Malecha MA. Limbal stem cell deficiency following topical mitomycin C treatment of conjunctival-corneal intraephitelial neoplasia. Am J Ophthalmol 2004; 137: 950-51.

Eng HL, Lin TM, Chen SY, et al. Failure to detect human papillomavirus DNA in malignant epithelial neoplasms of conjunctiva by polymerase chain reaction. Am J Clin Pathol 2002;117: 429–36.

Erie JC, Campbell RJ, Leisegang TJ. Conjunctival and corneal intraepithelial and invasive neoplasia. Ophtalmology 1986; 93:176-83.

Espana EM, Chodosh J, Mateo AJ, Di Pascuale MA, Tseng SC. Topical retinoids as a noninvasive treatment of conjunctival intraepithelial neoplasia. Microcirugía Ocular 2003, nº4: 1-7.

Frucht-Pery J, Rozenmam Y. Mitomycin C therapy for corneal intraepithelial neoplasia. Am J Ophthalmol 1994; 117: 164-68.

Frucht-Pery J, Sugar J, Baum J et al. Mitomycin C treatment for conjunctival-corneal intraepithelial neoplasia: a multicenter experience. Ophthalmology 1997; 104: 2085-93.

Frucht-Pery J, Rozenman Y, Pe'er J. Topical mitomycin-C for partially excised conjunctival squamous cell carcinoma. Ophthalmology 2002; 109: 548-52.

Giaconi JA, Karp CL. Current treatment options for conjunctival and corneal intraepithelial neoplasia. Ocul Surf 2003;1: 66-73.

Grossniklaus HE, Green WR, Luckenbach M, Chan CC. Conjunctival lesions in adults. A clinical and histopathologic review. Cornea 1987; 6:78-116.

Gündüz K, Uçakhan OO, Kanpolat A, Günalp I. Nonpreserved human amniotic membrane transplantation for conjunctival reconstruction after excision of extensive ocular surface neoplasia. Eye 2006; 20: 351-57.

Hawkins AS, Yu J, Hamming NA, Rubenstein JB. Treatment of recurrent conjunctival papillomatosis with mytomycin C. Am J Ophthalmol 1999; 128:638-40.

Herle RW, Durso F, Metzler JP, Varsa EW. Epibulbar squmaous cell carcinomas in brothers with xeroderma pigmantosa. J Pediatr Ophthalmol Strabismus 1991; 28. 350-53.

Hirst LW, Axelsen RA, Schwab I. Pterygium and associated ocular surface squamous neoplasia. Arch Ophthalmol 2009; 127: 31-2.

Huerva V, Mateo AJ, Mangues I, Jurjo C. Short-term mitomycin C followed by long-term interferon alpha 2β for conjunctiva-cornea intraepithelial neoplasia. Cornea 2006; 25:1220-23.

Huerva V, Sánchez MC, Mangues I. Tumor-volume increase at beginning of primary treatment with topical interferon alpha 2-beta in a case of conjunctiva-cornea intraepithelial neoplasia. J Ocul Pharmacol Ther 2007;23:143-45.

Huerva V, Mangues I. Treatment of conjunctival squamous neoplasias with interferon alpha 2b. J Fr Ophtalmol 2008; 31: 317-25.

Huerva V. Topical interferon alfa-2b or surgical excision for primary treatment of conjunctiva-cornea intraepithelial neoplasia. Arch Soc Esp Oftalmol 2009; 84: 5-6

Huerva V, Mangues I, Schoenenberger JA. Interferon alpha 2b eyedrops as treatment of conjunctival intraepithelial neoplasia. Farm Hosp 2009; 33: 335-36.

Kaimbo WA, Kaimbo D, Parys-Van Ginderdeuren R, Missotten L. Conjunctival squamous cell carcinoma and intraepithelial neoplasia in AIDS patients in Congo Kinshasa. Bull Soc Belge Ophtalmol 1998; 268: 135–41.

Karcioglu ZA, Issa TM. Human papillomavirus in neoplastic and non-neoplastic conditions of the external eye. Br J Ophthalmol 1997; 81: 595-598.

Karp CL, Scott IU, Chang TS, Pflugfelder SC. Conjunctival intraepithelial neoplasia: a possible marker for human immunodeficiency virus infection ?. Arch Ophthalmol 1996; 114: 257-61.

Karp CL, Moor JK, Rosa RH Jr. Treatment of conjunctival and corneal intraepithelial neoplasia with topical interpferon alpha-2b. Ophthalmology 2001; 108: 1093-98.

Karp CL, Galor A, Chhabra S, Barnes SD, Alfonso EC. Subconjunctival/Perilesional recombinant interferon α2b for ocular surface squamous neoplasia. Ophthalmology 2010; 117: 2241-46.

Karp CL, Galor A, Lee Y, Yoo SH. Pegylated interferon alpha 2b for treatment of ocular surface squamous neoplasia: a pilot study. Ocul Immunol Inflamm 2010;18:254-60.

Kheirkhah A, Mahbod M, Farzbod F, Zavareh MK, Behrouz MJ, Hashemi H. Repeated applications of impression cytology to increase sensitivity for diagnosis of conjunctival intraepithelial neoplasia. Br J Ophthalmol. 2011 Apr 15. [Epub ahead of print].

Khong JJ, Muecke J. Complications of mitomycin C therapy in 100 eyes with ocular surface neoplasia. Br J Ophthalmol 2006; 90: 819-22.

Lee GA, Hirst LW. Ocular surface squamous neoplasia. Surv Ophthalmol 1995;39: 429-50.

Lee GA, Williams G, Hirst LW, Green AC. Risk factors in the development of ocular surface epithelial dysplasia. Ophthalmology 1994; 101: 360-64.

Lee GA, Hirst LW. Retrospective study of ocular surface squamous neoplasia. Austr New Zealand J Ophthalmol 1997; 25: 269-76.

López García JS, Elosúa de Juan I, González Morales ML, de Pablo Martín C, Alvarez Lledo J, Martínez Garchitorena J. Squamous cell carcinoma of the conjunctiva with orbital invasion. Arch Soc Esp Oftalmol. 2000;75: 637-41.

Mahomed A, Chetty R. Human immunodeficiency virus infection, Bcl-2, p53 protein, and Ki-67 analysis in ocular surface squamous neoplasia. Arch Ophthalmol 2002; 12: 554-8.

Mauriello JA Jr, Napolitano J, McLean IW. Actinic keratosis and dysplasia of the conjunctiva: a clinicopatological study of 45 cases. Can J Ophthalnmol 1996; 30: 312-16.

Mauriello JA, Abdelsalam A, McLean IW. Adenoid aquamous carcinoma of the conjunctiva – a clinicapathological study of 14 cases. Br J Ophthalmol 1997; 81: 1001-05.

McGowan HD. Squamous Neoplasia of the Conjunctiva: The New TNM Classification by the American Joint Committee on Cancer (AJCC). Ophthalmology Rounds 2009; 7 (1) : 130-38E.

McKelvie PA, Daniell M, McNab A, Loughnan M, Santamaria JD. Squamous cell carcinoma of the conjunctiva: a series of 26 cases. British J Ophthalmol 2002; 86: 168-73.

Midena E, Angeli CD, Valenti M, de Belvis V, Boccato P. Treatment of conjunctival squamous cell carcinoma with topical 5-fluorouracil. Br J Ophthalmol 2000; 84: 68-72.

Napora C, Cohen EJ, Genvert GI, et al. Factors associated with conjunctival intraepithelial neoplasia: a case control study. Ophthalmic Surg 1990; 21: 27-30.

Nolan GR, Hirst LW, Wright RG, et al. Application of impression cytology to the diagnosis of conjunctival neoplasms. Diagn Cytopathol 1994;11: 246–49.

Palefsky JM. Human papillomavirus- associated anogenital neoplasia and other solid tumors in human immunodeficiency virus-infected individuals. Curr Opin Oncol 1991; 3. 881- 85.

Pizzarello ID, Jakobiec FA (1978). Bowen´s disease of the conjunctiva: a misnorner, In: *Ocular and adnexal tumors*, Jakobiec FA (ed), pp. (553-571), Al. Aesculapius, Birmingham.

Pola EC, Masanganise R, Rusakaniko S. The trend of ocular surface squamous neoplasia among ocular surface tumour biopsies submitted for histology from Sekuru Kaguvi Eye Unit, Harare between 1996 and 2000. Central African Journal of Medicine 2003 ;49:1-4.

Poole TR. Conjunctival squamous cell carcinoma in Tanzania. British J Ophthalmol 1999; 83: 177-9.

Porges Y, Groisman GM. Prevalence of HIV with conjunctival squamous cell neoplasia in an African provincial hospital. Cornea 2003;22: 1-4.

Ramos-Lopez JF, Martinez-Costa R, Cisneros-Lanuza AL, et. al. Treatment of conjunctival intraephitelial neoplasia with topical mitomycin C 0,02%. Arch Soc Esp Oftalmol 2004; 79: 375-78.

Rudkin AK, Dodd T, Muecke JS. The differential diagnosis of localised amelanotic limbal lesions: a review of 162 consecutive excisions. Br J Ophthalmol. 2011; 95: 350-54.

Schechter BA, Schrier A, Nagler RS, Smith EF, Velasquez GE. Regression of presumed primary conjunctival and corneal intraepithelial neoplasia with topical interferon alpha-2b. Cornea 2002; 21: 6-11.

Schechter BA, Nagler RS, Schrier A. Recurrent intraepithelial neoplasia treatment. Ophthalmology 2005, 112:1319.

Schechter BA, Koreishi AF, Karp CL, Feuer W. Long-term follow-up of conjunctival and corneal intraepithelial neoplasia treated with topical interferon alfa-2b. Ophthalmology 2008; 115: 1291-296.

Saornil MA, Becerra E, Méndez MC, Blanco G. Conjunctival tumors. Arch Soc Esp Oftalmol. 2009; 84: 7-22.

Scott IU, Karp CL, Nuovo GJ. Human papillomavirus 16 and 18 expression in conjunctival intraepithelial neoplasia. Ophthalmology 2002;109: 542-7.

Sears KS, Rundle PR, Mudhar HS, Rennie IG. The effects of photodynamic therapy on conjunctival in situ squamous cell carcinoma--a review of the histopathology. Br J Ophthalmol 2008; 92: 716-17.

Sherman MD, Feldman KA, Farahmand SM, Margolis TP. Treatment of conjunctival squamous cell carcinoma with topical cidofovir. Am J Ophthalmol 2002; 134: 432-33.

Shields CL, Naseripour M, Shields JA: Topical mitomycin C for extensive, recurrent conjunctival-corneal squamous cell carcinoma. Am J Ophthalmol 2002; 133: 601–06.

Shields CL, Shields JA. Tumors of the conjunctiva and cornea. Surv Ophthalmol 2004; 49: 3-24.

Shields CL, Demicri H, Karatza E, et al. Clinical survey of 1643 melanocytic and nonmelanocytic tumors of the conjunctiva. Ophthalmology 2004; 111: 1747-54.

Shields CL, Demirci H, Marr BP, et al. Chemoreduction with topical Mytomycin C prior to resection of extensive squamous cell carcinoma of the conjuntiva. Arch Ophthalmol 2005; 123: 109-13.

Shields CL, Manchandia A, Subbiah R, Eagle RC Jr, Shields JA. Pigmented squamous cell carcinoma in situ of the conjunctiva in 5 cases. Ophthalmology 2008; 115: 1673-78.

Shields JA, Shields CL. (2008). Eyelid, Conjunctival, and Orbital Tumors. Wolters Kluwer, Lippincott, Williams & Wilkins, Philadelphia.

Siganos CS, Kozobolis VP, Christodoulakis EV. The intraoperative use of mitomycin-C in excision of ocular surface neoplasia with or without limbal autograft transplantation. Cornea.2002; 21:12-16.

Stone DU, Butt AL, Chodosh J. Ocular surface squamous neoplasia. Cornea 2005; 24:297-300.

Sturges A, Butt AL, Lai JE, Chodosh J. Topical interferon or surgical excision for the management of primary ocular surface squamous neoplasia. Ophthalmology 2008; 115: 1297-1302.

Sun EC, Fears TR, Goedert JJ. Epidemiology of squamous cell conjunctival cancer. Cancer Epidemiol Biomarkers Prev. 1997;6: 73-77.

Tabin G, Levin S, Snibson G, Loughnan M, Taylor H. Late recurrences and the necessity for long-term follow-up in corneal and conjunctival intraepithelial neoplasia. Ophthalmology 1997; 104:485-92.

Tananuvat N, Lertprasertsuk N, Mahanupap P, Noppanakeepong P. Role of Impression Cytology in Diagnosis of Ocular Surface Neoplasia. Cornea 2008; 27: 269–74.

Timm A, Stropahl G, Schittowski M, et al. Association of malignant tumors of the conjunctiva and HIV infection in Kinshasa (D.R. Congo). First results. Ophthalmologe 2004; 101:1011–16.

Tole D, MecKelvie P, Daniell M. Reliability of impression cytology for the diagnosis of ocular surface squamous neoplasia employing the Biopore membrane. Br J Ophthalmol. 2001;85: 154–58.

Tseng SH. Conjuctival papilloma. Ophthalmology 2009; 116: 1013.

Tulwatana W, Bhattarakosol P, Sansopha L, et al. Risk factors for conjunctival squamous cell neoplasia: a matched case-control study. Br J Ohthalmol 2003; 87: 396-98.

Tunc M, Erbilen E. Topical Cyclosporine-A combined with Mitomycin C for conjunctival and corneal squamous cell carcinoma. Am J Ophthalmol 2006; 142: 673-75.

Tuppurainen K, Raninen A, Kosunen O, et al. Squamous cell carcinoma of the conjunctiva. Failure to demonstrate HPV DNA by in situ hybridization and polymerase chain reaction. Acta Ophthalmol (Copenh) 1992; 70: 248–54.

Verma V, Defan, S, Sieving P, Chan CC. The role of infectious agents in the etiology of ocular adnexal neoplasia. Surv Ophthalmol. 2008; 53: 312-331.

Waddell KM, Lewallen S, Lucas SB, Atenyi-Agaba C, Herrington CS, Liomba G. Carcinoma
        of the conjunctiva and HIV infection in Uganda and Malawi.. Br J Ophthalmol
        1996;80: 503-8.

Wilson MW, Hungerford JL, George SM, Madreperla SA. Topical Mitomycin C for the
        treatment of conjunctival and corneal epithelial dysplasia and neoplasia. Am J
        Ophthalmol 1997; 124: 303-11.

Yeatts RP, Engelbrecht NE, Curry CD, Ford JG, Walter KA. 5-Fluorouracil for the treatment
        of intraepithelial neoplasia of the conjunctiva and cornea. Ophthalmology 2000;
        107:2190-95.

4

Excess Fibroblast Growth Factor-7 (FGF-7) Activates β-Catenin and Leads to Ocular Surface Squamous Neoplasia in Mice

Chia-Yang Liu and Winston W.-Y. Kao
Edith J. Crawley Vision Research Center/Department of Ophthalmology,
College of Medicine/University of Cincinnati. Cincinnati OH
USA

1. Ocular surface squamous neoplasia

Human ocular surface squamous neoplasia (OSSN) is the most common ocular surface pre-cancerous and cancerous lesion previously known by various names such as conjunctival intraepithelial neoplasia, corneal intraepithelial neoplasia (CIN), or both together (CCIN) (Grossniklaus et al., 1987). Clinically, OSSN manifests in different grades ranging from simple dysplasia to squamous cell carcinoma (Grossniklaus et al., 1987). Because of the high incidence of OSSN in the limbal area, where the corneal epithelial stem cells are located, the limbal transition zone/stem cell theory has been proposed for the development of CIN by Lee and Hirst (Lee and Hirst, 1995). Tseng and co-investigators have suggested that the slow cycling limbal stem cells may become hyper-proliferative by stimulations such as alterations in this anatomic site influenced by other factors, e.g., carcinogens, irradiation (eg, UVB), and the phorbol ester tumor promoter, 12-O-tetradecanoylphorbol 13-acetate (TPA) (Tseng, 1989), which can cause abnormal proliferation of the conjunctival and corneal epithelium and lead to the formation of CIN. Nevertheless, the etiology and pathogenesis of CIN and ocular surface carcinoma remain elusive. To date, there is no appropriate animal model available to study the molecular and cellular mechanisms of this disease. Therefore, the availability of such animal model will not only aid to understand the pathogenesis but also yield a more effective treatment for OSSN.

2. Generation *Krt12*^{rtTA}/*tet-O-FGF-7* bi-transgenic mice and induction of FGF-7 overexpression by doxycycline

It has been well documented that the mouse *Krt12* gene expression is restricted to the differentiated corneal epithelium (Liu et al. 1993, 1994). To generate a corneal epithelium-specific and Dox-inducible transgenic driver mouse line, we have genetically introduced ("knock-in") an ires-rtTA (internal ribosome entry site-reverse tetracycline transactivator) cDNA into the 3'-untranslated region of the mouse *Krt12* gene locus via conventional gene-targeting techniques (Chikama et al., 2005). The resulting transgenic mouse line was designated as *Krt12*^{rtTA}, in which like K12 expression pattern, the rtTA is constitutively and specifically expressed by the corneal epithelium. The *Krt12*^{rtTA} mouse line was then crossed with the *tetO-FGF-7* mouse line (a gift from Dr. Jeffrey Whitsett, Cincinnati Children's Hospital Medical Center, Tichelaar et al., 2000) to generate the *Krt12^{rtTA/rtTA}/tet-O-FGF-7* bi-transgenic mouse strain. To induce FGF-7 expression, mice were injected once intra-peritoneally with Dox (80 µg/g body weight; Clontech Laboratories) dissolved in PBS (pH 7.4) at a concentration of 10 mg/ml and fed Dox-chow (1 g/Kg chow, Bioserv, Frenchtown, NJ). Control animals were fed regular chow. As shown in Figure 1, FGF-7 over-expression can be induced by doxycycline (Dox) induction and results in a phenotype resembling OSSN (Chikama et al., 2008). This bi-transgenic mouse line may serve as an animal model for understanding the relationship between signaling pathway and the pathological progression of this disease.

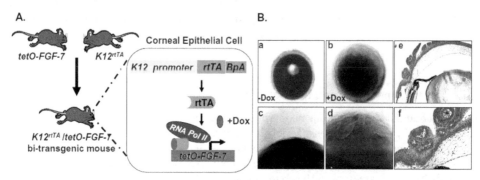

Fig. 1. **Over-expression of FGF-7 resulted in OSSN in Cornea.** A). Diagram showing that corneal epithelium-specific induction of FGF-7 by Dox in *Krt12^{rtTA}/rtTA/tetO-FGF-7* bi-transgenic mice. B) *K12^{rtTA}/tetO-FGF-7* bi-transgenic mice exposed to Dox through mother fed doxycycline chow in the dam since post-nantal day1 (P1) showed corneal intra-epithelial neoplasia at P21 (Bb, Bd) compared to age-matched non-induced mice (Ba, Bc). Papilloma-like epithelial lesion with mesenchymal invasion was found mainly in the peripheral/limbal region (Be, Bf).

3. FGF-7 over-expression and ocular surface carcinogenesis

FGF-7 is a potent mitogen for epithelial cells (Panos et al., 1993; Rubin et al., 1989). The pattern of expression of FGF-7 and its receptor suggest that FGF-7 serves as a paracrine produced by mesenchymal cells in modulating epithelial cells during embryonic development and the maintenance of homeostasis in adults (Finch et al., 1995). FGF-7 enhances epithelial cell proliferation in various organs (Finch et al., 1989, Rubin et al., 1995). Interestingly, aberrant up-regulation of FGF-7 has been reported to be associated with many human neoplastic tumors of epithelial cell origin (Cho et al., 2007; Manavi et al., 2007; Hishikawa et al., 2004; Mehta et al., 2000; Kovacs et al., 2006; Niu et al., 2007). Human papilloma virus 16 (HPV16) and long-term UV irradiation are the major risk factors for corneal intraepithelial neoplasia (Napora et al., 1990). Interestingly, it has been demonstrated that FGF-7 level within the cancer lesion was elevated throughout the progression of multi-stage epidermal carcinogenesis in *K14-HPV16* transgenic mice (Arbeit et al., 1996; Pietras et al., 2008). It has been reported that the exposure to UVB irradiation can induce a rapid intracellular production of ROS (reactive oxidative stress), which in turn is capable of triggering phosphorylation and activation of the FGF-7 receptor, FGFR2-IIIb, similar to those induced by FGF-7 (Marchese et al., 2003). These results lead to our hypothesis that aberrant activation of FGF-7 signaling pathway(s) may be accountable for tumorigenesis derived from limbal stem cells that undergo oncogenic transformation by insults such as long-term UVB exposure and/or infection of HPV etc, which exhibit the characteristic phenotypes of OSSN (Scott et al., 2002; Karp et al., 1996; Kiire et al., 2006). The FGF-7/FGFR signaling is likely the hub that integrates the input through UVB and HPV with the genesis and formation of OSSN (Figure 2). This may explain why excess FGF-7 caused OSSN phenotype in the Dox-treated *Krt12^{rtTA/rtTA}/tet-O-FGF-7* bi-transgenic mouse model (Chikama et al., 2008).

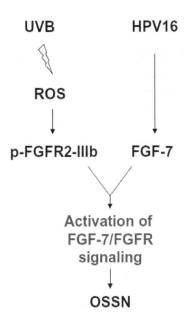

Fig. 2. Hypothetical schema showing that FGF-7/FGFR2 signaling can be activated by UVB and/or HPV type 16, two major risk factors for human OSSN. HPV16 transgene was known to be able to up-regulate FGF-7 (Artbeit et al., 1996). On the other hand, UVB can induced intracellular ROS which in turn phosphorylated and activated FGFR2-IIIb (Marchese et al., 2003).

The precise spatio-temporal expression of FGF-7 is important for ocular surface tissue morphogenesis. FGF-7 are secreted by mesenchymal cells, which bind with high affinity to the same FGF receptor 2 (FGFR2-IIIb) isoform expressed mainly by the epithelial cells (Igarashi et al., 1998). In cornea, expression of FGF-7 and its cognate receptor FGFR2-IIIb is higher in limbal stroma and epithelium, respectively, than in the central cornea, implicating that FGF-7 may promote limbal stem cell proliferation and participate in modulation of corneal epithelium renewal and homeostasis (Li and Tseng, 1996, 1997). However, excess FGF-7 are capable of altering epithelial fates during embryonic development. For example, over-expression of FGF-7 driven by αA-crystalline promoter, which is activated in mouse lens at E11.5, resulted in the suppression of cornea-type epithelial differentiation and the induction of ectopic lacrimal gland formation in the corneas of the transgenic mice (Makarenkova et al., 2000; Govindarajan et al., 2000; Lovicu et al., 1999). In order to understand such an influence at a later stage when epithelial cells have undergone corneal type epithelial differentiation, we developed a $Krt12^{rtTA/rtTA}/tetO$-FGF-7 bi-transgenic mouse line in which over-expression of FGF-7 by Dox induction caused squamous cell carcinoma of the cornea resembling OSSN in human (Chikama et al., 2008). Less is known, however, about the signaling pathways by which FGF-7 mediates control of corneal epithelial cell proliferation.

4. FGF signaling pathway and its action in mammalian cells

FGF signaling, which is involved in the control of cell proliferation, differentiation, migration, survival and polarity, is transduced through FGF receptors (FGFR). FGFR1, FGFR2, FGFR3 and FGFR4 are FGF receptors, consisting of an extracellular immunoglobulin-like (Ig-like) domain and a cytoplasmic tyrosine kinase domain (Lee et al., 1989; Dionne et al., 1990; Partanen et al., 1990, 1991; Powers et al., 2000; Katoh and Katoh, 2003). FGF receptor isoforms with distinct ligand affinity are generated by alternative splicing of mutually exclusive exons in the latter half of the third Ig-like domain. FGF dimers associated with heparan sulfate proteoglycan bind to FGF receptors to induce receptor dimerization and receptor auto-phosphorylation. FGF signals are transduced via multiple signaling pathways such as the mitogen-activated protein kinases (MAPK), the phospholipase-C gamma (PLCγ), and the PI3K-PKB/AKT (Eswarakumar et al., 2005; Dailey et al., 2005; Katoh and Katoh, 2006; Kouhara et al., 1997; Ong et al., 2000) (Figure 3). A key component of FGF signaling is the docking protein called FGFR substrate 2 (FRS2) which is phosphorylated on tyrosine residues upon FGF stimulation. FRS2 consists of N-terminal myristylation signal, phosphotyrosine binding (PTB) domain and C-terminal region with multiple Src homology-2 (SH2) binding sites. FRS2 is recruited to the auto-phosphorylated FGF receptors through the interaction with phospho-tyrosine residues. FRS2, bound to auto-phosphorylated FGFRs, is tyrosine phosphorylated in the C-terminal region, which in turns recruits growth factor receptor-bound protein 2 (GRB2) and a SH2-containing tyrosine phosphatase, SHP2. In most cell types, FRS2-GRB2-SHP2 signaling complex recruits the guanine nucleotide exchange factor, Son of sevenless (SOS), which activates Ras and downstream effectors of MAPK (Figure 3, highlighted in grey). The MAPKs which include ERKs, p38, and Jun N-terminal kinases (JNKs) regulate the activity of downstream kinases or transcription factors. Although MAPK family shares many structural similarities, the ERK1/2 kinases are generally considered responsible for the mitogenic response, while the p38 and JNK are usually associated with inflammatory or stress-responses (Johnson et al., 2002).

Alternatively upon FGF stimulation, FRS2-SHP2-GRB2 complex may recruit GRB2-associated binding protein 1 (GAB1) to activate PI3K (Figure 3, highlighted in yellow), which phosphorylates PIP2 to generate phosphatidylinositol-3,4,5-triphosphate (PIP3) (Ong et al., 2002; Cantley, 2002). PIP3 induces the translocation of PKB/AKT to plasma membrane, where PKB/AKT is phosphorylated and activated by phosphoinositide-dependent kinase (PDK) (Luo et al., 2003). The ability of FGFs to protect cells from apoptosis is primarily due to the activation of the PKB/AKT survival pathway, a PI3K dependent activation of PDK, leads to the activation of PKB/AKT which in turn attenuates the activity and/or suppression of pro-apoptotic factors (Schlessinger, 2000 Schlessinger. PKB/AKT phosphorylates a variety of substrates, such as glycogen synthase kinase (GSK-3β), forkhead transcription factor (FKHR), FOXO1 (Burgering and Medema, 2003). It is of interest to note that inhibition of GSK-3β via phosphorylation its Ser-9 by activated PKB/AKT and provides additional anti-apoptotic protection (Jope and Johnson, 2004) and possibly the accumulation and nuclear translocation of β-catenin. Furthermore, a wide variety of agents, e.g., SB-216763, SB-415286, and LiCl, is known to attenuate GSK-3β activities via phosphorylation at Ser-9 by PKB/AKT and allow the accumulation and nuclear translocation of transcription factor β-catenin (Frame and Cohen, 2000).

FGF signal also activates phosphatidyl inositol (Pt Ins) hydrolysis, release of intracellular calcium and activation of protein kinase C (PKC) via recruitment of the SH2 domain of PLCγ to the FGFR. Activated PLCγ hydrolyzes Pt Ins [4,5] P2 to form diacylglygerol (DAG) and Ins [1,4,5] P3 which stimulates calcium release and activation of calcium/calmodulin dependent protein kinases (Figure 3, highlighted in blue). The loss of this pathway does notappear to impair the proliferative response (Mohammadi et al., 1992). It remains unknown which individual aforementioned signaling pathways and/or their combination is responsible for the patho-physiology of OSSN caused by excess FGF-7 in our *Krt12^{rtTA/rtTA}/tet-O-FGF-7* mice.

5. Cross-talk between FGF and Wnt pathways may be mediated by common target proteins

It has been shown that FGF and Wnt signaling pathways crosstalk takes place during a variety of cellular processes, such as embryogenesis (Loebel et al., 2003; McGrew et al., 1997; Tickle, 1995; Ng, 2002; Villanueva et al., 2002; Gunhaga et al., 2003; Shackleford et al., 1993) and carcinogenesis (MacArthur et al., 1995; Katoh, 2002; McWhirter, 1997, 1999; Kirikoshi et al., 2000; Shimokawa et al., 2003; Chamorro et al., 2005; Katoh and Katoh, 2005). A key event in the canonical Wnt pathway is the activation of β-catenin that subsequently regulates transcription of specific target genes that modulate cell fate, proliferation, migration, and apoptosis. In the absence of Wnt signals, cytosolic β-catenin is phosphorylated by GSK-3β, which targets β-catenin for proteasome-mediated degradation. However, in the presence of a Wnt signal, GSK-3β is inactivated by phosphorylation, and results in accumulation and nuclear translocation of stable β-catenin that binds DNA elements of members of the lymphoid enhancer factor/T-cell factor (LEF/TCF) family of transcription factors. Thus, it activates transcription of Wnt-target genes (Moon et al., 2002) (Figure 4). β-catenin also functions at the cell membrane where, as a component of the adherent junction, it links cadherin to the cytoskeleton (Kemler et al., 1993).

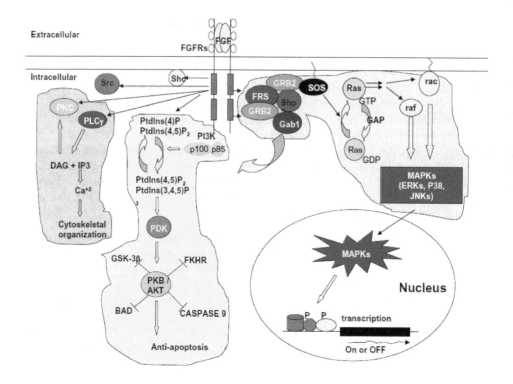

Fig. 3. **FGF signal transduction pathways.** Activated FGFRs (red rectangles) stimulate PLCγ pathway (blue highlight), the PI3K/PKB-AKT pathway (yellow highlight), and the FRS2-Ras-MAP kinase pathway (grey highlight). The activated MAP kinases (ERKs, p38, or JNKs) are translocated to the nucleus where they phosphorylate (P) transcription factors, thereby regulating target genes. In some cell types, FGF signaling also phosphorylates the Shc and Src proteins. Abbreviations: PKB/AKT, protein kinase b; DAG, diacyl glycerol; ERKs, Extracellular signal-regulated kinases; FKHR, forkhead homolog 1; FRS2, Fibroblast growth factor receptor substrate 2; GAP, GTPase-activating protein; Gab1, Grb2-associated binding protein 1; GRB2, Growth factor receptor-bound protein 2; GSK-3β, Glycogen synthase kinase 3β;MAPK, Mitogen-activated protein kinase; JNKs, Jun N-terminal Kinases; PDK, 3-phosophoinositide-dependent protein kinase;PI3K, Phosphoinositide 3-kinase; PKC, protein kinase C; PLCγ: phospholipase C gamma; Rac, Ras-related C3 botulinum toxin substrate; Shp2, Src homology 2-containing tyrosine phosphatase; SOS, Son of sevenless, guanine nucleotide exchange factor.

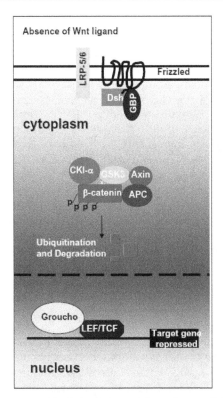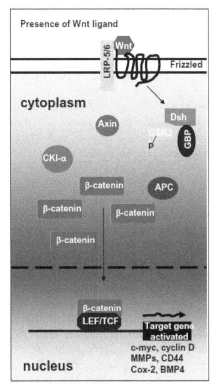

Fig. 4. **Graphical depiction of the schema of the canonical Wnt signaling pathway.** (Left panel) In the absence of Wnt ligand, LEF/TCF transcription factors are inert due to their association with a repressor (groucho protein) on their nuclear DNA binding sites. Beta-catenin, which is needed for release of the repressor protein, is captured by a degradation complex containing casein kinase I alpha (CKI-α), GSK-3β, Axin, and adenomatosis polyposis coli (APC) before it can enter the nucleus and is phosphorylated by GSK-3β and subsequently ubiquitinylated and degraded by the proteasome (Right panel). When Wnt ligand binds to its receptors, Frizzled and LRP5/6, dishevelled (Dsh) protein becomes activated. Repression of GSK-3β by an alternative complex containing Dsh and GSK-3β binding protein (GBP) releases β-catenin from the aforementioned degradation complex. Unbound β-catenin enters the nucleus to complex with LEF/TCF, displacing groucho and enabling formation of an active transcription complex.

A possible candidate that can mediate both Wnt and FGF signalings is GSK-3β, which can also be phosphorylated by PKB/AKT besides Dishevelled protein (Jope and Johnson, 2004). Evidence for this assertion has been provided during FGF-2 treatment of neuronal cells, which increases GSK-3β phosphorylation and nuclear localization of β-catenin. Over-expression of β-catenin maintains neural progenitor cells in a proliferative state in the presence of FGF-2, but enhances neuronal differentiation in the absence of FGF-2, suggesting that FGF signaling regulates neural progenitor cell proliferation and also affects lineage commitment during neural differentiation, in part, via β-catenin signaling (Israsena et al., 2004). Similarly, in neural cells, the neuroprotective effects of FGF-1 may utilize GSK-3β inactivation via activation of the PI3K-PKB/AKT cascade (Hashimoto et al., 2002). In addition, FGF-2-treated human endothelial cells show an increase in nuclear β-catenin by a reduction in GSK-3β activity (Holnthoner et al., 2002). Although GSK-3β has been shown to be a phosphorylation target of FGF as well as other signaling pathways such as IGF, the consequences of altered GSK-3β activity on β-catenin/LEF/TCF-dependent transcription seems to be cell-type dependent. For example in 293 embryonic kidney cells, IGF-mediated phosphorylation of GSK-3β does not lead to induction of LEF/TCF-dependent transcription (Playford et al., 2000). During early Xenopus development, FGF signaling leads to the inhibition of endogenous GSK-3β via a PKB/AKT independent p90RSK pathway (Torres et al., 1999). p90RSK over-expression increases the level of membrane-associated β-catenin without fluctuation in nuclear levels. The fact that both FGF and Wnt signaling target a common protein implicates that the assignment of particular proteins such as GSK-3β to a specific "pathway" is arbitrary since their activity can be influenced by a number of different routes. Furthermore, since proteins such as GSK-3β are targeted by several signal transduction pathways, they have the potential to act as molecular switches between these pathways, and thus serve as nodal points for pathway cross-talk (Figure 5). In addition, GSK-3β also binds to and phosphorylates the SNAIL transcriptional repressor, inducing its cytoplasmic translocation and degradation (Zhou et al., 2004). SNAIL represses E-cadherin transcription, a key tight junction molecule which establishes and maintains cell-cell adhesion (Batlle et al., 2000; Katoh and Katoh, 2005; Thiery and Sleeman, 2006). Taken together, these data support a model whereby the inhibition of GSK-3β activity via FGF-dependent PI3K-PKB/AKT signaling leads to an epithelial-mesenchymal transition (EMT) through the down-regulation of E-cadherin. This results in the release of β-catenin from the E-cadherin complex to promote nuclear accumulation of β-catenin. It remains unknown which aforementioned pathway is employed or in combination in the corneal epithelium elicited by excess FGF-7. Further investigation should lead to a clear mechanistic event of FGF-7 signaling in corneal epithelium and tumor formation.

6. Nuclear accumulation of the β-catenin protein in tumors

In human cancers, exon 3 of the *CTNNB1* gene, which encodes β-catenin, is a mutational "hot spot" for gain-of-function isoforms. This exon encodes the critical Ser/Thr residues, which are sites for priming by casein kinase1 (CK1) (Ser 45) and phosphorylation by GSK-3β (Ser 33, 37 and Thr 41). As a result, this β-transducin repeat-containing protein (β-TrCP) recognition site marks β-catenin for degradation. Therefore mutations within this exon increase the stability and nuclear accumulation of the β-catenin protein. Indeed, somatic mutations in *CTNNB1* gene are strongly associated with a wide variety of human tumors including colorectal carcinoma, desmoid tumor, endometrial carcinoma, hepatocellular carcinoma, hepatoblastoma, intestinal carcinoma gastric, medulloblastoma, melanoma, ovarian carcinoma, pancreatic carcinoma, pilomatricoma, prostate carcinoma, squamous cell carcinoma of the head and neck, thyroid carcinoma, and Wilms' tumor (Polakis, 2000). To test the hypothesis that β-catenin accumulation in corneal epithelial cell nucleus plays a pivotal role in mediating FGF-7 signaling networks leading to the formation of corneal neoplasia in *Krt12rtTA/rtTA/tetO-FGF-7* bi-transgenic mice induced by Dox, we have crossed *K12rtTA/rtTA/tetO-FGF-7/tetO-Cre* mice with two different floxed *Ctnnb1* mouse strains for the analysis the role of β-catenin in corneal intraepithelial neoplasia model. In the first mouse line, loxP sites flank an essential part (exon 2 through exon 6) of the *Ctnnb1*gene (*Ctnnb1^loxEx2-6*) (Brault et al., 2001). Upon Cre-mediated recombination, the region containing exon 2 to 6 of the *Ctnnb1* gene is deleted. This deletion causes a frame shift and premature termination of β-catenin translation. As a result, no β-catenin gene-product is produced from this allele. Interestingly, *K12^rtTA/rtTA/tetO-FGF-7/tetO-Cre/Ctnnb1^loxEx2-6/loxEx2-6* tetra-transgenic mice treated with Dox from P7 to P21 (i.e., short term induction for 14 days) failed to develop corneal epithelial hyperplasia, suggesting that this phenotype caused by excess FGF-7 is dependent on β-catenin (Chia-Yang Liu et al., in preparation for publication).

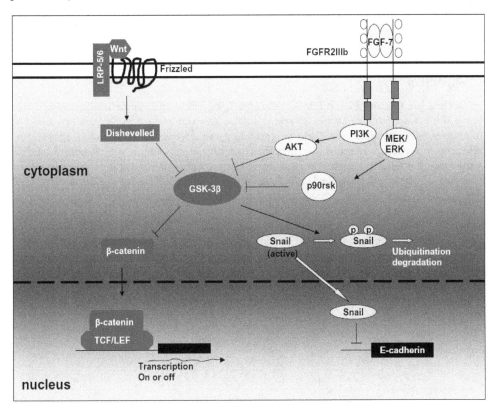

Fig. 5. **Cross-talk between FGF and Wnt pathways.** Elements of the FGF (yellow and red) and Wnt/β-catenin (blue) pathways are shown. Depicted is how activation of the FGF receptor on the cell membrane can lead to activation of β-catenin through different mechanisms. In some cell types, activation by FGF of either PI3K/AKT or MAP kinase/p90 ribosomal S6 kinase (p90rsk) can promote GSK-3β phosphorylation, which is associated with translocation of β-catenin to the nucleus. Inhibition of GSK-3b by Wnt, PI3K/AKt, or MAPK pathways suppressed the phosphorylation of Snail (green) and thus induces the nuclear localization and protein stabilization of Snail, which suppress E-cadherin and lead to cell migration.

In the second mouse line, loxP sites flank only exon 3 of the *Ctnnb1*gene (*Ctnnb1floxedEx3*) (Harada et al., 1999). Upon Cre-mediated recombination, exon 3 of the β-catenin gene is deleted. Exon 2 is spliced in frame to exon 4 and thus a mutant protein designated as ΔE3 β-catenin is produced. However, this mutant protein lacks its phosphorylation sites which are necessary for the subsequent ubiquitinylation and degradation of β-catenin. Therefore, this allele (*Ctnnb1floxedEx3*) encodes a more stable, constitutively dominant active β-catenin protein mimicking the canonical Wnt signaling pathway. Indeed, Dox-treated *K12rtTA/wt/tetO-Cre/Ctnnb1floxedEx3/Wt* tri-transgenic heterozygous mice all exhibited corneal epithelial hyperplasia at different developmental stages from P1 to P30. Moreover, X-gal positive cells completely correlated with the hyperplastic transformation and stromal invasion in the cornea of Dox-induced *Krt12rtTA/wt/tetO-Cre/Ctnnb1floxedEx3/Wt/TOPgal* quadruple transgenic heterozygous mice (Zhang et al., 2010). These results strongly indicate that regulation of β-catenin signaling plays a pivotal role in the maintenance of normal corneal epithelial homeostasis.

7. Pespective

Cornea is a unique and idea model for study angiogenesis and tumor progression *in vivo* because of its transparency and easy accessibility. Cornea consists of three cellular layers i.e., epithelium, stroma, and endothelium providing an idea model to study epithelium-mesenchym transition. In addition, corneal epithelium is also unique as a stratified but not keratinized epithelium, which allows us to study the epithelial metaplasia. Human OSSN consists of a spectrum of dysplasia and/or neoplasia which is relatively common in clinical ophthalmology practice. Primary surgical excision and various adjunctive therapies such as mitomycin-C (MMC), 5-fluorouracil (5-FU), or interferon alpha-2b (INF-α2b) remain as the mainstay of treatment but the recurrence rate is high. To improve the clinical outcome of OSSN treatment, agents aiming at the down regulation of signal transducing molecules that modulate β-catenin are required. It is anticipated that signal transduction molecule(s), e.g., ERK→p90RSK→GSK-3β→β-catenin and/or PI3K→PKB/AKT→GSK-3β→β-catenin, which could mediate FGF-7 signaling networks responsible for the formation of corneal epithelial neoplasia will aid in the design of novel regimens for the treatment of human OSSN and other human cancers.

8. Acknowledgment

This work is supported by RO1 grants EY12486 (CYL), EY13755 (WWYK), and Research to Prevent Blindness

9. References

Arbeit, J.M., Olson, D.C. & Hanahan, D. (1996). Upregulation of fibroblast growth factors and their receptors during multi-stage epidermal carcinogenesis in K14-HPV16 transgenic mice. Oncogene; 13:1847-1857

Batlle, E., Sancho, E., Franci, C., Dominguez, D., Monfar, M., Baulida, J. & Garcia, De Herreros A. (2000) The transcription factor snail is a repressor of *E-cadherin* gene expression in epithelial tumour cells. Nat Cell Biol; 2:84-89.

Brault, V., Moore, R., Kutsch, S., Ishibashi, M., Rowitch, D.H., McMahon, A.P., Sommer, L., Boussadia, O. & Kemler, R. (2001) Inactivation of the beta-catenin gene by Wnt1-Cre-mediated deletion results in dramatic brain malformation and failure of craniofacial development. Development; 128:1253-1264.

Burgering, B.M. & Medema, R.H. (2003) Decisions on life and death: FOXO Forkhead transcription factors are in command when PKB/Akt is off duty. J Leukoc Biol ; 73:689-701.

Chamorro, M.N., Schwartz, D.R., Vonica, A., Brivanlou, A.H., Cho, K.R. & Varmus, H.E. (2005) FGF20 and DKK1 are transcriptional target of b-catenin and FGF20 is implicated in cancer and development. EMBO J; 24:73-84.

Cantley, L.C. (2002) The phosphoinositide 3-kinase pathway. Science; 296:1655-1657.

Chikama, T., Hayashi, Y., Liu, C.Y., Terai, N., Terai, K., Kao, C.W., Wang, L., Hayashi, M., Nishida, T., Sanford, P., Doestchman, T. & Kao W. (2005) Characterization of tetracycline-inducible double transgenic Krt12rtTA/+/tet-O-LacZ mice. Invest Ophthalmol Vis Sci. 46:1966-1972.

Chikama, T., Liu, C.Y., Meij, J.T.A., Hayashi, Y., Wang, I.J., Liu, Y., Nishida, T. & Kao, W.W.Y. (2008) Excess FGF-7 in corneal epithelium causes corneal intraepithelial neoplasia in young mice and epithelium hyperplasia in adult mice. Am J of Pathol; 172:638-649.

Cho, K., Ishiwata, T., Uchida, E., Nakazawa, N., Korc, M., Naito, Z. & Tajiri, T. (2007) Enhanced expression of keratinocyte growth factor and its receptor correlates with venous invasion in pancreatic cancer. Am J Pathol.; 170:1964-1974.

Dailey, L., Ambrosetti, D., Mansukhani, A. & Basilico, C. (2005) Mechanisms underlying differential responses to FGF signaling. Cytokine Growth Factor Rev; 16:233-247.

Dionne, C.A., Crumley, G., Bellot, F., Kaplow, J.M., Searfoss, G., Ruta, M., Burgess, W.H., Jaye, M. & Schlessinger, J. (1990) Cloning and expression of two distinct high-affinity receptors cross-reacting with acidic and basic fibroblast growth factors. EMBO J; 9:2685-2692.

Eswarakumar, V.P., Lax, I. & Schlessinger, J. (2005) Cellular signaling by fibroblast growth factor receptors. Cytokine Growth Factor Rev; 16:139-149.

Finch, P.W., Rubin, J.S., Miki, T. & Ron, D. & Aaronson, S.A. (1989) Human KGF is FGF-related with properties of a paracrine effector of epithelial cell growth. Science; 245:752-755.

Finch, P.W., Cunha, G.R., Rubin, J.S., Wong, J. & Ron, D. (1995) Pattern of keratinocyte growth factor and keratinocyte growth factor receptor expression during mouse fetal development suggests a role in mediating morphogenetic mesenchymal-epithelial interactions. Dev Dyn; 203:223-240.

Frame, S. & Cohen P. (2000) GSK3 takes centre stage more than 20 years after its discovery. Biochem J; 359:1-16.

Govindarajan, V., Ito, M., Makarenkova, H.P., Lang, R.A. & Overbeek, P.A. (2000) Endogenous and ectopic gland induction by FGF-10. Dev. Biol.; 225:188-200.

Grossniklaus, H.E., Green, W.R., Lukenback, M. & Chan, C.C. (1987) Conjunctival lesions in adults: a clinical and histopathologic review. Cornea; 6:78-116.

Gunhaga, L., Marklund, M., Sjodal, M., Hsieh, J.C., Jessell, T.M. & Edlund, T. (2003) Specification of dorsal telencephalic character by sequential Wnt and FGF signaling. Nat Neurosci; 6:701-707.

Harada, N., Tamai, Y., Ishikawa, T., Sauer, B., Takaku, K., Oshima, M. & Taketo, M.M. (1999) Intestinal polyposis in mice with a dominant stable mutation of the beta-catenin gene. EMBO J.; 18:5931-5942.

Hashimoto, M., Sagara, Y., Langford, D., Everall, I.P., Mallory, M., Everson, A., Digicaylioglu, M. & Masliah, E. (2002) Fibroblast growth factor 1 regulates signaling via the glycogen synthase kinase-3beta pathway. Implications for neuroprotection. J Biol Chem; 277:32985-32991.

Hishikawa, Y., Tamaru, N., Ejima, K., Hayashi, T. & Koji, T. (2004) Expression of keratinocyte growth factor and its receptor in human breast cancer: its inhibitory role in the induction of apoptosis possibly through the over-expression of Bcl-2. Arch Histol Cytol.; 67:455-464.

Holnthoner, W., Pillinger, M., Groger, M., Wolff, K., Ashton, A.W., Albanese, C., Neumeister, P., Pestell, R.G. & Petzelbauer, P. (2002) Fibroblast growth factor-2 induces Lef/Tcf-dependent transcription in human endothelial cells. J Biol Chem; 277:45847-45853.

Igarashi, M., Finch, P.W. & Aaronson, S.A. (1998) Characterization of recombinant human fibroblast growth factor (FGF)-10 reveals functional similarities with keratinocyte growth factor (FGF-7). J Biol Chem.; 273:13230-13235.

Israsena, N., Hu, M., Fu, W., Kan, L. & Kessler, J,A. (2004) The presence of FGF2 signaling determines whether beta-catenin exerts effects on proliferation or neuronal differentiation of neural stem cells. Dev Biol; 268:220-231.

Johnson, G.L. & Lapadat, R. (2002) Mitogen-activated protein kinase pathways mediated by ERK, JNK, and p38 protein kinases. Science; 298:1911-1912.

Jope, R.S. & Johnson, G.V.W. (2004) The glamour and gloom of glycogen synthase kinase-3. Trends Biochem Sci; 29:95-102

Karp, C.L., Scott, I.U., Chang, T.S. & Pflugfelder, S.C. (1996) Conjunctival intraepithelial neoplasia. A possible marker for human immunodeficiency virus infection? Arch Ophthalmol.;114:257-261.

Katoh, M. & Katoh, M. (2003) *FGFR2* and *WDR11* are neighboring oncogene and tumor suppressor gene on human chromosome *10q26*. Int J Oncol;22:1155-1159.

Katoh, M. (2002) WNT and FGF gene clusters. Int J Oncol ; 21:1269-1273.

Katoh, M. & Katoh, M. (2005) Comparative genomics on FGF20 orthologs. Oncol Rep; 14:287-290.

Katoh, M. & Katoh, M. Comparative genomics on *SNAI1*, *SNAI2*, and *SNAI3* orthologs. Oncol Rep;14:1083-1086.

Katoh, M. & Katoh, M. (2005) Comparative genomics on FGF8, FGF17, and FGF18 orthologs. In J Mol Med; 16:493-496.

Katoh, M. & Katoh, M. (2005) Comparative genomics on FGF16 orthologs. Int J Mol Med; 16:959-963.

Katoh, M. & Katoh, M. (2006) FGF signaling network in the gastrointestinal tract. Int J Oncol; 29:163-168.

Kemler, R. (1993) From cadherins to catenins: cytoplasmic protein interactions and regulation of cell adhesion. Trends Genet; 9:317-321.

Kiire, C.A. & Dhillon, B. (2006) The aetiology and associations of conjunctival intraepithelial neoplasia. Br J Ophthalmol.; 90:109-113.

Kirikoshi, H., Sagara, N., Saitoh, T., Tanaka, K., Sekihara, H., Shiokawa, K. & Katoh, M. (2000) Molecular cloning and characterization of human FGF20 on chromosome 8p21.3-p22. Biochem Biophys Res Commun; 274:337-343.

Kouhara, H., Hadari, Y.R., Spivak-Kroizman, T., Schilling, J., Bar-Sagi, D., Lax, I. & Schlessinger, J. (1997) A lipid-anchored Grb2-binding protein that links FGF-receptor activation to the Ras/MAPK signaling pathway. Cell ; 89:693-702.

Kovacs, D., Cota, C., Cardinali, G., Aspite, N., Bolasco, G., Amantea, A., Torrisi, M.R. & Picardo, M. (2006) Expression of keratinocyte growth factor and its receptor in clear cell acanthoma. Exp Dermatol.; 15:762-768.

Lee, F.S., Lane, T.F., Kuo, A., Shackleford, G.M. & Leder, P. (1995) Insertional mutagenesis identifies a member of the Wnt gene family as a candidate oncogene in the mammary epithelium of int-2/Fgf-3 transgenic mice. Proc Natl Acad Sci USA; 92:2268-2272.

Lee, G.A. & Hirst, L.W. (1995) Ocular surface squamous neoplasia. Surv Ophthalmol; 39:429-450.

Lee, P.L., Johnson, D.E., Cousens, L.S., Fried, V.A. & Williams, L.T. (1989) Purification and complementary DNA cloning of a receptor for basic fibroblast growth factor. Science; 245:57-60.

Li, D.Q. & Tseng, S.C. (1996) Differential regulation of cytokine and receptor transcript expression in human corneal and limbal fibroblasts by epidermal growth factor, transforming growth factor-alpha, platelet-derived growth factor B, and interleukin-1 beta. Invest Ophthalmol Vis Sci; 37:2068-2080.

Li, D.Q. & Tseng, S.C. (1997) Differential regulation of keratinocyte growth factor and hepatocyte growth factor/scatter factor by different cytokines in human corneal and limbal fibroblasts. J Cell Physiol; 172:361-372.

Liu, C.Y., Zhu, G., Westerhausen-Larson, A., Converse, R.L., Kao, C.W.C., Sun, T.T. & Kao, W.W.Y. (1993) Cornea-specific expression of K12 keratin during mouse development. Curr Eye Res.; 12:963-974.

Liu, C.Y., Zhu, G., Converse, R.L., Kao, C.W.C., Nakamura, H., Tseng, S.C.G., Mui, M.M., Seyer, J., Justice, M.J., Stech, M.E., Hansen, G.M. & Kao, W.W.Y. (1994) Characterization and chromosomal localization of the cornea- specific murine keratin gene Krt1.12. J Biol Chem.; 269:24627-24636.

Loebel, D.A., Watson, C.M., DeYoung, R.A. & Tam, P.P. (2003) Lineage choice and differentiation in mouse embryos and embryonic stem cells. Dev Biol; 264:1-14.

Lovicu, F.J., Kao, W.W. & Overbeek, P.A . (1999) Ectopic gland induction by lens-specific expression of keratinocyte growth factor (FGF-7) in transgenic mice. Mech. Dev; 88:43-53.

Luo, J., Manning, B.D. & Cantley, L.C. (2003) Targeting the PI3K-Akt pathway in human cancer: Rationale and promise. Cancer Cell; 4:257-262.

MacArthur, C.A., Shankar, D.B. & Shackleford, G.M. (1995) Fgf-8, activated by proviral insertion, cooperates with the Wnt-1 transgene in murine mammary tumorigenesis. J Virol ; 69:2501-2507.

Makarenkova, H.P., Ito, M., Govindarajan, V., Faber, S.C., Sun, L., McMahon, G., Overbeek, P.A. & Lang, R.A. (2000) FGF10 is an inducer and Pax6 a competence factor for lacrimal gland development. Development; 127:2563-2572.

Manavi, M., Hudelist, G., Fink-Retter, A., Gschwandtler-Kaulich, D., Pischinger, K. & Czerwenka, K. (2007) Gene profiling in Pap-cell smears of high-risk human papillomavirus-positive squamous cervical carcinoma. Gynecol Oncol.; 105:418-426.

Marchese, C., Maresca, V., Cardinali, G., Belleudi, F., Ceccarelli, S., Bellocci, M., Frati, L., Torrisi, M.R. & Picardo, M. (2003) UVB-induced activation and internalization of keratinocyte growth factor receptor. Oncogene; 22:2422-24231.

McGrew, L.L., Hoppler, S. & Moon, R.T. (1997) Wnt and FGF pathways cooperatively pattern antero-posterior neural ectoderm in Xenopus. Mech Dev; 69:105-114.

McWhirter, J.R., Goulding, M., Weiner, J.A., Chun, J. & Murre, C. (1997) A novel fibroblast growth factor gene expressed in the developing nervous system is a downstream target of the chimeric homeodomain oncoprotein E2A-Pbx1. Development; 124:3221-3232.

McWhirter, J.R., Neuteboom, S.T., Wancewicz, E.V., Monia, B.P., Downing, J.R. & Murre, C. (1999) Oncogenic homeodomain transcription factor E2A-Pbx1 activates a novel WNT gene in pre-B acute lymphoblastoid leukemia. Proc Natl Acad Sci USA; 96:11464-11469.

Mehta, P.B., Robson, C.N., Neal, D.E. & Leung, H.Y. (2000) Serum keratinocyte growth factor measurement in patients with prostate cancer. J Urol; 164:2151-2155.

Mohammadi, M., Dionnem, C.A., Li, W., Li, N., Spivak, T. & Honegger, A.M. (1992) Point mutation in FGF receptor eliminates phosphatidylinositol hydrolysis without affecting mitogenesis. Nature; 358:681-684.

Moon, R.T., Bowerman, B., Boutros, M. & Perrimon, N. (2002) The promise and perils of Wnt signaling through beta-catenin. Science; 296:1644-1646.

Napora, C., Cohen, E.J., Genvert, G.I., Presson, A.C., Arentsen, J.J., Eagle, R.C. & Laibson, P.R. (1990) Factors associated with conjunctival intraepithelial neoplasia: a case control study. Ophthalmic Surg.; 21:27-30.

Ng, J.K., Kawakami, Y., Buscher, D., Raya, A., Itoh, T., Koth, C.M., Rodriguez, Esteban. C., Rodriguez-Leon, J., Garrity, D.M., Fishman, M.C., Izpisua Belmonte, J.C. (2002) The limb identity gene Tbx5 promotes limb initiation by interacting with Wnt2b and Fgf10. Development; 129:5161-5170.

Niu, J., Chang, Z., Peng, B., Xia, Q., Lu, W., Huang, P., Tsao, M.S. & Chiao, P.J. (2007) Keratinocyte growth factor/fibroblast growth factor-7-regulated cell migration and invasion through activation of NF-kappaB transcription factors. J Biol Chem.; 282:6001-6011.

Ong, S.H., Guy, G.R., Hadari, Y.R., Laks, S., Gotoh, N., Schlessinger, J. & Lax, I. (2000) FRS2 proteins recruit intracellular signaling pathways by binding to diverse targets on fibroblast growth factor and nerve growth factor receptors. Mol Cell Biol; 20:979-989.

Ong, S.H., Hadari, Y.R., Gotoh, N., Guy, G.R. & Schlessinger, J. & Lax, I. (2001) Stimulation of phosphatidylinositol 3-kinase by fibroblast growth factor receptors is mediated by coordinated recruitment of multiple docking proteins. Proc Natl Acad Sci USA; 98:6074-6079.

Panos, R.J., Rubin, J.S., Csaky, K.G., Aaronson, S.A. & Mason, R.J. (1993) Keratinocyte growth factor and hepatocyte growth factor/scatter factor are heparin-binding

growth factors for alveolar type II cells in fibroblast-conditioned medium. J Clin
Invest; 92:969-977

Partanen, J., Makela, T.P., Alitalo, R., Lehvaslaiho, H. & Alitalo, K. (1990) Putative tyrosine
kinases expressed in K-562 human leukemia cells. Proc Natl Acad Sci USA; 87:8913-
8917.

Partanen, J., Makela, T.P., Eerola, E., Korhonen, J., Hirvonen, H., Claesson-Welsh, L. &
Alitalo, K. (1991) FGFR-4, a novel acidic fibroblast growth factor receptor with a
distinct expression pattern. EMBO J; 10:1347-354.

Pietras, K., Pahler, J., Bergers, G. &, Hanahan, D. (2008) Functions of paracrine PDGF
signaling in the proangiogenic tumor stroma revealed by pharmacological
targeting. PLoS Med.; 5:e19

Playford, M.P., Bicknell, D., Bodmer, W.F. & Macaulay, V.M. (2000) Insulin-like growth
factor 1 regulates the location, stability, and transcriptional activity of beta-catenin.
Proc Natl Acad Sci USA; 97:12103-12108.

Polakis, P. (2000) Wnt signaling and cancer. Genes Dev. 2000; 14:1837-1851.

Powers, C.J., McLeskey, S.W. & Wellstein, A. (2000) Fibroblast growth factors, their
receptors and signaling. Endocr Relat Cancer; 7:165-197.

Rubin, J.S., Osada, H., Finch, P.W., Taylor, W.G., Rudikoff, S. & Aaronson, S.A. (1989)
Purification and characterization of a newly identified growth factor specific for
epithelial cells. Proc Natl Acad Sci USA; 86:802-806.

Rubin, J.S., Bottaro, D.P., Chedid, M., Miki, T., Ron, D., Cheon, G., Taylor, W.G., Fortney, E.,
Sakata, H., Finch, P.W. & LaRochelle, W.J. (1995) Keratinocyte growth factor. Cell
Biol. Intern.; 19:399-411.

Schlessinger, J. (2000) Cell signaling by receptor tyrosine kinases. Cell; 103:211-225.

Scott, I.U., Karp, C.L. & Nuovo, G.J. (2000) Human papillomavirus 16 and 18 expression in
conjunctival intraepithelial neoplasia. Ophthalmology.; 109:542-547.

Shackleford, G.M., MacArthur, C.A., Kwan, H.C. & Varmus, H.E. (1993) Mouse mammary
tumor virus infection accelerates mammary carcinogenesis in Wnt-1 transgenic
mice by insertional activation of int-2/Fgf-3 and Hst/Fgf-4. Proc Natl Acad Sci
USA; 90:740-744.

Shimokawa, T., Furukawa ,Y., Sakai, M., Li, M., Miwa, N., Lin, Y.M. & Nakamura, Y. (2003)
Involvement of the FGF18 gene in colorectal carcinogenesis, as a novel downstream
target of the β-catenin/T-cell factor complex. Cancer Res; 63:6116-6120.

Thiery, J.P. & Sleeman, J.P. (2006) Complex networks orchestrate epithelial-mesenchymal
transitions. Nat Rev Mol Cell Biol ; 7:131-142.

Tichelaar, J., Lu, W.W. & Whitsett, J.A. (2000) Conditional expression of fibroblast growth
factor-7 in the developing and mature lung, J. Biol. Chem. 275: 11858-11864.

Tickle, C. (1995) Vertebrate limb development. Curr Opin Genet Dev; 5:478-484.

Torres, M.A., Eldar-Finkelman. H., Krebs, E.G. & Moon, R.T. (1999) Regulation of ribosomal
S6 protein kinase-p90(rsk), glycogen synthase kinase 3, and beta-catenin in early
Xenopus development. Mol Cell Biol; 19:1427-1437.

Tseng, S.C.G. (1989) Concept and application of limbal stem cells. Eye; 3:141-157.

Villanueva, S., Glavic, A., Ruiz, P. & Mayor, R. (2002) Posteriorization by FGF, Wnt, and
retinoic acid is required for neural crest induction. Dev Biol; 241:289-301.

Zhang, Y., Call, M.K., Yeh, L.K., Liu, H., Kochel, T., Wang, I.J., Chu, P.H., Taketo, M.M.,
Jester, J.V., Kao, W.W. & Liu, C.Y. (2010) Aberrant expression of a beta-catenin

gain-of-function mutant induces hyperplastic transformation in the mouse cornea. J Cell Sci.; 123:1285-1294.

Zhou, B.P., Deng, J., Xia, W., Xu, J., Li, Y.M., Gunduz, M. & Hung, M.C. (2004) Dual regulation of Snail by GSK-3b-mediated phosphorylation in control of epithelial - mesenchymal transition. Nat Cell Biol; 6:931-940.

Part 3

Intraepithelial Neoplasia of Breast

5

Intraepithelial Neoplasia of Breast

Simonetta Monti and Andres Del Castillo
Senology Division, European Institute of Oncology, Milan,
Italy

1. Introduction

The early proliferative lesions of the breast have taken on greater significance as a result of mammographic screening and the use of even more sensitive imaging technologies. The intraductal proliferative lesions of the breast are a group of cytologically and architecturally diverse proliferations, typically originating from the terminal duct-lobular unit and confined to the mammary duct-lobular system. In this chapter we considere the most important of these lesions, lobular carcinoma in situ and ductal carcinoma in situ.

Lobular Carcinoma in situ

Foote and Stewart (Foote & Stewart, 1941) first described lobular carcinoma in situ (LCIS) in detail in 1941, emphasizing the morphologic similarity of LCIS cells to those of invasive lobular carcinoma. The multicentricity and high frequency of bilaterality of LCIS were recognized early on. LCIS generally occurs in younger women, and microcalcifications are not a feature (in contrast to the high frequency of microcalcifications seen with ductal proliferative lesions). Microscopically, LCIS is usually characterized by a solid, occlusive proliferation of loosely cohesive uniform cells, some of which may contain intracytoplasmic lumens. The incidence of LCIS is otherwise benign breast biopsy is reported as between 0,5% and 3,8%

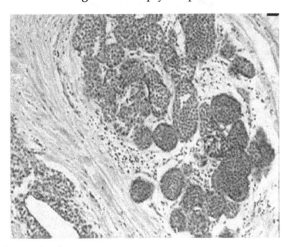

Fig. 1. Lobular carcinoma in situ

Ductal carcinoma in situ

Ductal carcinoma in situ is a heterogeneous group of lesions with diverse malignant potential and range of treatment options. It was infrequently diagnosed in the past, when it accounted for only 1% to 5% of all breast cancers. It usually presented as a palpable lesion, Paget disease, or bloody nipple discharge (Intra et al. 2003). The clinical presentation of Ductal Carcinoma In Situ shifted from a palpable lesion in the pre-mammographic era to a non-palpable lesion detected on the basis of mammographic microcalcifications or density (Tavassoli, 2008).

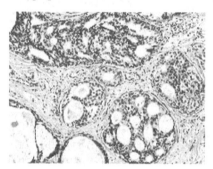

Fig. 2. Ductal carcinoma in situ

2. New classification

As mammographic screening became widespread, the frequency of diagnosis of intraepithelial proliferative lesions increased markedly, highlighting deficiencies in their classification as well a lack of data on natural history, and making clinical management a challenge. The new classification of these entities (DCIS and LCIS), principally due to Tavassoli (Tavassoli et al. 2003), was based on the concept of intraepithelial neoplasia

Ductal intra-epithelial neoplasia (DIN) terminology	Conventional terminology
Low-risk DIN	Intraductal hyper-plasia (IDH)
Flat DIN 1	Flat epithelial atypia
DIN 1	<2 mm: atypical ductal hyperplasia (ADH)
	>2 mm: ductal carcinoma in situ, low grade (DCIS, grade 1)
DIN 2	Ductal carcinoma in situ, intermediate grade (DCIS, grade 2)
DIN 3	Ductal carcinoma in situ, high grade (DCIS, grade 3)

Table 1. DIN translational table (Tavassoli FA 1997)

developed for cervix, vagina, vulva, prostate and pancreas. It does not use the term "cancer" diminishing the likelihood of overtreatment, and perhaps reduces also the level of anxiety and emotional stress a patient feels when told she has a cancer, even if it is only in situ.

3. Diagnosis and treatment

3.1 Diagnosis

The role of clinical examination in the ongoing surveillance of women with these high risk lesions is limited. While many DINs are detected through microcalcifications at mammography, the detection of LIN can be occasionally occur after a biopsy, for example in plastic surgery a histological evaluation of tissue excised in a breast reduction. In this case there is therefore a clear indication for meticulous assessment of both breasts ongoing surveillance of the breast at a relatively short interval. So standard core biopsy (14-18 gauge) is probably the most prevalent method for initial diagnosis of LIN. In the case of DIN it is diagnosed primarily via mammography plus ultrasound followed by stereotactic needle biopsy. However, new techniques such as magnetic resonance imaging (MRI) and analysis of ductal cytology aim to improve DIN detection (Sickles, 1983).

The use of breast MRI for patients with DIN is not yet established. MRI may be more sensitive for DIN detection than mammography but can lack specificity. The potential benefits of MRI include fewer re-excisions after BCS, decreased local recurrence rates after excision, and earlier detection and treatment of contralateral breast cancer (Leonard & Swain, 2004).

At present, therefore, mammography plus ultrasound and SCNB remain the standard diagnostic approaches for DIN (Leonard & Swain 2004).

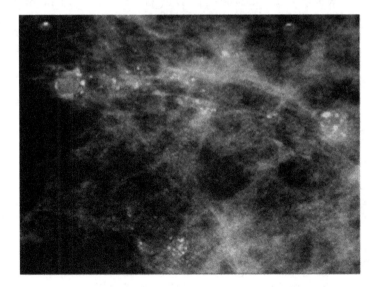

Fig. 3. A typical mammographic view of DIN

3.2 Treatment

Where the surgical biopsy reveals LIN , the margins of excision are paramount in determining further treatment. If the margins are clear, that is, the lesion appears to have been completely excised, no further treatment is required. Where surgical excision reveals a high risk lesion with an involved surgical margin or margins, re-excision should be considered. The goal of course is complete excision, but it must be recognized that LIN can be very extensive, sometimes occupying an entire quadrant or more of the breast. If a generous diagnostic excision suggests this, usually because several margins are involved, it may be clear that a substantially wider excision would have an adverse cosmetic outcome. In such cases observation and surveillance as detailed may be the preferred option. But the patient will be fully informed of the risk of subsequent cancer and involved in this decision. So, while for the LIN management is not yet clear the standard of treatment and in some institutions the decision is made on an individual, case by case basis, for DIN.

The standard of care consists of (a) breast conservative surgery (BCS) (mastectomy is still indicated in large lesions - masses or microcalcifications), axillary dissection is not indicated because of the low prevalence of nodal metastases and the significant morbidity associated with lymph node dissection (b) radiotherapy (RT) after conservative surgery, and (c) medical treatment in estrogen receptor-positive patients (Tavassoli ,2008; Farante et al. 2010; Cox et al. 2001; Intra et al.2003).

Surgical treatment

The main goal of surgical treatment for women with DIN is BCS plus RT, particularly for those patients with small solid masses, mammographically detected lesions, or limited microcalcification areas (Fisher et al. 1991; Schwartz et al. 2000; Silverstein et al. 1992).

However, mastectomy is still indicated in DIN patients with multicentricity, diffuse microcalcification, large palpable masses, when there is an inability to obtain negative margins as well other contraindications to breast conservation or a personal preference for mastectomy (Farante et al. 2010). At present, mastectomy is performed in about 30% of DIN patients, BCS without RT in about 30% of DIN patients, and BCS followed by RT in about 40% of DIN patients (Guerrieri-Gonzaga et al. 2009; Schmidt et al. 2006; Kuerer et al. 2008).

One important issue in breast cancer surgery is that of surgical margins, mainly in DIN patients, since these lesions are typically vague masses, which often cannot be adequately seen or felt and, thus, the pathological sampling of margins is fairly random (Fisher at al. 1999; Allred 2005). The current treatment of positive or focally involved margins in DIN patients is re-excision (Farante et al. 2010).

There is considerable debate regarding whether width of a negative margin is (width of a margin negative for tumor cells) associated with a decreased risk of recurrence, and classification of the margins makes summary statements difficult. About 10 years ago, the "Consensus on DCIS of Philadelphia" and Silverstein et al. proposed 10 mm of width margin as a limit of oncological safety (Schwartz et al. 2000). Since then other authors have proposed progressively smaller measures, down to 1 mm (Mansell, 2003).

Another important topic in DIN patients is the management of the axilla. Before the sentinel lymph node biopsy (SLNB) era, axillary dissection (AD) was a part of the standard surgical treatment for these patients (Farante et al. 2010). SLNB is recommended for patients with invasive breast cancer to determine prognosis and to guide adjuvant treatment decisions. In general, SLNB is not recommended for patients with a final or definitive diagnosis of DIN because the preinvasive cells do not metastasize (Virnig et al. 2010).

So not only is AD not necessary, neither should SLNB always be required because of the low prevalence of metastatic involvement, so an SLN biopsy should not be considered a standard procedure in the treatment of all patients with DCIS. The sole criteria for proposing SLN biopsy in DIN should be when there exists any uncertainty regarding the presence of invasive foci at definitive histology (Intra et al. 2008).

Major cancer centers agree that SLNB should be performed (a) always when mastectomy is performed (Cody 2007; Yi et al. 2008; Intra et al. 2007). (b)with large lesions (masses or micro-calcifications) and G3 tumors. (Cody 2007; Julian et. al 2007) and (c) after performing core or mammotome biopsies.

In cases of DIN patients with a positive SLNB, AD should not be immediately performed except for only those cases that present mammary invasion on final pathologic evaluation (Farante et al. 2010).

Radiation therapy

External RT

The standard course of current external RT after BCS for DIN, delivers a total dose of 50 Gy in daily fractions of 1.8/2 Gy without boost (Farante et al. 2010). The role of RT in DIN patients with conservative treatment has been mainly defined by four randomized trials (Bryan et al. 2003; Bijker et al. 2006; Houghton et al. 2003; Emdin et al. 2006). Additional radio-therapy reduced the LR rate by about 50%, with no effect on survival. The controversy is, instead, related if all DIN patients have to undergo RT. According to the 2009 St Gallen Consensus (Goldhirsch et al. 2009) RT could be avoided in elderly patients and in those with G1 DIN and clearly negative margins.

At the IEO in Milan, RT is not administered to DIN patients with G1 or G2 without comedo-necrosis. On the other hand, at least four significant papers from the Saint Gallen Consensus (Goldhirsch et al. 2009), the ECOG trial (Hughes et al. 2009), the Newport Consensus Conference III (Silverstein et al. 2009) and from the National Consensus Cancer Network (National Comprehensive cancer Network, 2009), suggested that some DIN patient sub-groups (i.e., G1 or G2 tumors without comedo-necrosis, and other low-risksub-groups) could not be candidates to receive RT after BCS.

4. Biomarkers

To guide such optimal treatment, histological classification is not sufficient and additional biological factors are being investigated for their ability to predict outcome for individual patients with intraepitalial neoplasia of the breast. As the molecular and genetic understanding of breast cancer has increased, new biological characteristics have been identified as prognostic indicators, as new adjuvant treatments have been developed. This has resulted in an increasingly personalized approach to breast cancer treatment that takes into account the diverse biological characteristics of the individual and their disease. A biomarker is an objectively measured feature that indicates a normal biological response, a pathogenic process, or the likelihood of response to a pharmacologic therapy. In oncology, biomarkers may be used to detect or stage disease, monitor response to therapy, and predict outcome (Madu & Lu 2010). A biomarker may be DNA, RNA, or a protein, measured directly in tissue, serum or other body fluids. An optimal biomarker for intraepithelial neoplasias of the breast would provide information additional to that provided by factors such as grade, lesion size, age and margin status (already established as related to the risk of local recurrence) so as to make it possible to predict which cases are unlikely to ever progress to invasive breast cancer and thus would require no further treatment after lesion removal, and also to predict which cases should receive local excision or mastectomy, or would benefit from adjuvant RT (Barker et al. 2003).

4.1 Estrogen receptors (ER)

Estrogens play a central role in the growth and differentiation of normal breast epithelium, stimulating cell proliferation and regulating the expression of genes, including that coding for the progesterone receptor (PgR) (Henderson et al. 1998; Peterson et al. 1987). In the normal pre-menopausal breast, ER-positive cells are luminal cells constituting about 7% of the total epithelial cell population (Peterson et al. 1987). They seem to secrete factors which influence, in paracrine manner, the proliferation of adjacent ER-negative cells (Peterson et al. 1987; Clarke et al. 1997). ER positivity and proliferation activity (as measured by Ki-67) are almost mutually exclusive in normal breast epithelium (Clarke et al. 1997). The proportion of ER-positive cells increases with age to reach a plateau after menopause (Shoker et al. 1999)

High ER expression in normal epithelium is a risk factor for breast cancer, conferring a 3-fold increase in risk compared to minimal expression (Khan et al. 1998). ER positivity together with KI-67 expression, may correlate with progression to more severe lesions in non-atypical epithelial hyperplasia (Iqbal et al. 2001). It has also been suggested that an increased percentage of ER-positive cells in adjacent normal lobules is associated with increased risk of invasive breast cancer rather than ER-positivity within the non-atypical epithelial hyperplasia *per se* (Gobbi et al. 2005)

In other benign breast lesions, such as sclerosing adenosis, radial scar, papilloma, fibroadenoma, and phylloides tumor, the percentage of ER-positive cells is higher than in normal breast tissue (Shoker et al. 2000). Similarly, ER-alpha expression is significantly elevated in hyperplastic enlarged lobular units (HELU), which are the earliest histologically identifiable lesions with premalignant potential. By contrast, intense ER-alpha staining in enlarged lobular units with columnar alteration (ELUCA) seems associated with reduced risk of subsequent invasive carcinoma (McLaren et al. 2005).

Unlike in normal breast, in ADH (atypical ductal hyperplasia), LN (lobular neoplasia) and DCIS (ductal carcinoma in situ), ER-positive cells are surrounded by contiguous cells which are also characterized by ER-positivity (Clarke et al. 1997). Furthermore, in DCIS, cells that are both ER-positive and Ki-67-positive are a characteristic finding (Shoker et al. 2000). In general, non-comedo carcinomas more frequently exhibit ER positivity (Bose et al. 1996; Page et al. 1982).

The expression of ER-beta by breast epithelium is the inverse of that of ER-alpha, declining progressively from normal breast tissue to ADH, DCIS, and IDC (Intraductal Carcinoma)

(Roger at al. 2001; Shaaban et al. 2003). According to a recent article, a high ER-alpha/ER-beta ratio in non-atypical epithelial hyperplasia predicts progression to carcinoma (Shaaban et al. 2005). For the optimal envisagement of the ER network, it should be kept in mind that important regulators exist, such as hsp-27 (O'Neill et al. 2004) or AIB1 (Hudelist et al. 2003), exhibiting more intense expression in breast cancer.

4.2 Progesterone receptors (PgR)

As is the case with ER levels, PgR levels are elevated in early premalignant breast lesions (Lee et al. 2006) and PgR expression decreases with progression to malignancy (Ariga et al. 2001). In DCIS, PgR positivity is associated with ER positivity and lack of comedo necrosis (Barnes et al. 2005; Claus et al. 2001). Studies on the relation of PgR expression to tumor grade (Barnes et al. 2005; Ringberg et al. 2001, Rody et al. 2004; Lebrecht et al. 2002) and recurrence rate (Kepple et al. 2006; Provenzano et al. 2003) have provided contrasting results (reviewed in Provenzano et al. 2003). In ductal carcinoma, PgR expression has been associated with histological grade, but not with lymph node involvement, tumor size, or prognosis (Ariga et al. 2001). Data on PgR expression in lobular neoplasia is scarce but it seems to be expressed in most cases (Fadare et al. 2006; Fisher et al. 1996).

The ratio of PgR-A to PgR-B appears important. In normal breast tissue and non-atypical hyperplasia, PgR-A and PgR-B are expressed in approximately equal quantities, but at an early stage of progression, one receptor (usually PgR-A in advanced lesions) predominates (Mote et al. 2002). In vitro studies indicate that PgR-A exerts modulating effects on cell morphology and adhesion (McGowan et al. 1999; Grahal et al. 2005). In the normal tissue of BRCA mutation carriers, PgR-B is absent (Mote et al. 2004).

4.3 HER2

HER2 or human epidermal growth factor receptor 2 (c-ErbB-2) is a tyrosine kinase receptor and oncoprotein encoded by the ERBB2 gene on chromosome 17q. Alterations in ERBB2 expression are important in malignant transformation (De Potter et al. 1989; Ross et al. 1999). Some studies have found that HER2 was not overexpressed in benign proliferative breast disease or ADH (Gusterson et al. 1988; Heffelfinger et al. 2000) while another used fluorescence in situ hybridization to demonstrate that the extent of HER2 amplification increased with progression to invasive carcinoma (Xu et al. 2002). Patients with benign breast lesions showing low levels of HER2 amplification were found in one study to have a two-fold increased risk of developing breast cancer (Stark et al. 2000); however another study found that HER2 overexpression in benign lesions was not a significant risk factor for developing cancer (Rohan et al. 1998).

A quarter of LCIS cases have been found to be HER2 positive, irrespective of the coexistence of an invasive component (Mohsin et al. 2005). Occasional positivity has also been found in pleomorphic ductal-lobular carcinomas in situ (Sneige et al. 2002).

As far as the role of HER2 in DCIS is concerned, HER2 immunoreactivity has been primarily associated with DCIS of higher grade, in the absence or presence (Tsuda et al. 1998) of IDC , and with comedo type (Albonico et al. 1996). Interestingly, given the association of higher grade with HER2 amplification, the latter has been regarded as an independent prognostic factor (Tsuda et al. 1993). Allred et al (Allred et al. 1992) documented that the percentage of

HER2 immunoreactivity is significantly higher in DCIS than IDC: one of the possible explanations proposed by the authors was that HER2 may be more important for the initiation than the progression of breast cancer, or that HER2 may be downregulated during breast cancer progression.

4.4 P53

P53 is a tumour suppressor gene located on 17p. p53 protein mediates its tumor suppressor functions via the transcriptional regulation or repression of a variety of genes (Toledo et al. 2006; Vogelstein et al. 2000) and is an important component of breast cancer pathophysiology (Gasco et al. 2002). Regarding the role of p53 as a risk factor in benign breast lesions, there data is controversial: the immunohistochemical detection of p53 in benign breast lesions has been associated with elevated cancer risk (Rohan et al. 1998), although there are studies with conflicting results (Younes et al. 1995).

Considering the various types of lesions in the continuum between benign lesions and breast cancer, various studies have assessed the role of p53. In epithelial hyperplasia without atypia, p53 mutations have not been detected (Done et al. 1998). In ADH, the presence and role of p53 mutations is still an open field: p53 mutations were initially not documented (Chitemerere et al. 1996); subsequently studies pointing to p53 mutations appeared (Kang et al. 2001), and, more recently, the presence of mutated p53 in ADH has been demonstrated with the use of laser capture microdissection microscope, single-stranded conformational polymorphism (SSCP) and sequencing (Keohavong et al. 2004). Regarding LN, there is scarcity of data: in two studies, no p53 immunoreactivity was demonstrated in LN lesions (Siziopikou et al. 1996; Sapino et al. 2000), whereas a more recent study on LCIS reported p53 immunoreactivity in one fifth of cases (Mohson et al. 2005).

p53 mutations/accumulation are present in a significant percentage of DCIS cases (Lebeau et al. 2003; Poller et al. 1993), especially in the comedo type (O'Malley et al. 1994). However, the clinical significance of p53 accumulation remains still elusive; although it has been found to influence the proliferation rate (Rudas et al. 1997), a recent study showed that it does not affect the proliferation rate of the DCIS lesion *per se* (Lebeau et al. 2003). Is worth noting that the coexistence of DCIS with IDC is not associated with a different degree of p53 immunostaining (Myonlas et al. 2005).

4.5 Ki-67

Ki-67 is a cell cycle-associated nuclear protein, which is expressed in all cycle phases, with the exception of G0 and early G1, and reacts with MIB-1 antibody (Gerdes et al. 1984). Protein Ki-67 is extensively used as a proliferative index and is linked with malignancy, even in FNA (fine needle aspiration) specimens (Midulla et al. 2002). Moreover, its intrinsic association with apoptosis (bcl-2 status, see below) and p53 expression (see above) seems to be of importance in the diagnosis and prognosis of precursors and pre-invasive breast lesions: low Ki-67 expression/bcl-2 positivity and p53 negativity are a trait of ADH and, subsequently, well-differentiated carcinomas. Conversely, high Ki-67 expression/bcl-2 negativity within the lobules implicate lesions with a potential of poorly differentiated carcinoma (Viacava P et al. 1999). As mentioned above, also in the context of non-atypical hyperplasia, high Ki-67 and ER-alpha expression seem to predict progression to cancer (Ariga et al. 2001; Shaaban et al. 2002).

Interestingly enough, a clinical application of Ki-67 expression intensity seems to emerge. In non-atypical ductal hyperplasia, lesions with high Ki-67 expression can be clinically detected scintimammographically, since high (99m)Tc-(V)**DMSA** uptake seems to be their characteristic feature. According to the authors, this could prove useful in identifying women with benign but high-risk breast disease (Papantoniou et al. 2006).

4.6 Bcl-2

The bcl-2 gene is located on 18q. Bcl-2 protein, and belongs to a family of proteins playing a central role in the regulation of apoptosis (reviewed in van Delft et al. 2006; Reed et al. 1994; Hockenbery et al. 1994) and other pathways (reviewed by Kim (Kim, 2005)). With respect to the overall role of apoptosis in breast cancer pathogenesis, there seems to be an intriguing pattern incorporating the proliferation of the lesion. Growth imbalance in favour of proliferation seems crucial in the transition from normal epithelium to hyperplasia and later, from pre-invasive lesions to IDC. Conversely, apoptosis becomes more important at an intermediate stage: in the transition from hyperplasia to preinvasive lesions, the imbalance is in favour of apoptosis (Bai et al. 2001). Bcl-2 is present in the whole spectrum of breast lesions: predominantly in benign lesions, ADH, LN, and well-differentiated DCIS (Sizioupikou et al. 1996; Kapucuoglu et al. 1997; Meteoglu et al. 2005). More specifically, there is a gradual increase in the extent of apoptosis (Bai et al. 2001; Mustonen et al. 1997) and a parallel decrease in bcl-2 expression in benign/precursors/preinvasive/invasive lesions as they become histologically more aggressive (Mustonen et al. 1997) . Bcl-2 positivity tends to coincide with p53 negativity in normal breast tissue, non-atypical ductal hyperplasia, ADH, LN and in the majority of the DCIS (Sizioupikou et al. 1996). The role of Bcl-2 expression as a risk factor for breast cancer is described above, together with Ki-67 (see above).

4.7 Vascular endothelial growth factor (VEGF) and angiogenesis

VEGF is a potent angiogenic growth factor, commonly involved in tumor-induced angiogenesis, with a putative therapeutic significance in the context of breast cancer (Lebeau et al. 2003). Interestingly, VEGF gene polymorphisms have been associated with modified breast cancer risk in various populations (Jacobs et al. 2006)

Viacava et al (Viacava et al. 2004) have thoroughly examined the angiogenesis in precursor and preinvasive lesions. Increased vascularization is present in all preinvasive lesions and increases with lesion severity. In ductal lesions, angiogenesis is more intense in poorly/intermediately differentiated intraductal carcinomas than in non-atypical ductal hyperplasia and ADH. Similarly, LCIS, showing microvascular density similar to that of poorly/intermediately differentiated intraductal carcinoma, is more vascularized than ALH. In the same study, VEGF expression in normal glandular structures was lower than in lesions, with the highest levels found in ductal lesions. Interestingly, no correlation was found between VEGF expression and the degree of vascularization in that study. On the other hand, Hieken TJ et al. suggested **that VEGF** expression may help predict the biologic aggressiveness of DCIS (Hieken et al. 2001). Additionally, in the context of DCIS, Vogl et al provide evidence to support the idea that VEGF expression is not regulated by the HER2 pathway (Vogl et al. 2005).

4.8 E-cadherin

E-cadherin, a tumor suppressor gene located on 17q, has been implicated especially in lobular breast cancer molecular pathogenesis (Berx et al. 1995). In clinical practice, immunohistochemistry for E-cadherin is a helpful marker for differential diagnosis, since most cases of low-grade DCIS exhibit E-cadherin positivity, whereas LN is almost always E-cadherin negative (Bratthauer et al. 2002, reviewed in Lerwill et al. 2004 and Putti et al. 2005). This implies that E-cadherin disruption is an early event, prior to progression, in lobular carcinogenesis (Vos et al. 1997; Mastracci et al. 2005); more specifically, DNA alterations accompanying the loss of protein expression pertain to LCIS but not to ALH (Mastracci et al. 2005). As expected according to the above, only few studies have focused on E-cadherin in ductal lesions. In the context of DCIS, hypermethylation of E-cadherin 5' CpG islands has been demonstrated (Nass et al. 2000) , and, at the protein level, E-cadherin has been linked to better differentiation (Gupta et al. 1997). Moreover, mutational analysis of E-cadherin provided evidence to support that DCIS is the precursor of invasive ductal carcinoma in cases where LCIS coexists (Rieger-Christ et al. 2001).

4.9 TGF-beta

The transforming growth factor-beta (TGF-β) pathway has ambivalent importance in the pathogenesis of breast cancer (reviewed in Wakefield et al. 2001). Serum TGF-beta levels do not differ between patients with breast cancer, DCIS and benign lesions (Lebrecht et al. 2004); however, TGF-beta expression becomes more accentuated in IDC, compared with DCIS (Walker et al. 1992). Surprisingly enough, an interesting study recently showed that loss of TGF-beta-RII expression in epithelial cells of hyperplasia without atypia is associated with increased risk of IDC (Gobbi et al. 1999). No reports exist on ADH and LN, to our knowledge.

4.10 P16 (INK4a)

p16 is an inhibitor of cyclin-dependent kinases 4 and 6 (reviewed in Rocco & Sidransky 2001). With respect to the role of p16, controversial results exist. According to some authors, aberrant methylation of p16 is not demonstrated in benign conditions, epithelial hyperplasia and intraductal papillomas, but is restricted in cancerous epithelium (Lehmann et al. 2004). Conversely, another study showed that IDC demonstrates hypomethylation of p16 and hyperactivity of the p16 gene (enhanced expression of p16 mRNA), contrary to the hypermethylated, inactive state in the normal epithelium (Van Zee et al. 1998). Independently, Di Vinci et al. distinguish between p16 hypermethylation and p16 protein overexpression; the former seems not to be specifically associated with malignancy and to occur both in benign and malignant lesions, whereas the latter, together with cytoplasmic sequestration, is a feature of breast carcinoma (Di Vinci et al. 2005). In the context of such controversy, no studies exist with respect to p16 as a risk factor, with the exception of a study in Poland envisaging p16 as a low penetrance breast cancer susceptibility gene (Debniak et al. 2005)

4.11 p27(Kip1)

The p27 gene encodes for an inhibitor of the cyclin – CDK (cyclin-dependent kinase) active complex. Although numerous studies exist with respect to the role of p27 in breast cancer (reviewed in Colozza et al. 2005; Alkarain et al. 2004 and Musgrove et al. 2004), there is a lack of data regarding precursors, pre-invasive lesions and other predisposing conditions. p27 expression has been documented in DCIS, but its clinicopathological significance is still uncertain (Oh et al. 2001).

4.12 P21 (Waf1)

p21 is a cell cycle regulator, implicated in a variety of pathways (Dotto 2000). p21 immunoreactivity has been detected both in benign and malignant epithelium, and thus its role is hard to interpret (Krogerus et al. 2000). Studies focusing especially on ADH or LN do not exist. As far as DCIS is concerned, p21 positivity has been independently associated with clinical recurrence (Provenzano et al. 2003) . On the other hand, Oh YL et al. found a significant correlation between positive p21 immunoreactivity (67.3% of the cases) and well-differentiated histologic grade, non-comedo type, ER-positive and p53-negativity. According to these authors, DCIS with p21+/p53- is likely to be the non-comedo type (Oh et al. 2001).

4.13 14-3-3 sigma

Umbricht and coworkers identified 14-3-3 sigma as a gene whose expression is lost in breast carcinomas, primarily by methylation-mediated silencing. Importantly, the hypermethylation of the locus was absent in hyperplasia without atypia, but was detectable with increasing frequency as the breast lesions progressed from atypical hyperplasia to DCIS, and finally to invasive carcinoma (Umbricht et al. 2001); interestingly, methylated alleles existed in the periductal stromal breast tissue. Subsequently, a parallel, stepwise reduction at the 14-3-3 sigma protein level was documented (Simooka et al. 2004).

Despite the emerging role of 14-3-3 sigma in breast carcinogenesis, to date no studies exist assessing its role as a risk factor for breast cancer development.

5. Genetic events

Complex and heterogeneous sets of genetic alterations are involved in the etiology of breast cancer. However, some of these genetic events occur more often early, or late, in carcinogenesis. Rather, breast cancer to be viewed as the result of accumulation of various major and minor genetic events in a fairly, random order, which is referred to as the "bingo principle" analogous to winning the "prize" (in this case cancer) in this popular game. With the establishment of new global genetic screening techniques such as comparative genomic hybridization (CGH), a pattern of genetic alterations has emerged. More recently other methods have been used for the characterization of pre-invasive breast lesions, such as cDNA microarray and proteonomics analysis. Numerous studies have documented differences in the copy number, sequence and expression level of specific genes in cohorts of invasive breast carcinomas, but relatively little is known of the events that mediate the transition of normal human breast epithelial cells to premalignant and early tumorigenic states. Non neoplastic breast tissue often harbors genetic changes that can be important to understanding the local breast environment within which cancer develops. In fact, most pre invasive lesions of the breast are thought to derive from the transition zone between the duct and the functional unit of the breast, the lobule, which is composed of acini that are lined by an outer myoepithelial layer and a inner luminal or glandular layer containing a putative stem or progenitor cell component, which gives rise to the above- mentioned cells. These cells have recently been described and characterized in more detail. It is noteworthy that many characteristics of these cells are shared in mouse and human cells. At present, the relationship between these cells and breast cancer specific stem cells is unclear. However, these cells can serve as a tool to explain the presence of monoclonal patches within a breast lobule or parts of the ductal tree. In addition, the description of non-recurrent genetic changes within the morphologically normal breast tissue, requiring a large subset of affected cells, favors the idea of long living cells as targets of the initial starting of the genetic cascade towards an overt malignancy. The finding of genetic changes within morphologically normal beast tissue is nowadays not only associated with an increased local recurrence risk, but also exerts a tremendous influence on the validity of progression models of breast cancer and especially the relationship toward proposed precursor lesions.

A recent study (Hannafon, et al. 2011) hypothesized that micro RNA expression might be dysregulated prior to invasive breast carcinoma. This study provides the first report of a microRNA expression profile in normal breast epithelium and the first integrated analysis of microRNA and microRNA expression in paired samples of histologically normal epithelia and preinvasive breast cancer. They further demonstrated, by modulating the expression of several microRNA samples, that the expression of their predicted target genes is affected. Taken together, these findings support their hypothesis that changes in microRNA expression in early breast cancer may control many of the parallel changes in gene expression in this stage. This work also implicates the loss of the tumor suppressor miR-125b and the gain of the oncogenic miRNA miR-182 and miR-183 as major contributors to early breast cancer development. Additionally this study has revealed novel candidate markers of preinvasive breast cancer, which could contribute to the identification of new diagnostic and therapeutic targets.

Another study (Kretschmer et al., 2011) has identified, using transgenic mouse model of DCIS (mice were transgenic for the WAP-SV40 early genome region, so that expression of the SV40 oncogene is activated by lactation) and identified seven genes that are significantly up regulated in DCIS: DEPDC1, NUSAP1, EXO1,RRM2,FOXM1,MUC1 and SPP1. A similar upregulation of homologues of these murine genes was observed in human DCIS samples.So, comparing murine markers for the DCIS of the mammary gland with genes up-regulated in human DCIS samples it is possible to identify a set of genes which might allow early detection of DCIS and invasive carcinoma in the future.

Cichon and her co-workers (Cichon et al., 2010) identified alterations in stromal cell function that may be critical for disease progression from benign disease to invasive cancer: key functions of myoepithelial cells that maintain tissue structure are lost, while tissue fibroblasts become activated to produce proteases that degrade the extracellular matrix and trigger the invasive cellular phenotype. Gene expression profiling of stromal alterations associated with disease progression has also identified key transcriptional changes that occur early in disease development. This study suggests approaches to identify processes that control earlier stages of disease progression.

Future studies aimed at studying post-translational modifications of histone proteins of the different stages of breast cancer promise to shed new light on the epigenetic regulatory control of gene expression during tumorigenesis (Fiegl et al., 2006).

6. Conclusions

Intraepithelial neoplasias of the breast are non-obligate precursor lesions with an increased risk of invasive carcinoma. The evolution to invasive carcinoma may not however be linear and may involve multiple pathways. Genomic instability drives tumorigenic process in invasive carcinoma and premalignant breast lesions and might promote the accumulation of genetic alterations in apparently normal tissue before histological abnormalities are detectable. Evidence suggests that genomic changes in breast parenchyma affect the behavior of epithelial cells and, ultimately, might affect tumor growth and progression. Inherent instability in genes that maintain genomic integrity, as well as exogenous chemical and environmental pollutants, have been implicated in breast cancer development. Although molecular mechanisms of tumorigenesis are unclear at present, carcinogenetic agents could contribute to field of genomic instability localized to specific areas of the breast. The use of molecular profiling technologies to identify distinct features that predict the future behavior of invasive disease is well documented. However, the application of such approaches to the identification of molecular predictors of clinical behavior of normal breast tissue and pre-invasive disease has been hampered by several problems. First, because pre-invasive disease is frequently microscopic in size, all of the tissue is processed through the use of standard pathological formalin fixed paraffin embedded (FFPE) processes and utilized for clinical diagnostic purposes. Second, standard FFPE processes pose a significant technical challenge for high throughput array CGH and gene expression microarray profiling. Third, and most importantly, large clinical cohorts and clinical trials of pre-invasive disease with well-annotated clinical samples and long (10-20 years) clinical follow up are lacking. Understanding the functional importance of genomic instability in early carcinogenesis is important for improving diagnostic and treatment strategies (Ellsworth et al., 2004)

7. Future directions

Despite many molecular studies, breast carcinogenesis is still not well understood. Our knowledge of the genetic and molecular biology of intraepithelial breast lesions is increasing at a remarkably rapid rate. In addition, more and more data are now available on the morphology and immunophenotype of the different precursor lesions, allowing the pathologists to recognize them. Epidemiologic studies have yielded information on the progression risk of several lesions. Future studies are likely to identify markers at a very early stage indeed that can play a role in the development of these precursor lesions from normal breast tissue. Clearly, prospective studies based on larger patient cohorts representing the whole spectrum of breast cancer are needed before the full power of gene expression profiling will be realized in clinical medicine. Results from studies so far are encouraging for the future.

8. References

Albonico, G., Querzoli, P., Ferretti, S., Magri, E. & Nenci, I. (1996)., Biophenotypes of breast carcinoma in situ defined by image analysis of biological parameters., *Pathology, Research & Practice*, vol. 192, no. 2, pp. 117-123.

Alkarain, A., Jordan, R. & Slingerland, J. (2004). p27 deregulation in breast cancer: prognostic significance and implications for therapy, *Journal of Mammary Gland Biology & Neoplasia*, vol. 9, no. 1, pp. 67-80.

Allred, D.C., Clark, G.M., Molina, R., Tandon, A.K., Schnitt, S.J., Gilchrist, K.W., Osborne, C.K., Tormey, D.C. & McGuire, W.L. (1992). Overexpression of HER-2/neu and its relationship with other prognostic factors change during the progression of in situ to invasive breast cancer., *Human pathology*, vol. 23, no. 9, pp. 974-979.

Allred, DC (2005). Ductal carcinoma in situ of the breast: pathologic and biologic perspectives. American Society of Clinical Oncology 2005 educational book 41st annual meeting, May 13-17, pp 75-79.

Ariga, N., Suzuki, T., Moriya, T., Kimura, M., Inoue, T., Ohuchi, N. & Sasano, H. (2001). Progesterone receptor A and B isoforms in the human breast and its disorders., *Japanese Journal of Cancer Research*, vol. 92, no. 3, pp. 302-308.

Aubele, M., Werner, M. & Hofler, H. (2002). Genetic alterations in presumptive precursor lesions of breast carcinomas, *Analytical Cellular Pathology*, vol. 24, no. 2-3, pp. 69-76.

Bai, M., Agnantis, N.J., Kamina, S., Demou, A., Zagorianakou, P., Katsaraki, A. & Kanavaros, P. (2001). In vivo cell kinetics in breast carcinogenesis., *Breast Cancer Research*, vol. 3, no. 4, pp. 276-283.

Barker, P.E. (2003). Cancer biomarker validation: standards and process: roles for the National Institute of Standards and Technology (NIST), *Annals of the New York Academy of Sciences*, vol. 983, pp. 142-150.

Barnes, N.L., Boland, G.P., Davenport, A., Knox, W.F. & Bundred, N.J. (2005). Relationship between hormone receptor status and tumour size, grade and comedo necrosis in ductal carcinoma in situ., *British Journal of Surgery*, vol. 92, no. 4, pp. 429-434.

Berx, G., Cleton-Jansen, A.M., Nollet, F., de Leeuw, W.J., van de Vijver, M., Cornelisse, C. & van Roy, F. (1995). E-cadherin is a tumour/invasion suppressor gene mutated in human lobular breast cancers., *EMBO Journal*, vol. 14, no. 24, pp. 6107-6115.

Bjurstam, N., Bjorneld, L., Warwick, J., Sala, E., Duffy, S.W., Nystrom, L., Walker, N., Cahlin, E., Eriksson, O., Hafstrom, L.O., Lingaas, H., Mattsson, J., Persson, S., Rudenstam, C.M., Salander, H., Save-Soderbergh, J. & Wahlin, T. (2003). The Gothenburg Breast Screening Trial., *Cancer*, vol. 97, no. 10, pp. 2387-2396.

Bose, S., Lesser, M.L., Norton, L. & Rosen, P.P. (1996). Immunophenotype of intraductal carcinoma., *Archives of Pathology & Laboratory Medicine*, vol. 120, no. 1, pp. 81-85.

Bratthauer, G.L., Moinfar, F., Stamatakos, M.D., Mezzetti, T.P., Shekitka, K.M., Man, Y.G. & Tavassoli, F.A. (2002). Combined E-cadherin and high molecular weight cytokeratin immunoprofile differentiates lobular, ductal, and hybrid mammary intraepithelial neoplasias., *Human pathology*, vol. 33, no. 6, pp. 620-627.

Bryan J., Land S., Allred C et al. (2003). DCIS: evidence from randomized trials. Proceedings of the 8th international conference March 12-15, 2003, St Gallen, Switzerland. Breast 12(Suppl. 1):S9 (S24) Breast Cancer Research and Treatment 123.

Buchberger, W., DeKoekkoek-Doll, P., Springer, P., Obrist, P. & Dunser, M. (1999). Incidental findings on sonography of the breast: clinical significance and diagnostic workup., *AJR.American Journal of Roentgenology*, vol. 173, no. 4, pp. 921-927.

Buchberger, W., Niehoff, A., Obrist, P., DeKoekkoek-Doll, P. & Dunser, M. (2000). Clinically and mammographically occult breast lesions: detection and classification with high-resolution sonography., *Seminars in Ultrasound, CT & MR,* vol. 21, no. 4, pp. 325-336.

Burbank, F. (1997). Stereotactic breast biopsy of atypical ductal hyperplasia and ductal carcinoma in situ lesions: improved accuracy with directional, vacuum-assisted biopsy., *Radiology,* vol. 202, no. 3, pp. 843-847.

Burstein, H.J., Polyak, K., Wong, J.S., Lester, S.C. & Kaelin, C.M. (2004). Ductal carcinoma in situ of the breast, *New England Journal of Medicine*, vol. 350, no. 14, pp. 1430-1441.

Chitemerere, M., Andersen, T.I., Holm, R., Karlsen, F., Borresen, A.L. & Nesland, J.M. (1996). TP53 alterations in atypical ductal hyperplasia and ductal carcinoma in situ of the breast., *Breast Cancer Research & Treatment,* vol. 41, no. 2, pp. 103-109.

Cichon, M.A., Degnim A.C, Visscher DW, Radisky DC. (2010). Microenvironmental influences that drive progression from benign breast disease to invasive breast cancer. *Journal of Mammary Gland Biology and Neoplasia*, vol. 15, no. 4, pp. 389-397.

Clarke, R.B., Howell, A., Potten, C.S. & Anderson, E. 1997, Dissociation between steroid receptor expression and cell proliferation in the human breast., *Cancer research,* vol. 57, no. 22, pp. 4987-4991.

Claus, E.B., Chu, P., Howe, C.L., Davison, T.L., Stern, D.F., Carter, D. & DiGiovanna, M.P. (2001). Pathobiologic findings in DCIS of the breast: morphologic features, angiogenesis, HER-2/neu and hormone receptors., *Experimental & Molecular Pathology,* vol. 70, no. 3, pp. 303-316.

Cody, H.S.,3rd (2007). Sentinel lymph node biopsy for breast cancer: indications, contraindications, and new directions., *Journal of surgical oncology,* vol. 95, no. 6, pp. 440-442.

Colozza, M., Azambuja, E., Cardoso, F., Sotiriou, C., Larsimont, D. & Piccart, M.J. (2005). Proliferative markers as prognostic and predictive tools in early breast cancer: where are we now?, *Annals of Oncology,* vol. 16, no. 11, pp. 1723-1739.

Cornfield, D.B., Palazzo, J.P., Schwartz, G.F., Goonewardene, S.A., Kovatich, A.J., Chervoneva, I., Hyslop, T. & Schwarting, R. (2004). The prognostic significance of multiple morphologic features and biologic markers in ductal carcinoma in situ of the breast: a study of a large cohort of patients treated with surgery alone., *Cancer,* vol. 100, no. 11, pp. 2317-2327.

Cox, C.E., Nguyen, K., Gray, R.J., Salud, C., Ku, N.N., Dupont, E., Hutson, L., Peltz, E., Whitehead, G., Reintgen, D. & Cantor, A. (2001). Importance of lymphatic mapping in ductal carcinoma in situ (DCIS): why map DCIS?., *American Surgeon,* vol. 67, no. 6, pp. 513-519.

De Potter, C.R., Van Daele, S., Van de Vijver, M.J., Pauwels, C., Maertens, G., De Boever, J., Vandekerckhove, D. & Roels, H. (1989). The expression of the neu oncogene product in breast lesions and in normal fetal and adult human tissues., *Histopathology,* vol. 15, no. 4, pp. 351-362.

Debniak, T., Gorski, B., Huzarski, T., Byrski, T., Cybulski, C., Mackiewicz, A., Gozdecka-Grodecka, S., Gronwald, J., Kowalska, E., Haus, O., Grzybowska, E., Stawicka, M., Swiec, M., Urbanski, K., Niepsuj, S., Wasko, B., Gozdz, S., Wandzel, P., Szczylik, C., Surdyka, D., Rozmiarek, A., Zambrano, O., Posmyk, M., Narod, S.A. & Lubinski, J. (2005). A common variant of CDKN2A (p16) predisposes to breast cancer., *Journal of medical genetics*, vol. 42, no. 10, pp. 763-765.

Dershaw, D.D., Abramson, A. & Kinne, D.W. (1989). Ductal carcinoma in situ: mammographic findings and clinical implications., *Radiology*, vol. 170, no. 2, pp. 411-415.

Di Vinci, A., Perdelli, L., Banelli, B., Salvi, S., Casciano, I., Gelvi, I., Allemanni, G., Margallo, E., Gatteschi, B. & Romani, M. (2005). p16(INK4a) promoter methylation and protein expression in breast fibroadenoma and carcinoma., *International Journal of Cancer*, vol. 114, no. 3, pp. 414-421.

Done, S.J., Arneson, N.C., Ozcelik, H., Redston, M. & Andrulis, I.L. (1998). p53 mutations in mammary ductal carcinoma in situ but not in epithelial hyperplasias., *Cancer research*, vol. 58, no. 4, pp. 785-789.

Dooley, W.C., Ljung, B.M., Veronesi, U., Cazzaniga, M., Elledge, R.M., O'Shaughnessy, J.A., Kuerer, H.M., Hung, D.T., Khan, S.A., Phillips, R.F., Ganz, P.A., Euhus, D.M., Esserman, L.J., Haffty, B.G., King, B.L., Kelley, M.C., Anderson, M.M., Schmit, P.J., Clark, R.R., Kass, F.C., Anderson, B.O., Troyan, S.L., Arias, R.D., Quiring, J.N., Love, S.M., Page, D.L. & King, E.B. (2001). Ductal lavage for detection of cellular atypia in women at high risk for breast cancer., *Journal of the National Cancer Institute*, vol. 93, no. 21, pp. 1624-1632.

Dotto, G.P. (2000). p21(WAF1/Cip1): more than a break to the cell cycle?, *Biochimica et biophysica acta*, vol. 1471, no. 1, pp. M43-56.

Edorh, A., Leroux, A., N'sossani, B., Parache, R.M. & Rihn, B. (1999). Detection by immunohistochemistry of c-erbB2 oncoprotein in breast carcinomas and benign mammary lesions., *Cellular & Molecular Biology*, vol. 45, no. 6, pp. 831-840

Ellsworth, D.L., Ellsworth R.E., Liebman M.N., Hooke J.A., & Shriver C. D. (2004). Genomic instability in histologically normal breast tissues: implications for carcinogenesis., *Lancet Oncology*, vol.5, no.12, pp.753-8

Emdin, S.O., Granstrand, B., Ringberg, A., Sandelin, K., Arnesson, L.G., Nordgren, H., Anderson, H., Garmo, H., Holmberg, L., Wallgren, A. & Swedish Breast Cancer, G. (2006). SweDCIS: Radiotherapy after sector resection for ductal carcinoma in situ of the breast. Results of a randomised trial in a population offered mammography screening., *Acta Oncologica*, vol. 45, no. 5, pp. 536-543.

EORTC Breast Cancer Cooperative, G., EORTC Radiotherapy, G., Bijker, N., Meijnen, P., Peterse, J.L., Bogaerts, J., Van Hoorebeeck, I., Julien, J.P., Gennaro, M., Rouanet, P., Avril, A., Fentiman, I.S., Bartelink, H. & Rutgers, E.J. (2006). Breast-conserving treatment with or without radiotherapy in ductal carcinoma-in-situ: ten-year results of European Organisation for Research and Treatment of Cancer randomized phase III trial 10853--a study by the EORTC Breast Cancer Cooperative Group and EORTC Radiotherapy Group., *Journal of Clinical Oncology*, vol. 24, no. 21, pp. 3381-3387.

Ernster, V.L., Ballard-Barbash, R., Barlow, W.E., Zheng, Y., Weaver, D.L., Cutter, G., Yankaskas, B.C., Rosenberg, R., Carney, P.A., Kerlikowske, K., Taplin, S.H., Urban, N. & Geller, B.M. (2002). Detection of ductal carcinoma in situ in women undergoing screening mammography., *Journal of the National Cancer Institute*, vol. 94, no. 20, pp. 1546-1554.

Fabian, C.J., Kimler, B.F., Zalles, C.M., Klemp, J.R., Kamel, S., Zeiger, S. & Mayo, M.S. (2000). Short-term breast cancer prediction by random periareolar fine-needle aspiration cytology and the Gail risk model., *Journal of the National Cancer Institute*, vol. 92, no. 15, pp. 1217-1227.

Fadare, O., Dadmanesh, F., Alvarado-Cabrero, I., Snyder, R., Stephen Mitchell, J., Tot, T., Wang, S.A., Ghofrani, M., Eusebi, V., Martel, M. & Tavassoli, F.A. (2006). Lobular intraepithelial neoplasia [lobular carcinoma in situ] with comedo-type necrosis: A clinicopathologic study of 18 cases., *American Journal of Surgical Pathology*, vol. 30, no. 11, pp. 1445-1453.

Farante, G., Zurrida, S., Galimberti, V., Veronesi, P., Curigliano, G., Luini, A., Goldhirsch, A. & Veronesi, U. (2011). The management of ductal intraepithelial neoplasia (DIN): open controversies and guidelines of the Istituto Europeo di Oncologia (IEO), Milan, Italy, *Breast Cancer Research & Treatment*, vol. 128, no. 2, pp. 369-378.

Fiegl H., Millinger S., Goebel G., Müller-Holzner E., Marth C., Laird PW., & Widschwendter M.(2006). Breast cancer DNA methylation profiles in cancer cells and tumor stroma: association with HER-2/neu status in primary breast cancer., *Cancer Research*, vol. 66, no. 1, pp. 29-33.

Fisher, B., Land, S., Mamounas, E., Dignam, J., Fisher, E.R. & Wolmark, N. (2001). Prevention of invasive breast cancer in women with ductal carcinoma in situ: an update of the National Surgical Adjuvant Breast and Bowel Project experience., *Seminars in oncology*, vol. 28, no. 4, pp. 400-418.

Fisher, E.R., Costantino, J., Fisher, B., Palekar, A.S., Paik, S.M., Suarez, C.M. & Wolmark, N. (1996). Pathologic findings from the National Surgical Adjuvant Breast Project (NSABP) Protocol B-17. Five-year observations concerning lobular carcinoma in situ., *Cancer*, vol. 78, no. 7, pp. 1403-1416.

Fisher, E.R., Dignam, J., Tan-Chiu, E., Costantino, J., Fisher, B., Paik, S. & Wolmark, N. (1999). Pathologic findings from the National Surgical Adjuvant Breast Project (NSABP) eight-year update of Protocol B-17: intraductal carcinoma., *Cancer*, vol. 86, no. 3, pp. 429-438.

Fisher, E.R., Leeming, R., Anderson, S., Redmond, C. & Fisher, B. (1991). Conservative management of intraductal carcinoma (DCIS) of the breast. Collaborating NSABP investigators., *Journal of surgical oncology*, vol. 47, no. 3, pp. 139-147.

Foote, F.W. & Stewart, F,W. (1941). Lobular carcinoma in situ. A rare form of mammary cancer. *American Journal of Pathology*, vol. 17, pp. 491-496.

Frisell, J., Lidbrink, E., Hellstrom, L. & Rutqvist, L.E. (1997). Followup after 11 years--update of mortality results in the Stockholm mammographic screening trial., *Breast Cancer Research & Treatment*, vol. 45, no. 3, pp. 263-270.

Gasco, M., Shami, S. & Crook, T. (2002). The p53 pathway in breast cancer, *Breast Cancer Research*, vol. 4, no. 2, pp. 70-76.

Gerdes, J., Lemke, H., Baisch, H., Wacker, H.H., Schwab, U. & Stein, H. (1984). Cell cycle analysis of a cell proliferation-associated human nuclear antigen defined by the monoclonal antibody Ki-67., *Journal of Immunology*, vol. 133, no. 4, pp. 1710-1715.

Gobbi, H., Dupont, W.D., Parl, F.F., Schuyler, P.A., Plummer, W.D., Olson, S.J. & Page, D.L. (2005). Breast cancer risk associated with estrogen receptor expression in epithelial hyperplasia lacking atypia and adjacent lobular units., *International Journal of Cancer*, vol. 113, no. 5, pp. 857-859.

Gobbi, H., Dupont, W.D., Simpson, J.F., Plummer, W.D.,Jr, Schuyler, P.A., Olson, S.J., Arteaga, C.L. & Page, D.L. (1999). Transforming growth factor-beta and breast cancer risk in women with mammary epithelial hyperplasia., *Journal of the National Cancer Institute*, vol. 91, no. 24, pp. 2096-2101.

Goldhirsch, A., Ingle, J.N., Gelber, R.D., Coates, A.S., Thurlimann, B., Senn, H.J. & Panel, m. (2009). Thresholds for therapies: highlights of the St Gallen International Expert Consensus on the primary therapy of early breast cancer 2009, *Annals of Oncology*, vol. 20, no. 8, pp. 1319-1329.

Graham, J.D., Yager, M.L., Hill, H.D., Byth, K., O'Neill, G.M. & Clarke, C.L. (2005). Altered progesterone receptor isoform expression remodels progestin responsiveness of breast cancer cells., *Molecular Endocrinology*, vol. 19, no. 11, pp. 2713-2735.

Guerrieri-Gonzaga, A., Botteri, E., Rotmensz, N., Bassi, F., Intra, M., Serrano, D., Renne, G., Luini, A., Cazzaniga, M., Goldhirsch, A., Colleoni, M., Viale, G., Ivaldi, G., Bagnardi, V., Lazzeroni, M., Decensi, A., Veronesi, U. & Bonanni, B. (2009). Ductal intraepithelial neoplasia: postsurgical outcome for 1,267 women cared for in one single institution over 10 years., *Oncologist*, vol. 14, no. 3, pp. 201-212.

Gupta, S.K., Douglas-Jones, A.G., Jasani, B., Morgan, J.M., Pignatelli, M. & Mansel, R.E. (1997). E-cadherin (E-cad) expression in duct carcinoma in situ (DCIS) of the breast., *Virchows Archiv*, vol. 430, no. 1, pp. 23-28.

Gusterson, B.A., Machin, L.G., Gullick, W.J., Gibbs, N.M., Powles, T.J., Elliott, C., Ashley, S., Monaghan, P. & Harrison, S. (1988). c-erbB-2 expression in benign and malignant breast disease., *British journal of cancer*, vol. 58, no. 4, pp. 453-457.

Hannafon, B.N., Sebastiani, P., de Las Morenas, A., Lu, J.,& Rosenberg, C.L. (2011). Expression of microRNA and their gene targets are dysregulated in preinvasive breast cancer. *Breast Cancer Research & Treatment*, vol. 13, no. 2 , pp. R24

Heffelfinger, S.C., Yassin, R., Miller, M.A. & Lower, E.E. (2000). Cyclin D1, retinoblastoma, p53, and Her2/neu protein expression in preinvasive breast pathologies: correlation with vascularity., *Pathobiology*, vol. 68, no. 3, pp. 129-136.

Henderson, B.E., Ross, R. & Bernstein, L. (1988). Estrogens as a cause of human cancer: the Richard and Linda Rosenthal Foundation award lecture, *Cancer research*, vol. 48, no. 2, pp. 246-253.

Hieken, T.J., Farolan, M., D'Alessandro, S. & Velasco, J.M. (2001). Predicting the biologic behavior of ductal carcinoma in situ: an analysis of molecular markers., *Surgery*, vol. 130, no. 4, pp. 593-600.

Hockenbery, D.M. (1994). bcl-2 in cancer, development and apoptosis, *Journal of Cell Science - Supplement*, vol. 18, pp. 51-55.

Holland, R., Peterse, J.L., Millis, R.R., Eusebi, V., Faverly, D., van de Vijver, M.J. & Zafrani, B. (1994). Ductal carcinoma in situ: a proposal for a new classification, *Seminars in diagnostic pathology*, vol. 11, no. 3, pp. 167-180.

Houghton, J. George, W.D. Cuzick, J. Duggan, C. Fentiman, IS. Spittle, M. UK Coordinating Committee on Cancer Research. Ductal Carcinoma in situ Working Party. DCIS trialists in the UK, Australia,and New Zealand (2003). Radiotherapy and tamoxifen in women with completely excised ductal carcinoma in situ of the breast in the UK, Australia, and New Zealand: randomised controlled trial., *Lancet*, vol. 362, no. 9378, pp. 95-102.

Hudelist, G., Czerwenka, K., Kubista, E., Marton, E., Pischinger, K. & Singer, C.F. (2003). Expression of sex steroid receptors and their co-factors in normal and malignant breast tissue: AIB1 is a carcinoma-specific co-activator., *Breast Cancer Research & Treatment*, vol. 78, no. 2, pp. 193-204.

Hughes, L.L., Wang, M., Page, D.L., Gray, R., Solin, L.J., Davidson, N.E., Lowen, M.A., Ingle, J.N., Recht, A. & Wood, W.C. (2009). Local excision alone without irradiation for ductal carcinoma in situ of the breast: a trial of the Eastern Cooperative Oncology Group., *Journal of Clinical Oncology*, vol. 27, no. 32, pp. 5319-5324.

Hwang, E.S., Kinkel, K., Esserman, L.J., Lu, Y., Weidner, N. & Hylton, N.M. (2003). Magnetic resonance imaging in patients diagnosed with ductal carcinoma-in-situ: value in the diagnosis of residual disease, occult invasion, and multicentricity., *Annals of Surgical Oncology*, vol. 10, no. 4, pp. 381-388.

Ikeda, D.M. & Andersson, I. (1989). Ductal carcinoma in situ: atypical mammographic appearances., *Radiology*, vol. 172, no. 3, pp. 661-666.

Intra, M., Rotmensz, N., Mattar, D., Gentilini, O.D., Vento, A., Veronesi, P., Colleoni, M., De Cicco, C., Cassano, E., Luini, A. & Veronesi, U. (2007). Unnecessary axillary node dissections in the sentinel lymph node era., *European journal of cancer*, vol. 43, no. 18, pp. 2664-2668.

Intra, M., Rotmensz, N., Veronesi, P., Colleoni, M., Iodice, S., Paganelli, G., Viale, G. & Veronesi, U. (2008). Sentinel node biopsy is not a standard procedure in ductal carcinoma in situ of the breast: the experience of the European institute of oncology on 854 patients in 10 years., *Annals of Surgery*, vol. 247, no. 2, pp. 315-319.

Intra, M., Veronesi, P., Mazzarol, G., Galimberti, V., Luini, A., Sacchini, V., Trifiro, G., Gentilini, O., Pruneri, G., Naninato, P., Torres, F., Paganelli, G., Viale, G. & Veronesi, U. (2003). Axillary sentinel lymph node biopsy in patients with pure ductal carcinoma in situ of the breast., *Archives of Surgery*, vol. 138, no. 3, pp. 309-313.

Iqbal, M., Davies, M.P., Shoker, B.S., Jarvis, C., Sibson, D.R. & Sloane, J.P. (2001). Subgroups of non-atypical hyperplasia of breast defined by proliferation of oestrogen receptor-positive cells., *Journal of Pathology*, vol. 193, no. 3, pp. 333-338.

Jacobs, E.J., Feigelson, H.S., Bain, E.B., Brady, K.A., Rodriguez, C., Stevens, V.L., Patel, A.V., Thun, M.J. & Calle, E.E. (2006). Polymorphisms in the vascular endothelial growth factor gene and breast cancer in the Cancer Prevention Study II cohort., *Breast Cancer Research*, vol. 8, no. 2, pp. R22.

Julian, T.B., Land, S.R., Fourchotte, V., Haile, S.R., Fisher, E.R., Mamounas, E.P., Costantino, J.P. & Wolmark, N. (2007). Is sentinel node biopsy necessary in conservatively treated DCIS?, *Annals of Surgical Oncology,* vol. 14, no. 8, pp. 2202-2208.

Kang, J.H., Kim, S.J., Noh, D.Y., Choe, K.J., Lee, E.S. & Kang, H.S. (2001). The timing and characterization of p53 mutations in progression from atypical ductal hyperplasia to invasive lesions in the breast cancer., *Journal of Molecular Medicine,* vol. 79, no. 11, pp. 648-655.

Kapucuoglu, N., Losi, L. & Eusebi, V. (1997). Immunohistochemical localization of Bcl-2 and Bax proteins in in situ and invasive duct breast carcinomas., *Virchows Archiv,* vol. 430, no. 1, pp. 17-22.

Keohavong, P., Gao, W.M., Mady, H.H., Kanbour-Shakir, A. & Melhem, M.F. (2004). Analysis of p53 mutations in cells taken from paraffin-embedded tissue sections of ductal carcinoma in situ and atypical ductal hyperplasia of the breast., *Cancer letters,* vol. 212, no. 1, pp. 121-130.

Kepple, J., Henry-Tillman, R.S., Klimberg, V.S., Layeeque, R., Siegel, E., Westbrook, K. & Korourian, S. (2006). The receptor expression pattern in ductal carcinoma in situ predicts recurrence., *American Journal of Surgery,* vol. 192, no. 1, pp. 68-71.

Kerlikowske, K., Molinaro, A., Cha, I., Ljung, B.M., Ernster, V.L., Stewart, K., Chew, K., Moore, D.H. 2nd & Waldman, F. (2003). Characteristics associated with recurrence among women with ductal carcinoma in situ treated by lumpectomy., *Journal of the National Cancer Institute,* vol. 95, no. 22, pp. 1692-1702.

Khan, S.A., Rogers, M.A., Khurana, K.K., Meguid, M.M. & Numann, P.J. (1998). Estrogen receptor expression in benign breast epithelium and breast cancer risk., *Journal of the National Cancer Institute,* vol. 90, no. 1, pp. 37-42.

Kim, R. (2005). Unknotting the roles of Bcl-2 and Bcl-xL in cell death, *Biochemical & Biophysical Research Communications,* vol. 333, no. 2, pp. 336-343.

Krassenstein, R., Sauter, E., Dulaimi, E., Battagli, C., Ehya, H., Klein-Szanto, A. & Cairns, P. (2004). Detection of breast cancer in nipple aspirate fluid by CpG island hypermethylation., *Clinical Cancer Research,* vol. 10, no. 1 Pt 1, pp. 28-32.

Kretschmer ,C., Sterner-Kock, A., Siedentopf, F., Schoenegg, W., Schlag, P.M., & Kemmner, W., (2011). Identification of early molecular markers for breast cancer. *Molecular Cancer.* vol.10, no.1, pp. 15.

Krishnamurthy, S., Sneige, N., Thompson, P.A., Marcy, S.M., Singletary, S.E., Cristofanilli, M., Hunt, K.K. & Kuerer, H.M. (2003). Nipple aspirate fluid cytology in breast carcinoma., *Cancer,* vol. 99, no. 2, pp. 97-104.

Krogerus L.A., Leivonen M. & Häastö A.L. (2000). Expression patterns of biologic markers in small breast cancers and preneoplastic breast lesions. *Breast,* vol.9, no.5, pp.281-285.

Kuerer, H.M., Albarracin, C.T., Yang, W.T., Cardiff, R.D., Brewster, A.M., Symmans, W.F., Hylton, N.M., Middleton, L.P., Krishnamurthy, S., Perkins, G.H., Babiera, G., Edgerton, M.E., Czerniecki, B.J., Arun, B.K. & Hortobagyi, G.N. (2009). Ductal carcinoma in situ: state of the science and roadmap to advance the field, *Journal of Clinical Oncology,* vol. 27, no. 2, pp. 279-288.

Lagios, M.D. (1990). Duct carcinoma in situ. Pathology and treatment., *Surgical Clinics of North America*, vol. 70, no. 4, pp. 853-871.

Lebeau, A., Unholzer, A., Amann, G., Kronawitter, M., Bauerfeind, I., Sendelhofert, A., Iff, A. & Lohrs, U. (2003). EGFR, HER-2/neu, cyclin D1, p21 and p53 in correlation to cell proliferation and steroid hormone receptor status in ductal carcinoma in situ of the breast., *Breast Cancer Research & Treatment*, vol. 79, no. 2, pp. 187-198.

Lebrecht, A., Buchmann, J., Hefler, L., Lampe, D. & Koelbl, H. (2002). Histological category and expression of hormone receptors in ductal carcinoma in situ of the breast., *Anticancer Research*, vol. 22, no. 3, pp. 1909-1911.

Lebrecht, A., Grimm, C., Euller, G., Ludwig, E., Ulbrich, E., Lantzsch, T., Hefler, L. & Koelbl, H. (2004). Transforming growth factor beta 1 serum levels in patients with preinvasive and invasive lesions of the breast., *International Journal of Biological Markers*, vol. 19, no. 3, pp. 236-239.

Lee, C.H., Carter, D., Philpotts, L.E., Couce, M.E., Horvath, L.J., Lange, R.C. & Tocino, I. (2000). Ductal carcinoma in situ diagnosed with stereotactic core needle biopsy: can invasion be predicted?., *Radiology*, vol. 217, no. 2, pp. 466-470.

Lee, S., Mohsin, S.K., Mao, S., Hilsenbeck, S.G., Medina, D. & Allred, D.C. (2006). Hormones, receptors, and growth in hyperplastic enlarged lobular units: early potential precursors of breast cancer., *Breast Cancer Research*, vol. 8, no. 1, pp. R6.

Lehman, C.D., Gatsonis, C., Kuhl, C.K., Hendrick, R.E., Pisano, E.D., Hanna, L., Peacock, S., Smazal, S.F., Maki, D.D., Julian, T.B., DePeri, E.R., Bluemke, D.A., Schnall, M.D. & ACRIN Trial 6667 Investigators, G. (2007). MRI evaluation of the contralateral breast in women with recently diagnosed breast cancer., *New England Journal of Medicine*, vol. 356, no. 13, pp. 1295-1303.

Leonard, G.D. & Swain, S.M. (2004). Ductal carcinoma in situ, complexities and challenges, *Journal of the National Cancer Institute*, vol. 96, no. 12, pp. 906-920.

Lerwill, M.F. (2004). Current practical applications of diagnostic immunohistochemistry in breast pathology, *American Journal of Surgical Pathology*, vol. 28, no. 8, pp. 1076-1091.

Li, C.I., Malone, K.E., Saltzman, B.S. & Daling, J.R. (2006). Risk of invasive breast carcinoma among women diagnosed with ductal carcinoma in situ and lobular carcinoma in situ, 1988-2001., *Cancer*, vol. 106, no. 10, pp. 2104-2112.

Madu, C.O. & Lu, Y. (2010). Novel diagnostic biomarkers for prostate cancer., *Journal of Cancer*, vol. 1, pp. 150-177.

Mansel, R.E. (2003). Ductal carcinoma in situ: surgery and radiotherapy., *Breast*, vol. 12, no. 6, pp. 447-450.

Mastracci, T.L., Tjan, S., Bane, A.L., O'Malley, F.P. & Andrulis, I.L. (2005). E-cadherin alterations in atypical lobular hyperplasia and lobular carcinoma in situ of the breast., *Modern Pathology*, vol. 18, no. 6, pp. 741-751.

McGowan, E.M. & Clarke, C.L. (1999). Effect of overexpression of progesterone receptor A on endogenous progestin-sensitive endpoints in breast cancer cells., *Molecular Endocrinology*, vol. 13, no. 10, pp. 1657-1671.

McLaren, B.K., Gobbi, H., Schuyler, P.A., Olson, S.J., Parl, F.F., Dupont, W.D. & Page, D.L. (2005). Immunohistochemical expression of estrogen receptor in enlarged lobular units with columnar alteration in benign breast biopsies: a nested case-control study., *American Journal of Surgical Pathology*, vol. 29, no. 1, pp. 105-108.

Megha, T., Ferrari, F., Benvenuto, A., Bellan, C., Lalinga, A.V., Lazzi, S., Bartolommei, S., Cevenini, G., Leoncini, L. & Tosi, P. (2002). p53 mutation in breast cancer. Correlation with cell kinetics and cell of origin., *Journal of clinical pathology*, vol. 55, no. 6, pp. 461-466.

Menell, J.H., Morris, E.A., Dershaw, D.D., Abramson, A.F., Brogi, E. & Liberman, L. (2005). Determination of the presence and extent of pure ductal carcinoma in situ by mammography and magnetic resonance imaging., *Breast Journal*, vol. 11, no. 6, pp. 382-390.

Meteoglu, I., Dikicioglu, E., Erkus, M., Culhaci, N., Kacar, F., Ozkara, E. & Uyar, M. (2005). Breast carcinogenesis. Transition from hyperplasia to invasive lesions., *Saudi medical journal*, vol. 26, no. 12, pp. 1889-1896.

Midulla, C., Pisani, T., De Iorio, P., Cenci, M., Divizia, E., Nofroni, I. & Vecchione, A. (2002). Cytological analysis and immunocytochemical expression of Ki67 and Bcl-2 in breast proliferative lesions., *Anticancer Research*, vol. 22, no. 2B, pp. 1341-1345.

Miller, A.B., To, T., Baines, C.J. & Wall, C. (2002). The Canadian National Breast Screening Study-1: breast cancer mortality after 11 to 16 years of follow-up. A randomized screening trial of mammography in women age 40 to 49 years, *Annals of Internal Medicine*, vol. 137, no. 5 Part 1, pp. 305-312.

Miller, A.B., To, T., Baines, C.J. & Wall, C. (2000). Canadian National Breast Screening Study-2: 13-year results of a randomized trial in women aged 50-59 years., *Journal of the National Cancer Institute*, vol. 92, no. 18, pp. 1490-1499.

Mohsin, S.K., O'Connell, P., Allred, D.C. & Libby, A.L. (2005). Biomarker profile and genetic abnormalities in lobular carcinoma in situ., *Breast Cancer Research & Treatment*, vol. 90, no. 3, pp. 249-256.

Morrow, M., Strom, E.A., Bassett, L.W., Dershaw, D.D., Fowble, B., Harris, J.R., O'Malley, F., Schnitt, S.J., Singletary, S.E., Winchester, D.P., American College of, S., College of American, P., Society of Surgical, O. & American College of, R. (2002). Standard for the management of ductal carcinoma in situ of the breast (DCIS), *CA: a Cancer Journal for Clinicians*, vol. 52, no. 5, pp. 256-276.

Mote, P.A., Bartow, S., Tran, N. & Clarke, C.L. (2002). Loss of co-ordinate expression of progesterone receptors A and B is an early event in breast carcinogenesis., *Breast Cancer Research & Treatment*, vol. 72, no. 2, pp. 163-172.

Mote, P.A., Leary, J.A., Avery, K.A., Sandelin, K., Chenevix-Trench, G., Kirk, J.A., Clarke, C.L. & kConFab, I. (2004). Germ-line mutations in BRCA1 or BRCA2 in the normal breast are associated with altered expression of estrogen-responsive proteins and the predominance of progesterone receptor A., *Genes, chromosomes & cancer*, vol. 39, no. 3, pp. 236-248.

Musgrove, E.A., Davison, E.A. & Ormandy, C.J. (2004). Role of the CDK inhibitor p27 (Kip1) in mammary development and carcinogenesis: insights from knockout mice, *Journal of Mammary Gland Biology & Neoplasia*, vol. 9, no. 1, pp. 55-66.

Mustonen, M., Raunio, H., Paakko, P. & Soini, Y. (1997). The extent of apoptosis is inversely associated with bcl-2 expression in premalignant and malignant breast lesions., *Histopathology*, vol. 31, no. 4, pp. 347-354.

Mylonas, I., Makovitzky, J., Jeschke, U., Briese, V., Friese, K. & Gerber, B. (2005). Expression of Her2/neu, steroid receptors (ER and PR), Ki67 and p53 in invasive mammary ductal carcinoma associated with ductal carcinoma In Situ (DCIS) Versus invasive breast cancer alone., *Anticancer Research*, vol. 25, no. 3A, pp. 1719-1723.

Nass, S.J., Herman, J.G., Gabrielson, E., Iversen, P.W., Parl, F.F., Davidson, N.E. & Graff, J.R. (2000). Aberrant methylation of the estrogen receptor and E-cadherin 5' CpG islands increases with malignant progression in human breast cancer., *Cancer research*, vol. 60, no. 16, pp. 4346-4348.

National Comprehensive Cancer Network (NCCN) Guidelines for Treatment of Cancer by Site: Breast Cancer. (2009). Available from http://www.nccn.org/professionals/physician_gls/breast.pdf

Nofech-Mozes, S., Spayne, J., Rakovitch, E. & Hanna, W. (2005). Prognostic and predictive molecular markers in DCIS: a review, *Advances in Anatomic Pathology*, vol. 12, no. 5, pp. 256-264.

Nystrom, L., Andersson, I., Bjurstam, N., Frisell, J., Nordenskjold, B. & Rutqvist, L.E. (2002). Long-term effects of mammography screening: updated overview of the Swedish randomised trials, *Lancet*, vol. 359, no. 9310, pp. 909-919.

Oh, Y.L., Choi, J.S., Song, S.Y., Ko, Y.H., Han, B.K., Nam, S.J. & Yang, J.H. (2001). Expression of p21Waf1, p27Kip1 and cyclin D1 proteins in breast ductal carcinoma in situ: Relation with clinicopathologic characteristics and with p53 expression and estrogen receptor status., *Pathology international*, vol. 51, no. 2, pp. 94-99.

O'Malley, F.P., Vnencak-Jones, C.L., Dupont, W.D., Parl, F., Manning, S. & Page, D.L. (1994). p53 mutations are confined to the comedo type ductal carcinoma in situ of the breast. Immunohistochemical and sequencing data., *Laboratory Investigation*, vol. 71, no. 1, pp. 67-72.

O'Neill, P.A., Shaaban, A.M., West, C.R., Dodson, A., Jarvis, C., Moore, P., Davies, M.P., Sibson, D.R. & Foster, C.S. (2004). Increased risk of malignant progression in benign proliferating breast lesions defined by expression of heat shock protein 27., *British journal of cancer*, vol. 90, no. 1, pp. 182-188.

Orel, S.G., Mendonca, M.H., Reynolds, C., Schnall, M.D., Solin, L.J. & Sullivan, D.C. (1997). MR imaging of ductal carcinoma in situ., *Radiology*, vol. 202, no. 2, pp. 413-420.

Page, D.L., Dupont, W.D., Rogers, L.W. & Landenberger, M. (1982). Intraductal carcinoma of the breast: follow-up after biopsy only., *Cancer*, vol. 49, no. 4, pp. 751-758.

Papantoniou, V., Tsiouris, S., Koutsikos, J., Sotiropoulou, M., Mainta, E., Lazaris, D., Valsamaki, P., Melissinou, M., Zerva, C. & Antsaklis, A. (2006). Scintimammographic detection of usual ductal breast hyperplasia with increased proliferation rate at risk for malignancy., *Nuclear medicine communications*, vol. 27, no. 11, pp. 911-917.

Parker, S.H., Burbank, F., Jackman, R.J., Aucreman, C.J., Cardenosa, G., Cink, T.M., Coscia, J.L.,Jr, Eklund, G.W., Evans, W.P.,3rd & Garver, P.R. (1994). Percutaneous large-core breast biopsy: a multi-institutional study., *Radiology,* vol. 193, no. 2, pp. 359-364.

Petersen, O.W., Hoyer, P.E. & van Deurs, B. (1987). Frequency and distribution of estrogen receptor-positive cells in normal, nonlactating human breast tissue., *Cancer research,* vol. 47, no. 21, pp. 5748-5751.

Poller, D.N., Roberts, E.C., Bell, J.A., Elston, C.W., Blamey, R.W. & Ellis, I.O. (1993). p53 protein expression in mammary ductal carcinoma in situ: relationship to immunohistochemical expression of estrogen receptor and c-erbB-2 protein., *Human pathology,* vol. 24, no. 5, pp. 463-468.

Provenzano, E., Hopper, J.L., Giles, G.G., Marr, G., Venter, D.J. & Armes, J.E. (2003). Biological markers that predict clinical recurrence in ductal carcinoma in situ of the breast., *European journal of cancer,* vol. 39, no. 5, pp. 622-630.

Putti, T.C., Pinder, S.E., Elston, C.W., Lee, A.H. & Ellis, I.O. (2005). Breast pathology practice: most common problems in a consultation service, *Histopathology,* vol. 47, no. 5, pp. 445-457.

Reed, J.C. (1994). Bcl-2 and the regulation of programmed cell death, *Journal of Cell Biology,* vol. 124, no. 1-2, pp. 1-6.

Rieger-Christ, K.M., Pezza, J.A., Dugan, J.M., Braasch, J.W., Hughes, K.S. & Summerhayes, I.C. (2001). Disparate E-cadherin mutations in LCIS and associated invasive breast carcinomas., *Molecular Pathology,* vol. 54, no. 2, pp. 91-97.

Ringberg, A., Anagnostaki, L., Anderson, H., Idvall, I., Ferno, M. & South Sweden Breast Cancer, G. (2001). Cell biological factors in ductal carcinoma in situ (DCIS) of the breast-relationship to ipsilateral local recurrence and histopathological characteristics., *European journal of cancer,* vol. 37, no. 12, pp. 1514-1522.

Roberts, M.M., Alexander, F.E., Anderson, T.J., Forrest, A.P., Hepburn, W., Huggins, A., Kirkpatrick, A.E., Lamb, J., Lutz, W. & Muir, B.B. (1984). The Edinburgh randomised trial of screening for breast cancer: description of method., *British journal of cancer,* vol. 50, no. 1, pp. 1-6.

Rocco, J.W. & Sidransky, D. (2001). p16(MTS-1/CDKN2/INK4a) in cancer progression, *Experimental cell research,* vol. 264, no. 1, pp. 42-55.

Rody, A., Diallo, R., Poremba, C., Speich, R., Wuelfing, P., Kissler, S., Solbach, C., Kiesel, L. & Jackisch, C. (2004). Estrogen receptor alpha and beta, progesterone receptor, pS2 and HER-2/neu expression delineate different subgroups in ductal carcinoma in situ of the breast., *Oncology reports,* vol. 12, no. 4, pp. 695-699.

Roger, P., Sahla, M.E., Makela, S., Gustafsson, J.A., Baldet, P. & Rochefort, H. (2001). Decreased expression of estrogen receptor beta protein in proliferative preinvasive mammary tumors., *Cancer research,* vol. 61, no. 6, pp. 2537-2541.

Rohan, T.E., Hartwick, W., Miller, A.B. & Kandel, R.A. (1998). Immunohistochemical detection of c-erbB-2 and p53 in benign breast disease and breast cancer risk., *Journal of the National Cancer Institute,* vol. 90, no. 17, pp. 1262-1269.

Ross, J.S. & Fletcher, J.A. (1999). HER-2/neu (c-erb-B2) gene and protein in breast cancer, *American Journal of Clinical Pathology,* vol. 112, no. 1 Suppl 1, pp. S53-67.

Rudas, M., Neumayer, R., Gnant, M.F., Mittelbock, M., Jakesz, R. & Reiner, A. (1997). p53 protein expression, cell proliferation and steroid hormone receptors in ductal and lobular in situ carcinomas of the breast., *European journal of cancer*, vol. 33, no. 1, pp. 39-44.

Santamaria, G., Velasco, M., Farrus, B., Zanon, G. & Fernandez, P.L. (2008). Preoperative MRI of pure intraductal breast carcinoma--a valuable adjunct to mammography in assessing cancer extent., *Breast*, vol. 17, no. 2, pp. 186-194.

Sapino, A., Frigerio, A., Peterse, J.L., Arisio, R., Coluccia, C. & Bussolati, G. (2000). Mammographically detected in situ lobular carcinomas of the breast., *Virchows Archiv*, vol. 436, no. 5, pp. 421-430.

Schnitt, S.J., Silen, W., Sadowsky, N.L., Connolly, J.L. & Harris, J.R. (1988). Ductal carcinoma in situ (intraductal carcinoma) of the breast, *New England Journal of Medicine*, vol. 318, no. 14, pp. 898-903.

Schoonjans, J.M. & Brem, R.F. (2000). Sonographic appearance of ductal carcinoma in situ diagnosed with ultrasonographically guided large core needle biopsy: correlation with mammographic and pathologic findings., *Journal of Ultrasound in Medicine*, vol. 19, no. 7, pp. 449-457.

Schwartz, G.F., Solin, L.J., Olivotto, I.A., Ernster, V.L. & Pressman, P.I. (2000). Consensus Conference on the Treatment of In Situ Ductal Carcinoma of the Breast, April 22-25, 1999, *Cancer*, vol. 88, no. 4, pp. 946-954.

Shaaban, A.M., Jarvis, C., Moore, F., West, C., Dodson, A. & Foster, C.S. (2005). Prognostic significance of estrogen receptor Beta in epithelial hyperplasia of usual type with known outcome., *American Journal of Surgical Pathology*, vol. 29, no. 12, pp. 1593-1599.

Shaaban, A.M., O'Neill, P.A., Davies, M.P., Sibson, R., West, C.R., Smith, P.H. & Foster, C.S. (2003). Declining estrogen receptor-beta expression defines malignant progression of human breast neoplasia., *American Journal of Surgical Pathology*, vol. 27, no. 12, pp. 1502-1512.

Shaaban, A.M., Sloane, J.P., West, C.R. & Foster, C.S. (2002). Breast cancer risk in usual ductal hyperplasia is defined by estrogen receptor-alpha and Ki-67 expression., *American Journal of Pathology*, vol. 160, no. 2, pp. 597-604.

Shapiro, S. (1997). Periodic screening for breast cancer: the HIP Randomized Controlled Trial. Health Insurance Plan., *Journal of the National Cancer Institute*, vol. Monographs, no. 22, pp. 27-30.

Shoker, B.S., Jarvis, C., Clarke, R.B., Anderson, E., Hewlett, J., Davies, M.P., Sibson, D.R. & Sloane, J.P. (1999). Estrogen receptor-positive proliferating cells in the normal and precancerous breast., *American Journal of Pathology*, vol. 155, no. 6, pp. 1811-1815.

Shoker, B.S., Jarvis, C., Clarke, R.B., Anderson, E., Munro, C., Davies, M.P., Sibson, D.R. & Sloane, J.P. (2000). Abnormal regulation of the oestrogen receptor in benign breast lesions., *Journal of clinical pathology*, vol. 53, no. 10, pp. 778-783.

Sickles E.A. (1983). Sonographic detectability of breast calcifications., *Proceedings of SPIE*, vol. 419, pp. 51-52

Silverstein, M.J., Cohlan, B.F., Gierson, E.D., Furmanski, M., Gamagami, P., Colburn, W.J., Lewinsky, B.S. & Waisman, J.R. (1992). Duct carcinoma in situ: 227 cases without microinvasion., *European journal of cancer*, vol. 28, no. 2-3, pp. 630-634.

Silverstein, M.J., Recht, A., Lagios, M.D., Bleiweiss, I.J., Blumencranz, P.W., Gizienski, T., Harms, S.E., Harness, J., Jackman, R.J., Klimberg, V.S., Kuske, R., Levine, G.M., Linver, M.N., Rafferty, E.A., Rugo, H., Schilling, K., Tripathy, D., Vicini, F.A., Whitworth, P.W. & Willey, S.C. (2009). Special report: Consensus conference III. Image-detected breast cancer: state-of-the-art diagnosis and treatment, *Journal of the American College of Surgeons*, vol. 209, no. 4, pp. 504-520.

Simooka, H., Oyama, T., Sano, T., Horiguchi, J. & Nakajima, T. (2004). Immunohistochemical analysis of 14-3-3 sigma and related proteins in hyperplastic and neoplastic breast lesions, with particular reference to early carcinogenesis., *Pathology international*, vol. 54, no. 8, pp. 595-602.

Siziopikou, K.P., Prioleau, J.E., Harris, J.R. & Schnitt, S.J. (1996). bcl-2 expression in the spectrum of preinvasive breast lesions., *Cancer*, vol. 77, no. 3, pp. 499-506.

Slanetz, P.J., Giardino, A.A., Oyama, T., Koerner, F.C., Halpern, E.F., Moore, R.H. & Kopans, D.B. (2001). Mammographic appearance of ductal carcinoma in situ does not reliably predict histologic subtype., *Breast Journal*, vol. 7, no. 6, pp. 417-421.

Smith, G.L., Smith, B.D. & Haffty, B.G. (2006). Rationalization and regionalization of treatment for ductal carcinoma in situ of the breast., *International journal of radiation oncology, biology, physics*, vol. 65, no. 5, pp. 1397-1403.

Sneige, N., Wang, J., Baker, B.A., Krishnamurthy, S. & Middleton, L.P. (2002). Clinical, histopathologic, and biologic features of pleomorphic lobular (ductal-lobular) carcinoma in situ of the breast: a report of 24 cases., *Modern Pathology*, vol. 15, no. 10, pp. 1044-1050.

Stark, A., Hulka, B.S., Joens, S., Novotny, D., Thor, A.D., Wold, L.E., Schell, M.J., Melton, L.J.,3rd, Liu, E.T. & Conway, K. (2000). HER-2/neu amplification in benign breast disease and the risk of subsequent breast cancer., *Journal of Clinical Oncology*, vol. 18, no. 2, pp. 267-274.

Stomper, P.C., Connolly, J.L., Meyer, J.E. & Harris, J.R. (1989). Clinically occult ductal carcinoma in situ detected with mammography: analysis of 100 cases with radiologic-pathologic correlation., *Radiology*, vol. 172, no. 1, pp. 235-241.

Tabar, L., Vitak, B., Chen, H.H., Duffy, S.W., Yen, M.F., Chiang, C.F., Krusemo, U.B., Tot, T. & Smith, R.A. (2000). The Swedish Two-County Trial twenty years later. Updated mortality results and new insights from long-term follow-up., *Radiologic clinics of North America*, vol. 38, no. 4, pp. 625-651.

Tavassoli, F.A. (2008). Lobular and ductal intraepithelial neoplasia, *Pathologe*, vol. 29, no. Suppl 2, pp. 107-111.

Tavassoli, F.A., Hoeffler H, Rosai J et al. 2003 Intraductal proliferative lesions In: Tavassoli F.A,, Devilee P (eds) World Health Organization Classification of Tumors: Pathology and genetics of Tumor of the breast and the female genital organs IARC Press, Lyon pp 65-66

Toledo, F. & Wahl, G.M. (2006). Regulating the p53 pathway: in vitro hypotheses, in vivo veritas, *Nature Reviews.Cancer*, vol. 6, no. 12, pp. 909-923.

Tsuda, H. & Hirohashi, S. (1998). Multiple developmental pathways of highly aggressive breast cancers disclosed by comparison of histological grades and c-erbB-2 expression patterns in both the non-invasive and invasive portions., *Pathology international,* vol. 48, no. 7, pp. 518-525.

Tsuda, H., Iwaya, K., Fukutomi, T. & Hirohashi, S. (1993). p53 mutations and c-erbB-2 amplification in intraductal and invasive breast carcinomas of high histologic grade., *Japanese Journal of Cancer Research,* vol. 84, no. 4, pp. 394-401.

Umbricht, C.B., Evron, E., Gabrielson, E., Ferguson, A., Marks, J. & Sukumar, S. (2001). Hypermethylation of 14-3-3 sigma (stratifin) is an early event in breast cancer., *Oncogene,* vol. 20, no. 26, pp. 3348-3353.

van Delft, M.F. & Huang, D.C. (2006). How the Bcl-2 family of proteins interact to regulate apoptosis, *Cell research,* vol. 16, no. 2, pp. 203-213.

Van Zee, K.J., Calvano, J.E. & Bisogna, M. (1998). Hypomethylation and increased gene expression of p16INK4a in primary and metastatic breast carcinoma as compared to normal breast tissue., *Oncogene,* vol. 16, no. 21, pp. 2723-2727.

Viacava, P., Naccarato, A.G. & Bevilacqua, G. (1999). Different proliferative patterns characterize different preinvasive breast lesions., *Journal of Pathology,* vol. 188, no. 3, pp. 245-251.

Viacava, P., Naccarato, A.G., Bocci, G., Fanelli, G., Aretini, P., Lonobile, A., Evangelista, G., Montruccoli, G. & Bevilacqua, G. (2004). Angiogenesis and VEGF expression in pre-invasive lesions of the human breast., *Journal of Pathology,* vol. 204, no. 2, pp. 140-146.

Virnig, B.A., Tuttle, T.M., Shamliyan, T. & Kane, R.L. (2010). Ductal carcinoma in situ of the breast: a systematic review of incidence, treatment, and outcomes, *Journal of the National Cancer Institute,* vol. 102, no. 3, pp. 170-178.

Vogelstein, B., Lane, D. & Levine, A.J. (2000). Surfing the p53 network., *Nature,* vol. 408, no. 6810, pp. 307-310.

Vogl, G., Dietze, O. & Hauser-Kronberger, C. (2005). Angiogenic potential of ductal carcinoma in situ (DCIS) of human breast., *Histopathology,* vol. 47, no. 6, pp. 617-624.

Vos, C.B., Cleton-Jansen, A.M., Berx, G., de Leeuw, W.J., ter Haar, N.T., van Roy, F., Cornelisse, C.J., Peterse, J.L. & van de Vijver, M.J. (1997). E-cadherin inactivation in lobular carcinoma in situ of the breast: an early event in tumorigenesis., *British journal of cancer,* vol. 76, no. 9, pp. 1131-1133.

Wakefield, L.M., Piek, E. & Bottinger, E.P. (2001). TGF-beta signaling in mammary gland development and tumorigenesis, *Journal of Mammary Gland Biology & Neoplasia,* vol. 6, no. 1, pp. 67-82.

Walker, R.A. & Dearing, S.J. (1992). Transforming growth factor beta 1 in ductal carcinoma in situ and invasive carcinomas of the breast., *European journal of cancer,* vol. 28, no. 2-3, pp. 641-644.

Warren, J.L., Weaver, D.L., Bocklage, T., Key, C.R., Platz, C.E., Cronin, K.A., Ballard-Barbash, R., Willey, S.C. & Harlan, L.C. (2005). The frequency of ipsilateral second tumors after breast-conserving surgery for DCIS: a population based analysis., *Cancer,* vol. 104, no. 9, pp. 1840-1848.

Winchester, D.P. & Strom, E.A. (1998). Standards for diagnosis and management of ductal carcinoma in situ (DCIS) of the breast. American College of Radiology. American

College of Surgeons. College of American Pathologists. Society of Surgical Oncology, *CA: a Cancer Journal for Clinicians,* vol. 48, no. 2, pp. 108-128.

Wrensch, M.R., Petrakis, N.L., Miike, R., King, E.B., Chew, K., Neuhaus, J., Lee, M.M. & Rhys, M. (2001). Breast cancer risk in women with abnormal cytology in nipple aspirates of breast fluid., *Journal of the National Cancer Institute,* vol. 93, no. 23, pp. 1791-1798.

Xu, R., Perle, M.A., Inghirami, G., Chan, W., Delgado, Y. & Feiner, H. (2002). Amplification of Her-2/neu gene in Her-2/neu-overexpressing and -nonexpressing breast carcinomas and their synchronous benign, premalignant, and metastatic lesions detected by FISH in archival material., *Modern Pathology,* vol. 15, no. 2, pp. 116-124.

Yi, M., Krishnamurthy, S., Kuerer, H.M., Meric-Bernstam, F., Bedrosian, I., Ross, M.I., Ames, F.C., Lucci, A., Hwang, R.F. & Hunt, K.K. (2008). Role of primary tumor characteristics in predicting positive sentinel lymph nodes in patients with ductal carcinoma in situ or microinvasive breast cancer., *American Journal of Surgery,* vol. 196, no. 1, pp. 81-87.

Younes, M., Lebovitz, R.M., Bommer, K.E., Cagle, P.T., Morton, D., Khan, S. & Laucirica, R. (1995). p53 accumulation in benign breast biopsy specimens., *Human pathology,* vol. 26, no. 2, pp. 155-158.

Part 4

Intraepithelial Neoplasia of Prostate

6

Prostate Cancer Precursor Diseases

A.G. Papatsoris, C. Kostopoulos, V. Migdalis and M. Chrisofos
2nd Department of Urology, School of Medicine, University of Athens,
Sismanoglio General Hospital, Athens,
Greece

1. Introduction

Prostate cancer remains the most common non-cutaneous malignancy in the Western world and is the second leading cause of cancer death in males, after lung cancer (Nelson et al., 2003; Papatsoris & Anagnostopoulos, 2009). Based on experimental and preclinical findings, novel anti-prostate cancer strategies have been developed (Papatsoris & Papavassiliou, 2001; Papatsoris et al., 2005). However, the causes of prostate cancer, prostatic carcinogenesis, and the histological changes preceding and leading to the initiation of prostate cancer have yet to be elucidated. Several research groups are trying to solve the puzzle of prostatic carcinogenesis by focusing within the morphological continuum between benign glands at one end and premalignant lesions and invasive disease at the other (Vis & Van Der Kwast, 2001). In parallel, clinicians are frequently confronted with morphological features on the prostate needle biopsy that, although negative for cancer, raise suspicion of concomitant malignancy.

2. Definition of prostate cancer precursor lesions

There are several criteria that should be met in order to consider a prostatic lesion as premalignant (Vis & Van Der Kwast, 2001). An epidemiological relationship must be revealed, the precursor lesion should present at an earlier age than the cancer, and clear morphological similarities (e.g. cellular, histological, and architectural) should be present. Also, premalignant lesions should be close to their presumed malignant equivalents. The prostate has a greater frequency, severity, and extent of premalignant lesions in comparison with other organs. The definitive proof of a relationship between a premalignant lesion and malignancy is the clinical evidence of progression into invasive prostate cancer.

The earliest report on premalignant prostatic lesions dates back to 1926 (Orteil, 1926). In 1965, *McNeal* described lesions with possible premalignant features in prostatic epithelium (McNeal, 1965). In 1986, *McNeal and Bostwick* described the first criteria for the diagnosis of "intraductal neoplasia" which was classified into three grades (McNeal & Bostwick, 1986). In 1987, *Bostwick and Brawer* introduced the term "PIN" - prostatic intraepithelial neoplasia (Bostwick & Brawer, 1987). At an international conference in 1989, the term "PIN" was accepted as a replacement for various other terms (e.g. intraductal hyperplasia, hyperplasia with malignant change, large acinar atypical hyperplasia, marked atypia, ductal-acinar dysplasia). Initially, PIN was categorized into three grades with regard to architectural and cytological characteristics, taking into account that the alterations cover a continuum. However, in 1989, at the aforementioned workshop on premalignant prostatic lesions, the classification was altered to low-grade (formerly grade I) and high-grade (formerly grades II and III) PIN - LGPIN and HGPIN, respectively (Bostwick, 1989).

Conventional histopathology examination is used to differentiate precursor lesions of prostate cancer. Benign glands show a continuous basal cell layer, while in prostate cancer the basal cell layer is immunohistochemically absent. Immunostaining for p63 (a p53 homologue) was shown to be useful as a basal cell-specific marker (Signoretti et al., 2000; Shah, 2004). It is frequently used in addition to 34βE12 immunostaining in difficult diagnostic cases, where the main advantage of p63 over 34βE12 is that there is less variable staining.

A wide range of "atypical" epithelial proliferative processes with a variety of names, often with confusing and overlapping terminology, has been described. Several morphological lesions have been proposed that may act as potential precursor lesions of prostate cancer. These are the morphologically distinct entities of focal atrophy or post-atrophic hyperplasia (PAH), atypical adenomatous hyperplasia (AAH) or adenosis, and PIN.

3. Prostatic atrophy and PAH

Focal prostatic atrophy reportedly is present in up to 85% of prostates at autopsy (Amin, 1999; Billis, 1998). It should be distinguished from diffuse atrophy, as the latter is not considered premalignant. A role for focal atrophy in the pathogenesis of PIN and/or prostate cancer was proposed by Franks, over 50 years ago (Franks, 1954). De Marzo observed that focal atrophic lesions showed an increased proliferative activity of luminal cells and a decreased frequency of apoptosis (DeMarzo et al., 1999). Concerning the classification of focal prostatic atrophy, although there are distinct histologic variants, the terminology is currently non-standardized and no formal classification has been tested for interobserver reliability. According to the current classification focal atrophy lesions were categorized into 4 distinct subtypes as follows: (i) simple atrophy, (ii) simple atrophy with cyst formation, (iii) postatrophic hyperplasia and (iv) partial atrophy (De Marzo et al., 2006).

Simple atrophy consists of atrophic cells lining acini with relatively normal caliber that lack papillary fronds, where the number of glands per unit area does not appear to be increased relative to normal tissue. (De Marzo et al., 2006). Simple atrophy demonstrates strong basal cell-specific antikeratin immunoreactivity (Bostwick, 1996). Simple atrophy with cyst formation is a subtype of simple atrophy. Two general patterns are now encompassed: those containing very large diameter glands (> 1 mm) and those containing smaller, rounded glands. The amount of cytoplasm at times may be so attenuated as to be nearly invisible. When there is significant cytoplasm in the luminal cell, it tends to be clear. Atrophy with cyst formation tends to have less inflammation than the other sub-types (De Marzo et al., 2006). In sclerotic atrophy, the stroma is more extensively sclerosed, resulting in a wider separation of the acinar elements; these continue to have the cytological features described above (DeMarzo et al., 1999; De Marzo et al., 2006). Post-Atrophic Hyperplasia (PAH) consists of acini that are smaller, round and appear in a lobular distribution, often surrounding a somewhat dilated duct with an apparent increase in the number of small glands compared to normal tissue. Some authors tend to refer to some of these lesions as "lobular atrophy" or "lobular hyperplasia" (De Marzo et al., 2006). In lobular atrophy, the lesion is circumscribed with a central duct surrounded by small acini (Grignon & Sakr, 1996). The acini frequently have ectatic lumens and are lined by a flattened epithelium having scant cytoplasm and hyperchromatic nuclei with inconspicuous nucleoli. Basal cells are present but are difficult to recognize; however, they are readily identified with immunohistochemical stains for 34βE12 cytokeratin. The stroma is sclerotic and compressed, particularly around the central duct.

PAH, which may closely mimic the histology of prostate cancer, may represent a diagnostic pitfall (Bostwick, 1996). Recent studies have reported that the frequency of PAH in radical prostatectomy specimens was remarkably similar to that in cystoprostatectomy specimens, implying that the simultaneous finding of PAH with prostate cancer is coincidental (Amin, 1999; Grignon & Sak, 1996). PAH develops in a background of lobular or sclerotic atrophy and so retains many features of these lesions (Anton et al., 1999; Cheville & Bostwick, 1995; Grignon & Sakr, 1996). In PAH, there is an apparent secondary proliferation of small acini. The secretory cells have more abundant pale or clear cytoplasm than in usual atrophy, though generally not as much as in adenocarcinoma or AAH. The nuclei also become less hyperchromatic, and small chromocenters or nucleoli may be seen. The double cell layer is maintained and can be confirmed with basal cell-specific anticytokeratin antibodies. Despite the observation that focal atrophic lesions and PAH consist of flattened and dispersed acini, immunostaining with 34βE12 cytokeratin is almost always positive and continuous, as it is for benign epithelial glands (Anton et al., 1999; Cheville & Bostwick, 1995).

4. AAH and sclerosing adenosis

The prevalence of AAH in transurethral prostatectomy (TURP) specimens without cancer ranges from 1.6% to 7.3% (Gaudi & Epstein, 1994). In biopsy specimens, the prevalence is lower, for example 0.8% in one series (Gaudin & Epstein, 1995). The increase in frequency of AAH in needle biopsies is presumably related to ultrasound-guided biopsy of the transition zone. AAH can be diagnosed throughout the prostate, but it is most often located in the transition zone of the prostate in intimate association with benign nodular hyperplasia (Bostwick & Qian, 1995). It can also be found near the apex and in the periurethral area (Bostwick, 1996). In AAH, the basal cell layer is discontinuous and fragmented on 34βE12 cytokeratin immunostaining (Cheng et al., 1998).

There is considerable morphologic evidence suggesting that AAH is associated with low-grade adenocarcinoma arising in the transition zone (Cheville & Bostwick, 1995). AAH, a putative precursor of transition zone adenocarcinoma, has common features with low-grade adenocarcinoma and may cause problems in differential diagnosis, especially in the needle biopsy setting (Srigley, 2004). AAH is a lesion characterized by a proliferation of small acinar structures that mimics adenocarcinoma because of histological similarities (Grignon & Sakr, 1996). At low magnification, the lesion is circumscribed, although the small acini may show some infiltrative features. These acini are seen in association with a usually hyperplastic nodule and are most often at the periphery of the nodule. The nuclei tend to be uniform and round with inconspicuous or small nucleoli. There is limited data that AAH has a proliferation rate higher than hyperplasia but lower than adenocarcinoma (Bostwick & Qian, 1995; Cheng et al., 1998; Grignon & Sakr, 1996). AAH is diploid, as are most examples of low-grade adenocarcinoma, while a few markers (blood group antigens, peanut agglutinin) show similar patterns of expression in AAH and adenocarcinoma (Grignon & Sakr, 1996). Recent cytogenetic analyses have detected abnormalities of chromosome 8 in a very small proportion (4–7%) of AAH cases (Bostwick & Qian, 1995; Cheng et al., 1998).

Sclerosing adenosis is a circumscribed proliferation of small acinar structures in a cellular spindled sclerotic stroma (Grignon & Sakr, 1996; Srigley, 2004). The acini range from irregular in shape to small, round, and uniform. Usually, there is thickening of the tubular basement membrane, a valuable diagnostic clue (Grignon & Sakr, 1996). Nucleoli are generally inconspicuous but can be prominent in a few cells, while the lumens can contain basophilic mucin or crystalloids (Grignon & Sakr, 1996). Sclerosing adenosis is usually an incidental finding in about 2% of transurethral resection of the prostate (TURP) or radical prostatectomy specimens, (Bostwick et al., 2008; Grignon et al., 1992; Cheng & Bostwick, 2010; Sakamoto et al, 1991) and rarely is present in needle biopsy specimens (Cheng & Bostwick 2010; Srigley, 2004). Sclerosing adenosis may simulate adenocarcinoma and accounts for up to 10% of cases overdiagnosed as adenocarcinoma (Berney et al., 2007; Bostwick & Cheng, 1999; Cheng & Bostwick, 2010). Multiple light microscopic and immunohistochemical features separate typical sclerosing adenosis from adenocarcinoma, including: (i) intact basal cell layer, a finding that can be confirmed immunohistochemically with antibodies directed against high-molecular-weight cytokeratin 34βE12, (ii) unique immunophenotype of many of the basal and spinde cells in the stroma, including abundant S100 protein and smooth muscle actin (SMA) reactivity, as well as structural characteristics of myoepithelial cells, (iii) cellular spindle cell stroma, (iv) variably thickened basement membrane and (v) absence of significant cytological atypia. (Bostwick et al., 1994; Collina et al., 1992; Grignon et al., 1992; JonesC et al., 1991; Cheng & Bostwick, 2010; Sakamoto et al. 1991; Young & Clement, 1987 ; Young & Clement, 1990). Sclerosing adenosis differs from AAH by displaying myoepithelial features of the basal cells and an exuberant stroma of fibroblasts and loose ground substance (Bostwick et al., 1994).

5. Lesion Suspicious for Cancer (LSC)

As a result of the limited quantity of tissue sampled in prostate biopsies, there is the probability of finding a lesion that raises diagnostic confusion. There may be lesions suspicious for but not conclusive of malignancy. These lesions are small and have a wide diversity of architectural and morphological features. *Vis* proposed the terminology "prostate biopsy suspicious for malignancy" to classify these lesions (Vis et al., 2001). However, histology reports should be unequivocal and as concise as possible and vague diagnoses should not lead to unnecessary biopsy with its associated morbidities.

The controversial diagnostic term "atypical small acinar proliferation" (ASAP) is no longer considered acceptable. It is not considered a diagnostic entity as it can only be diagnosed on needle biopsies and not in prostatectomy specimens. The term ASAP has been replaced by the term "lesion suspicious for cancer" (LSC), as the prostate lesion lacks sufficient criteria to call it a carcinoma. LSC (fig. 1, 2) has gained acceptance as a legitimate way for pathologists to describe minute foci of small prostatic acini that raise the suspicion of carcinoma but that fail to attain the requisite diagnostic threshold for carcinoma (Fadare et al., 2004).

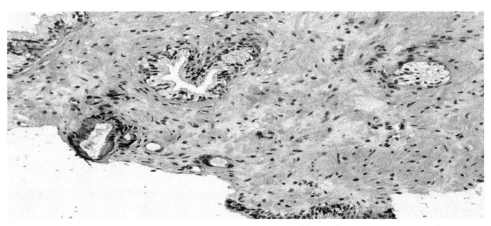

Fig. 1. An accumulation of a few atypical prostatic glands with amphophilic cytoplasm and nuclear enlargement. Compare with benign adjacent glands. This field is consistent with either atypical small prostate gland proliferation or limited prostate adenocarcinoma. (HE × 200)

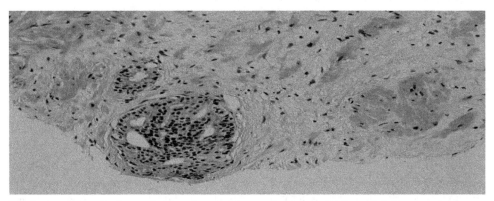

Fig. 2. Atypical cribriform gland at the periphery of a biopsy. Differential diagnosis between HGPIN and cribriform invasive adenocarcinoma may be impossible.

Studies have demonstrated that LSC is diagnosed in 1.5% to 9% of prostatic biopsies and that it predicts definite prostate cancer in about 45% of repeat biopsies (Iczkowski et al., 1997; Iczkowski et al., 1998; Iczkowski & Bostwick, 2000; Iczkowski et al., 2002). Since 1997, there have been efforts to stratify risk in cases of LSC into three categories: likely benign, uncertain, and likely carcinoma (Iczkowski et al., 1997). No single pathologic feature of LSC appeared to increase the likelihood for subsequent cancer (Iczkowski et al., 1998; Iczkowski & Bostwick, 2000; Iczkowski et al., 2002; Scattoni et al., 2005). The mean MIB-1 proliferation index of LSC was significantly higher than in benign prostatic tissue and did not differ from that of low-grade carcinoma.

LSC continues to be associated with a high risk of prostate cancer and requires a repeat biopsy with the extended peripheral zone biopsy scheme (Moore et al., 2005). Sampling should include multiple sites in the prostate, as in 40% of patients; cancer was in different sites from the initial LSC site. The detection rate was lower for patients with a larger prostate than those with a smaller prostate (Scattoni et al., 2005). Hence, patients with LSC should be followed up and undergo repeat biopsy. The role of radical prostatectomy for LSC is not clear, although *Brausi* advocated that prostatectomy could be the treatment of choice in young men with LSC (Brausi et al., 2004).

6. PIN

6.1 Epidemiology of PIN

Epidemiological studies have demonstrated the presence of PIN in men as early as the fourth decade of life and showed that the incidence and extent of PIN increased with age (Sakr et al., 1993). It has been postulated that PIN pre-dates the onset of prostate cancer by 5–10 years (Sakr et al., 1993). In several autopsy and surgical series, PIN was identified in 60% to 90% of prostates harboring carcinoma and was often close to its presumed invasive equivalent (Qian et al., 1997). Studies showed that PIN was present in 82% of step-sectioned autopsy prostates with cancers, but in only 43% of benign prostates from patients of similar age (Vis & Van Der Kwast, 2001). *Qian* found that 86% of a series of 195 whole-mount radical prostatectomies contained HGPIN, usually within 2 mm of the cancer (Qian et al., 1997).

In the United States, 1.300.000 prostate biopsies are performed annually to detect 230.000 new cases of prostate cancer (Joniau et al., 2007; Steiner, 2003). There are approximately 115.000 cases of isolated HGPIN diagnosed each year, representing an estimated 9% of prostate biopsies (Steiner, 2003). The incidence of HGPIN in biopsies ranges from 1.5% to 16.5%, with an average of 6% (Bostwick et al., 1995; Epstein, 2002). The different incidence of HGPIN in published studies derives from differences in defining HGPIN and in the number of patients.

The most likely explanation to account for the variation in incidence of PIN is interobserver reproducibility (Sakr, 1995). Those pathologists who use a lower threshold to define prominent nucleoli will have a higher incidence of HGPIN. Other plausible explanations for the variation in reported incidence of HGPIN relate to the fixative used and to differences in sampling. Furthermore, the variations from one institution to another can be attributed to variation in the population study, the indications for biopsy, and the biopsy compliance rates. The site of the prostate biopsied, the number of biopsies taken, and the quality and processing technique of the biopsy cores can also influence the incidence of PIN.

HGPIN starts in young individuals and increases with age in Caucasians and African-Americans, but is more prevalent in the latter (Epstein, 1995; Sakr et al., 1996). HGPIN in African-Americans precedes HGPIN in Caucasians by approximately a decade (Epstein, 1995). A more extensive form of HGPIN with multifocal or diffuse involvement of the glands appears at a younger age in African-Americans in comparison with Caucasians (Sakr et al., 1996). The replication of the association of chromosome 8q24 variants with increased prostate cancer risk in Tobago men and the higher frequency of the risk alleles in controls in populations of African ancestry further strengthens the possible role of this genomic region in the disproportionate higher burden of prostate cancer in men of African ancestry (Sakr et al., 1996). It has also been shown that prostate cancer grows more rapidly in black than in white men and/or earlier transformation from latent to aggressive prostate cancer occurs in black than in white men (Powell et al., 2010).

6.2 Molecular biology of PIN

The development of PIN is characterized by increased expression of several biomarkers that influence the proliferative potential of the dysplastic prostatic cells. Studies of potential biomarkers, such as growth factors, growth factor receptors, oncogene products, glycosylated tumor antigens, and other biomarkers in PIN, are difficult because these lesions are focal.

Unlike the premalignant polyps of the colon, it is difficult to obtain relatively pure preparations of PIN. One approach, to microdissect areas of PIN, is tedious and still may produce results contaminated by surrounding stroma and histologically normal epithelium. In addition, this technique does not allow differentiation of biomarker expression among the various components (basal versus luminal) of the dysplastic gland or duct.For these reasons, immuno - histochemical techniques as well as fluorescence *in situ* hybridization (FISH) is perhaps best suited for the assessment of biomarker expression in PIN.

FISH analysis has demonstrated strong expression of epidermal growth factor receptor (EGFr) mRNA in PIN (Myers & Grizzle, 1996). The c-erbB-2 gene product (p185erbB-2) is a transmembrane receptor that demonstrates significant homology to EGFr. Moderate-to-strong immunoreactivity for p185erbB-2 was noted in the luminal as well as the basal cells of PIN lesions. This immunostaining was frequently equivalent in pattern and intensity to that of adjacent malignant cells. The pattern of expression was typically coarse cytoplasmic immunoreactivity. Increased expression of the growth factor-related receptors p185erbB-2 and p180erbB-3, as well as the product of the *c-met* protooncogene (a transmembrane tyrosine kinase receptor that binds the mitogen hepatocyte growth factor/scatter factor), is frequently detected in the dysplastic luminal cells and in malignant cells of the prostate (Myers & Grizzle, 1996). Mutation of the p53 gene in PIN may precede the development of highly aggressive prostate cancer (Myers & Grizzle, 1996). The expression of the nm-23H1 gene product is strongly expressed in dysplastic and malignant prostatic cells (Myers & Grizzle, 1996).

It has been demonstrated that expression of the proliferative markers Ki-67 and proliferating cell nuclear antigen (PCNA) in PIN is increased as compared to benign prostatic epithelium (Myers & Grizzle, 1996). Increased PCNA expression also has been detected in the nuclei of stromal and endothelial cells adjacent to PIN (Myers & Grizzle, 1996). This may be associated with the observation of a higher density of blood vessels in the vicinity of PIN lesions. In contrast to the enhanced expression of the biomarkers associated with proliferation, decreased expression of prostate specific antigen (PSA), prostate acid phosphatase, and Leu 7 by dysplastic luminal cells is indicative of an impairment of the process of cellular differentiation (Myers & Grizzle, 1996). *Bostwick* demonstrated a decrease in the expression of neuroendocrine markers (neuron-specific enolase, serotonin, chromagranin, and human chorionic gonadotropin) in PIN. Aberrant glycosylation as well as inappropriate expression of glycosylated tumor antigens was demonstrated by enhanced binding of the lectin *Ulex europaeus* and by increased expression of tumor-associated glycoprotein 72 and the Lewis Y antigen (Myers & Grizzle, 1996). Enhanced expression of proteolytic enzymes, such as cathepsin D and the 72-kD form of collagenase IV, by dysplastic cells may represent an integral event in the development of invasive prostate cancer (Boag & Young, 1994). Moderate-to-strong immunoreactivity for fatty acid synthetase was also detected in PIN (Swinnen et al., 2002).

In HGPIN, studies have demonstrated notable loss of the three critical signaling components of the apoptotic action of transforming growth factor-β; that is, the transmembrane receptor II (TβRII), the key cell cycle inhibitor p27Kip1, and the protagonist downstream Smad4 receptor-activated protein (Zeng & Kyprianou, 2005). Quantitative evaluation of the apoptotic index revealed significantly less value in HGPIN when compared with adjacent areas of benign prostatic hyperplasia (Zeng & Kyprianou, 2005). Apoptotic profiling of HGPIN may contribute to a better understanding of factors that play a role in deregulated prostate growth (Zeng & Kyprianou, 2005).

Prostate carcinogenesis is the result of the accumulation of multiple genetic changes. The most frequently found chromosomal anomalies are overexpression on chromosomes 7p, 7q, and 8q, and inactivation on chromosomes 8p, 10q, 13q, 16q, and 18q (Joniau et al., 2005). Inactivation of tumor suppression genes, such as NKX3-1 (8p) and PTEN (10q), and overexpression of oncogenes, such as c-myc (8q), play an important role in PIN and the initiation of prostate cancer (Qian et al 1995). These findings support the multi-step theory in which PIN is considered a precursor lesion of prostate cancer.

6.3 Similarities between PIN and prostate cancer

The frequency and extent of PIN lesions increase with age, and this is similar to the increase in diagnosis of prostate cancer (Joniau et al., 2005). HGPIN is found significantly more frequently in prostates with cancer (McNeal & Bostwick, 1986). PIN is predominantly located in the peripheral zone of the prostate, the area in which most clinically important prostate cancers are found, and PIN, like prostate cancer, is often multifocal (Joniau et al., 2005). In an autopsy study, HGPIN was found in 63% of cases solely in the peripheral zone; in 36%, in the peripheral and transition zone; and in 1%, solely in the transition zone (Haggman et al., 1997). These findings are similar to the zonal distribution of prostate cancer.

Several genotypic and phenotypic studies have indicated that there are remarkable morphological, molecular, and biochemical similarities between PIN and prostate cancer (Vis & Van Der Kwast, 2001). Molecular abnormalities in PIN are mostly intermediate between benign gland and cancer, reflecting an impairment of cell-differentiation and regulatory control (Bostwick, 1999). PIN is characterized by cellular crowding and stratification. There is inequality in cell and nuclear size. Hyperchromatism is frequently seen with an enlarged nucleus, often containing prominent nucleoli lines. These changes are also seen in Gleason grade 1–4 prostate cancer (Bostwick et al., 1998). Biochemically, the cells of PIN show changes in the cytoskeletal proteins, secretory proteins, and nuclei that are shared with established prostate malignancies.

Prostate cancer and HGPIN have similar proliferative and apoptotic indices (Bostwick et al., 1998). Mitotic figures and apoptotic bodies increase progressively from nodular hyperplasia to HGPIN (Bostwick et al., 1998). During the malignant transformation of PIN, the basal cell layer loses its proliferative function, which is transferred to secretory luminal cell types, as demonstrated by Bonkhoff (Bonkhoff, 1996). Moreover, there is a progressive increase in the number of apoptotic bodies from nodular hyperplasia through PIN to prostate cancer (Bostwick et al., 1996). Greater cytoplasmic expression of bcl-2 is observed in PIN and cancer than in benign and hyperplastic epithelium (Bostwick et al., 1996). Two members of the platelet-derived growth factor (PDGF) peptide family, PDGF-A and PDGF-a, are up-regulated in PIN and prostate cancer compared with benign prostatic hyperplasia; BPH (Bostwick et al., 1996). Similarly, there is up-regulation of cathepsinD in PIN and prostate cancer; this autocrine mitogen, which has been studied extensively in other organs as a marker of invasion, correlates with tumor grade and DNA ploidy status in prostate cancer (Bostwick et al., 1996).

Histologically, the atypia observed in HGPIN is virtually indistinguishable from that of prostate cancer except that in HGPIN the basal membrane is still intact (Sakr et al., 1999). As HGPIN progresses, the likelihood of basal cell layer disruption increases. In HGPIN, the basal cell layer is disrupted or fragmented as demonstrated by high-molecular-weight cytokeratin immunolabeling. In prostate cancer, there is a complete loss of the basal cell layer. Both in PIN and prostate cancer, collagenase type IV expression is increased compared to normal prostate epithelium; this enzyme is responsible for basal membrane degradation and thus facilitates invasion (Bostwick et al., 1996). PIN and prostate cancer share several nuclear properties, such as amount of DNA, chromatin texture, chromatin distribution, nuclear perimeter, diameter, and nuclear abnormalities (Baretton et al., 1994).

Several genetic changes encountered in prostate cancer cells can be found in PIN lesions (Bostwick et al., 1996). Allelic loss is common in PIN and prostate cancer (Sakr et al., 1999). The frequent 8p12-21 allelic loss commonly found in prostate cancer is also found in microdissected PIN. Other examples of genetic changes found in prostate cancer that already exist in PIN include loss of heterozygosity at 8p22, 12pter-p12, and 10q11.2 and gain of chromosomes 7, 8, 10, and 12. Alterations in oncogene bcl2 expression and RER+ phenotype are similar for PIN and prostate cancer (Baltaci et al., 2000). As in prostate cancer, there is also evidence of aneuploidy and an increase in microinvascular density, both frequently regarded as evidence of aggressiveness in PIN (Montironi et al., 1993).

6.4 LGPIN

In LGPIN (Fig. 3), secretory cells of the lining epithelium proliferate and "pill up" with irregular spaces between them (Bostwick, 2000 ; Newling, 1990). The nuclei are enlarged, vary in size, have normal or slightly increased chromatin content, and possess small or inconspicuous nucleoli (Zeng & Kyprianou, 2005). More prominent nucleoli, when observable, comprise less than 10% of dysplastic cells. The basal cell layer normally surrounding secretory cells of ducts and acini remains intact. In LGPIN, only 0.7% of reported cases reveal evidence of basal cell layer disruption (Newling, 1999).

LGPIN is rather difficult to recognize, as it shares common features with normal and hyperplastic epithelium (Bostwick, 2000; Newling, 1999). The most common issue that may lead in some cases to discrepant diagnoses between LGPIN and HGPIN is the definition of "prominent" with regard to nucleolar enlargement and visibility.

It has been suggested that LGPIN should not be commented on in diagnostic reports (Epstein, 2002). Firstly, pathologists cannot reproducibly distinguish LGPIN from benign prostate tissue (Epstein et al., 1995). Secondly, when LGPIN is diagnosed on needle biopsy, these patients are not at greater risk of having prostate cancer on repeat biopsy (Keetch, 1995).

The distinction between HGPIN and LGPIN is based primarily on the extent of cytological abnormalities (prominence of the nucleoli) and secondarily on the degree of architectural complexity (Goeman et al., 2003; Weinstein & Epstein, 1993). Immunostaining studies of microvessel density may help to differentiate HGPIN from LGPIN (Sinha et al., 2004).

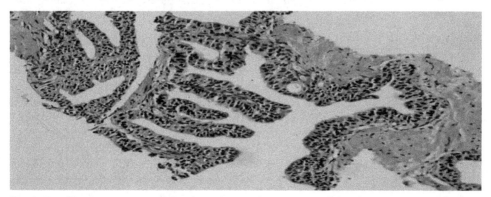

Fig. 3. Papillary structures within a large hyperplastic prostate gland. Minimal nuclear atypia. (HE × 200)

6.5 HGPIN

6.5.1 Why HGPIN?

PIN was initially divided into three different grades (I–III), which now are reduced to the abovementioned LGPIN for PIN I, and HGPIN for PIN II and III. HGPIN includes PIN II and III for two reasons. Firstly, there was a great deal of inter-observer variability in the distinction between PIN II and III (Epstein et al., 1995). Secondly, the finding of PIN II or III on needle biopsy was associated with the same risk of prostate cancer on subsequent biopsy (Weinstein & Epstein, 1993).

6.5.2 Histology of HGPIN

In HGPIN, uniform morphologic abnormalities are detectable (Vis & Van Der Kwast, 2001). Cells have large nuclei of relatively uniform size, and possess prominent nucleoli that are similar to those of cancer cells (fig. 4, 5). Regarding cytological features, the acini and ducts are lined by malignant cells which are uniformly enlarged with an increased nuclear/cytoplasmic ratio, and with less variation in nuclear size in comparison to LGPIN. In HGPIN, at least 10% of cells demonstrate prominent nucleoli similar to those of carcinoma cells, and the majority of cells show coarse clumping of the chromatin which may be accentuated along the nuclear membrane (Vis & Van Der Kwast, 2001). The expanded nuclear chromatin area probably explains the darker "blue" appearance of the lining which characterizes HGPIN at low power microscopic examination. Nuclei toward the centre of the gland tend to have blander cytology than peripherally located nuclei.

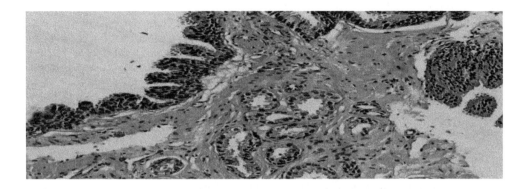

Fig. 4. HGPIN tufted pattern and adjacent carcinoma (with an adequate number of small malignant glands). Small atypical glands are too numerous to represent outpouchings of HGPIN. (HE × 200)

6.5.3 Patterns of HGPIN

There are four architectural patterns of HGPIN: tufting, micropapillary, flat and cribriform, based on the arrangement of the cells within pre-existing ducts or glands (Vis & Van Der Kwast, 2001). Tufting HGPIN is by far the most common pattern (present in 97% of all HGPINs) followed by micropapillary, flat, and cribriform patterns (Yamauchi et al., 2006). In the flat pattern, nuclear atypia is evident without significant architectural changes. In the tufting pattern, nuclei become more pilled up, and undulating mounds of cells are formed. Columns of atypical epithelial cells typically lacking fibrovascular cores characterize the micropapillary pattern. In the cribriform pattern, more complex architectural features, such as a "Roman bridge" and cribriform formation, are encountered.

Patients with HGPIN in only one initial biopsy or a predominant flat/tufting pattern clearly have less risk of cancer being found in subsequent biopsies compared to patients with HGPIN in more than one initial biopsy. Furthermore, a micropapillary and/or cribriform pattern are correlated with a greater risk for development of prostate cancer (Joniau et al., 2005).

Unusual subtypes of HGPIN include PIN with signet-ring morphology and neuroendocrine cells with either Paneth cell-like or small-cell morphology (Bostwick et al., 1993; Vis & Van Der Kwast, 2001). Intraductal HGPIN, in prostates with established cancer, has been associated with high tumor volumes, poorly differentiated tumor components, and a higher progression rate after radical prostatectomy than prostate cancers without these coexisting proliferations (Cohen et al., 2000; McNeal & Bostwick, 1986). Hence, a separate histological entity was proposed, namely, intraductal carcinoma of the prostate, which would be distinguished from HGPIN.

6.5.4 HGPIN and prostate cancer

HGPIN is the most likely precursor of prostatic adenocarcinoma, according to current literature (Dovey et al., 2005; Gaudin et al., 1997; Joniau et al., 2005; Lefkowitz et al., 2002; Pacelli & Bostwick, 1997; Powell et al., 2010; Singh et al., 2009; Vis & Van Der Kwast, 2001). The expression of various biomarkers in HGPIN is either the same as with prostate cancer or intermediate between prostate cancer and benign prostate tissue. The cytological changes are characterized by prominent nucleoli in a substantial proportion (≥5%) of cells, nuclear enlargement and crowding, increased density of cytoplasm, and anisonucleosis (Vis & Van Der Kwast, 2001). Ploidy seems not to discriminate between HGPIN and infiltrating cancer (Baretton et al., 1994). Also, studies reveal consistent down-regulation of epithelial cell adhesion molecules and transmembrane proteins in PIN (Vis & Van Der Kwast, 2001). This is accompanied by up-regulation of enzymes responsible for degradation of the extracellular matrix.

Unlike in prostate cancer, incomplete disruption of the basal cell layer can be shown by 34βE12 cytokeratin immunostaining (Vis & Van Der Kwast, 2001). In HGPIN, more than 50% of abnormal cells are seen to have a disrupted basal cell layer in the acini. Immunohistochemistry is usually not helpful since the lack of a basal cell layer in only a few cribriform or small glands is not sufficient for the diagnosis of cancer. However, in cases where many glands are totally immunonegative for high-molecular-weight cytokeratin, these foci may be diagnostic of cancer. Cases where some of the glands show the expected patchy basal cell layer of PIN and a few, morphologically identical glands are negative for high-molecular-weight keratin should still be diagnosed as HGPIN. In rare cases when sperm can be identified in the glandular lumen, the diagnosis of PIN is favored because only PIN glands are able to communicate with the main prostatic glands that contain sperm; malignant invasive glands cannot retain their continuity with main prostatic glands (Vis & Van Der Kwast, 2001).

In cases of HGPIN with neighboring small atypical glands, the possibility of coexistent invasive carcinoma should be examined (Vis & Van Der Kwast, 2001). When the latter are few, the issue is whether the small glands represent outpouchings or tangential sections of the adjacent HGPIN or whether they represent microinvasive cancer. When these small atypical glands are too many or too crowded to be outpouchings or tangential sections of the HGPIN glands, then the diagnosis of invasive carcinoma can be made (Bostwick et al., 1996).

6.6 Differential diagnosis of PIN

Histologically, PIN can be confused with several benign entities as well as with ductal and acinar adenocarcinoma (fig. 5) of the prostate (Epstein et al., 2002). Benign conditions include prostate central zone hyperplasia, since glands within the central zone at the base of the prostate are complex and large with many papillary infoldings and clear cell cribriform hyperplasia, which consists of crowded cribriform glands with clear cytoplasm (Vis & Van Der Kwast, 2001; Joniau et al., 2005). Both these entities lack significant nuclear atypia. The basal cell layer can display prominent nucleoli but secretory cells can be recognized. Cytologically atypical basal cell hyperplasia usually forms small solid nests of atypical basal cells, mainly in the central zone; these are inconsistent with PIN, which affects medium- or large-sized glands, mainly in the peripheral zone of the prostate (Vis & Van Der Kwast, 2001). In any case, basal cells can be easily identified by immunohistochemistry either with antibodies against high-molecular-weight cytokeratins (cytoplasmic staining pattern) or against p63 (nuclear staining pattern).

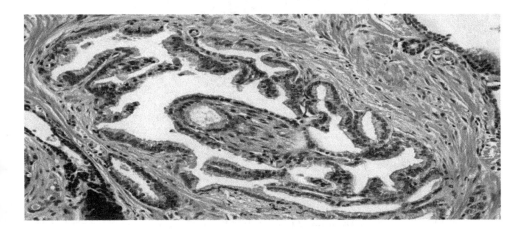

Fig. 5. Complex atypical gland with prominent nucleoli and perineural invasion. Gleason pattern 3 of cribriform adenocarcinoma. Note the coexistent microacinar cancerous pattern on the bottom right. (HE × 200)

With regard to malignant conditions, cribriform acinar adenocarcinoma can be discriminated from cribriform HGPIN when a sufficient number of cribriform glands totally lack basal cells (Vis & Van Der Kwast, 2001). Furthermore, in cribriform carcinoma, sometimes the appearance of foci of back-to-back glands, rather than true cribriform formations, is evident.

Ductal adenocarcinomas of the prostate may demonstrate a patchy basal cell layer (like PIN), but they develop in the transition zone. They may develop true papillary fronts with fibrovascular cores (in contrast to micropapillary PIN). Ductal adenocarcinoma glands are larger, may contain back-to-back glands, may show extensive comedonecrosis, and are usually fragmented in needle biopsy specimens.

Finally, the possibility of intraductal carcinoma (fig. 6) should be considered when multiple cribriform glands with prominent cytological atypia containing comedonecrosis are encountered (Cohen et al., 2000; McNeal & Bostwick, 1986). In these glands, basal cells can be identified, though this lesion should be distinguished from HGPIN since it appears to be a late event in prostate gland carcinogenesis and warrants immediate therapy.

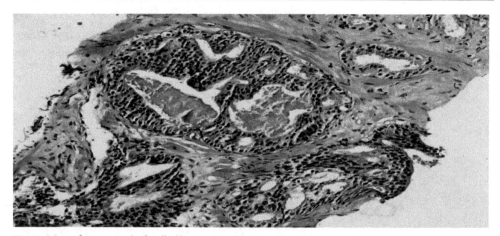

Fig. 6. Many large atypical cribriform glands with extensive comedonecrosis. Retention of basal cell layer remnants would be consistent with intraductal carcinoma rather than HGPIN. (HE × 200)

6.7 Clinical markers of PIN

6.7.1 PSA

PIN lesions do not contribute to an elevation of serum PSA, PSA density, or a decrease in the free-to-total PSA ratio (Alexander et al., 1996; Darson et al., 1999; Ronnett et al., 1993). PIN lesions show less expression of PSA in luminal cells, as determined by immunohistochemistry, than do benign epithelial glands (Ronnett et al., 1993; Alexander et al., 1996). An elevation of PSA should be attributed to the presence of prostate cancer, BPH, or concurrent prostatic inflammation rather than to the presence of PIN.

Immunohistochemical studies show a lower PSA expression in PIN lesions compared to benign tissue and prostate cancer (Darson et al., 1999). PSA produced by PIN lesions follows the route of least resistance and is excreted in the seminal fluid, whereas cancer forms tissue islands without a surrounding basal layer, and PSA diffuses into the blood.

6.7.2 Potential markers

Swinnen demonstrated that fatty acid synthetase immunostaining intensity tended to increase from LGPIN to HGPIN and prostate cancer (Swinnen et al., 2002). This key enzyme in the de novo production of fatty acids enables cancer progression and invasion. Another enzyme, alpha-methylacyl coenzyme A racemase (AMARC), which plays a key role in the beta-oxidation of fatty acids, is rarely expressed in benign prostatic tissue, in contrast to PIN and prostate cancer (Rubin et al., 2002). A statistically significant association of this biomarker with the risk of prostate cancer is yet to be revealed (Hailemariam et al., 2011). A80, a membrane-bound glycoprotein that is related to exocrine differentiation, may be useful in detecting residual and/or recurrent prostate carcinoma after radiation or hormonal therapy (Coogan et al., 2003). Benign glands are generally negative for A80 except for scattered positive cells in about 15% of glandular hyperplasia (Shin et al., 1989). *Coogan* demonstrated that A80 immunostaining in prostate cancer, HGPIN, and LGPIN, in 100%, 92%, and 73% of the examined specimens, respectively (Coogan et al., 2003). Markers including kallikrein-related peptidase 2 (KLK2), early prostate cancer antigen (EPCA), PCA3, hepsin, prostate stem cell antigen are under investigation for the early diagnosis and management of prostate cancer (Darson et al., 1997; Sardana et al., 2008). PCA3 is a prostate specific, non-protein coding RNA that is significantly over expressed in prostate cancer, without any correlation to prostatic volume and/or other prostatic diseases like prostatitis. Recent studies have shown the potential of PCA3, in correlation with other markers, to be used as a prognostic marker for prostate cancer (Bourdoumis et al., 2010).

6.8 Management of PIN

6.8.1 Repeat prostate biopsy

As a consequence of programs for the early detection of prostate cancer, the number of biopsies performed and specimens evaluated has increased substantially. False-positive results may strongly influence a man's quality of life through unnecessary psychological stress, unnecessary treatment, and treatment-associated morbidities. Furthermore, for medico-legal reasons, it is obvious that biopsy false-positive results should be minimized. Currently, the consensus is that the finding of focal atrophy, PAH, AAH, or LGPIN on needle biopsy or in TURP material for BPH should not lead to any diagnostic follow-up (Vis & Van Der Kwast, 2001). However, the finding of HGPIN on needle biopsy indicates a field effect by which the entire prostate is at higher risk of harboring cancer (Langer et al., 1996).

The decisions for diagnostic follow-up in men with PIN should take into account the patient's age, physical status, and co-morbidities. In men developing HGPIN in the eighth decade, knowing that the development of symptomatic prostate cancer will probably occur only after 10 years, a policy of watchful waiting should be recommended.(Ravery V , 2009 ; Vis AN & Van Der Kwast TH , 2001) Men who may not potentially benefit from curative treatment or early hormonal therapy should not undergo follow-up biopsy.

When more extensive repeat biopsy is performed, the likelihood of detecting prostate cancer is increased. If isolated HGPIN is detected in a 12-core biopsy protocol, the cancer incidence in the immediate 12-core repeat biopsy will be only 2% to 3% (Lefkowitz, 2002). In contrast, in repeat biopsies following initial sextant or octant biopsies, the cancer detection rate is 27–30% (Lefkowitz, 2002). However, taking too many biopsies can increase the risk of detecting too many clinically insignificant cancers and can lead to overtreatment (Joniau et al., 2005). Authors have proposed an 8-biopsy regimen, which clearly outperformed the sextant regimen in cancer detection (Joniau et al., 2005; Lefkowitz, 2002).

When follow-up biopsies are performed in men with foci of isolated HGPIN, the site of prostate cancer may not be the same site that raised the suspicion of concurrent carcinoma (Bostwick et al., 1995). The finding of HGPIN after TURP (2.8%-33%) also appears to place men at a higher risk of harboring cancer, although there are few studies on this topic (Gaudin et al., 1997; Pacelli & Bostwick, 1997). It is reasonable to perform needle biopsies on patients, especially younger men, who have HGPIN after TURP.

It has been demonstrated that patients with a flat or tufting HGPIN pattern on initial biopsy clearly have less risk of cancer being found in subsequent biopsy (20%), in comparison with patients with micropapillary or cribriform pattern, who have a relative risk of 70% (Chan & Epstein, 1999). The isolated finding of HGPIN in the cystoprostatectomy specimen has no clinical implications, and the prognosis of the patient is determined by the initial indication (e.g. invasive bladder cancer) for surgery.

The jury is still out concerning the best repeat biopsy strategy following the diagnosis of PIN on initial prostate biopsy. The length of the interval still needs to be established in large prospective studies (Joniau et al., 2005). Repeat biopsy six weeks after the initial biopsy has led to the diagnoses of prostate cancer in 9% of cases with isolated HGPIN (Chan & Epstein, 1999; Ellis & Brawer, 1995; Kronz et al., 2001; O'Dowd et al., 2000). The risk for finding prostate cancer in repeat biopsies seems to increase with the length of the biopsy interval. Age, PSA, and HGPIN were independent predictors for prostate cancer in repeat biopsies, with HGPIN providing the highest risk ratio (Chan & Epstein, 1999; Langer et al., 1996; Sakr et al., 1996). Most urologists recommend follow-up biopsy after 6–12 months, followed by regular PSA monitoring and repeat biopsies as indicated (Shepherd et al., 1996). Men within screening settings who are diagnosed with isolated HGPIN should be followed at regular intervals, and if clinical suspicion persists, the biopsy should be repeated (Ellis & Brawer, 1995; Kronz et al., 2001; Shepherd et al., 1996). The finding of intraductal HGPIN on initial biopsy needs further investigation with repeat biopsy, because this lesion is related to potentially aggressive cancer.

6.8.2 Chemoprevention

Examples of treated premalignant lesions include cervical intraepithelial neoplasia, ductal CIS (carcinoma *in situ*) of the breast, adenomatous polyps of the colon, and Barrett's esophagus (Sporn, 1999). The American Association for Cancer Research designates intraepithelial neoplasia an important target for chemoprevention (O'Shaughnessy et al., 2002). As HGPIN precedes the development of prostate cancer by several years and is easy identifiable, it is a candidate for chemoprevention. Chemoprevention means the administration of drugs or agents aimed at preventing the initiation and progression of cancer. A number of potential preventive agents have been investigated in patients with HG-PIN, including hormones (flutamide, finasteride, leuprolide acetate) and antioxidants such as lycopene, selenium, and catechins. An association beween the E-cadherin/catenin complex and high-grade prostate cancer has been proved and the therapeutic potential of integrin antagonists is being evaluated by ongoing clinical trials with promising results (Drivalos et al., 2011). One of the most promising chemoprevention drugs is the selective oestrogen receptor modulator toremifene citrate (Ravery, 2009). Recognizing the slow growth rate of prostate cancer and the considerable amount of time needed in animal and human studies for adequate follow-up, the noninvasive precursor lesion PIN is a suitable intermediate histological marker to indicate high likelihood of subsequent prostate cancer. HGPIN offers promise as an intermediate endpoint in studies of chemoprevention of prostate cancer (Montironi et al., 1999). Hence, HGPIN is a suitable intermediate histological marker to indicate subsequent likelihood of cancer and it may be worth monitoring young men with a high risk of developing HG-PIN in the future as potential targets for chemoprevention rather than focusing only on chemoprevention in the high-risk HG-PIN patient group (Ravery, 2009).

Anti-androgens (e.g. fluatamide) induce the regression of prostatic epithelium by enhancing apoptosis, suppressing proliferative activity, and inhibiting angiogenesis in BPH, PIN, and prostate cancer (Montironi et al., 1994). PIN is ablated by androgen deprivation therapy, as a result of accelerated apoptosis with subsequent exfoliation of cells into the glandular lumens (Montironi et al., 1994). Studies have documented that angiogenesis in the surrounding stroma of HGPIN glands is severely decreased via suppression of vascular endothelial growth factor (VEGF) production after androgen deprivation therapy (Papatsoris & Papavassiliou, 2001). A marked decrease in the extent and prevalence of HGPIN occurs in patients treated with anti-androgens in comparison to untreated patients (Bostwick & Qian, 1999). It has been suggested that anti-androgens might halt or reverse the process of carcinogenesis and prevent the transition of HGPIN to overt prostate cancer (Bostwick & Qian, 1999; Lieberman et al., 2001; Montironi et al., 1994; van der Kwast et al., 1999). The observed morphological changes (cytoplasmic clearing, prominent glandular atrophy, decreased ratio of glands to stroma) are reversible, and HGPIN lesions recover rapidly. However, it is unclear whether the histopathologic changes of anti-androgen treatment are clinically important (Lieberman et al., 2001; van der Kwast et al., 1999). *Yamauchi* demonstrated that the anti-androgen bicalutamide permitted the persistence of PIN after effective chemoprevention of microscopic prostate cancer in a rat model (Yamauchi et al., 2006). Moreover, there is a risk for amplification of the androgen-receptor (AR) gene in androgen-deficient conditions, as in cases of hormone-refractory prostate cancer (Koivisto et al., 1999). The blockage of 5-alpha reductase with finasteride does not seem to have any effect on the incidence of PIN (Yang et al., 1999). In addition to the above, the role of Ras/mitogen-activated protein kinase (MAPK) in prostate cancer, as well as the therapeutic potential of Ras/map inhibitors are currently under investigation (Papatsoris et al., 2007). Furthermore, it has been demonstrated that men with no evidence of prostate cancer on initial biopsy who were pretreated with finasteride had a significantly greater prostate cancer detection rate at one year than had men in the control group; 30% versus 4% (Cote et al., 1998).

Besides anti-androgens, other drugs (e.g. anti-angiogenics agents) and nutritional supplements (e.g. vitamin D, selenium) have been applied in ongoing chemoprevention trials (Montironi et al., 1994). In a prospective trial evaluating the effects of selenium-vitamin E-isoflavonoid supplement in 100 men with isolated HGPIN in octant biopsies, PSA decreased in a large subgroup (64%). In this subgroup, the overall risk of detecting cancer was 24.5%, compared to 55.6% in a smaller subgroup of patients in whom the PSA continued to rise under supplements (Joniau et al., 2005). *Bettuzzi* administered green tea catechin (GTC) in men with HGPIN and demonstrated that GTC is safe and very effective (Bettuzzi et al., 2006). In particular, after one year, only one prostate cancer was diagnosed among the 30 GTC-treated men (3%), whereas nine cancers were found among the 30 placebo-treated men (30%).

Studies suggest that administration of the nerve-growth factor (NGF) induces a reversion of the androgen-independent / androgen-receptor negative prostate cancer cell lines to a less malignant phenotype, which raises thoughts for a new perspective in prostate cancer therapy (Papatsoris et al., 2007). Moreover, deregulation of the IGF-1/IGF-1-receptor axis has been liked to progression of prostate cancer to androgen independenace and new therapeutic possibilities are currently under research (Papatsoris et al., 2005).

The ideal agent and duration of therapy remains to be defined. The selective alpha-estrogen receptor modulator toremifene was investigated in HGPIN. Studies using the transgenic adenocarcinoma of mouse prostate model (TRAMP) and this anti-estrogen demonstrated a reduction in the incidence of HGPIN and prostate cancer, along with an increase in animal survival (Raghow et al., 2002). The statistically significant reduction in the incidence of prostate cancer and the tolerability profile support toremifene's promise as a chemopreventive agent.

Although small, high-risk population trials will remain the key to the early evaluation of novel chemoprevention agents, large-scale, population-based clinical trials will still be necessary to ensure that valid recommendations are made to men regarding chemoprevention. Until the efficacy of chemopreventive agents is confirmed in well-conducted, randomized, controlled studies, there should be a reluctance to offer chemopreventive agents to men with isolated HGPIN on initial biopsy.

6.8.3 Radical prostatectomy and radiotherapy

HGPIN is sometimes associated with a PSA above normal levels; in other words, in these cases, HGPIN could be regarded as T1c prostate cancer (Newling, 1999). The firm evidence that within six months of the first biopsy showing HGPIN, invasive prostate cancer would be diagnosed in 60% of the cases has made some urologists offer radical prostatectomy to this group of patients (Newling, 1999). Nowadays, radical prostatectomy is not regarded as appropriate therapy for the management of patients with HGPIN (Davidson et al., 1995; Montironi et al., 2002; Newling, 1999). It seems logical that malignant histological changes should be seen before such radical therapy is offered. It has been recently shown that PSA and HGPIN focality at biopsy do not enhance cancer predictivity, thus patients who underwent prostate biopsy with a HGPIN diagnosis do not seem to need any different follow-up rebiopsy strategy than patients with a diagnosis of BPH (Gallo et al., 2008).

The prevalence and extent of PIN lesions decreases significantly after radiation therapy. Following such therapy, PIN retains the typical characteristics of untreated PIN and is readily recognized on histopathology (Cheng et al., 1999). The question remains if recurrence after radiation therapy is due to the growth of incompletely eradicated tumor or progression of incompletely eradicated PIN.

6.8.4 Potential anti-PIN agents

a. **Anti-Angiogenesis Agents**. The changes that occur in HGPIN leading to focal carcinoma include neo-angiogenesis; hence, the use of the anti-angiogenesis agent's thalidomide and platelet growth factor 4 could be important therapeutic interventions (Papatsoris et al., 2005).

b. **Differentiation Factors**. Retinoids and vitamin D analog are known to improve differentiation of epithelial cells, including prostate epithelium.The development of invasiveness, as seen in HGPIN, is characterized by loss of adhesion facility and dedifferentiation with aneuploid nuclear characteristics; these processes may be sensitive to retinoids or vitamin D analogs (Papatsoris et al., 2005; Banach-Petrosky et al, 2006; Kelloff et al., 1999). Gene therapy and immunotherapy are still experimental in prostate cancer and HGPIN. Serial examination of prostate biopsies and subsequent prostatectomy specimens may give an indication of the effectiveness of these agents.

c. **Epigenetic Therapeutics** *(Histone Deacetylase Inhibitors, Hypomethylating Agents).*
Epigenetic events, such as histone acetylation/deacetylation and aberrant DNA
methylation, represent crucial steps in prostate cancer development, which cause
alterations in gene expression (e.g. silencing tumor suppressor genes) without changes
in the DNA coding sequence (Kopelovich et al., 2003). Epigenetic changes can be
reversed by the use of small molecules, such as histone deacetylase (HDAC) inhibitors
and hypomethylating agents. Histones are core protein components of nucleosomes,
and their acetylation status regulates gene expression. Deacetylated histones are
generally associated with silencing gene expression (Marks et al., 2001). HDAC
inhibitors have been shown to induce expression of genes linked to growth inhibition
and cellular differentiation. Several phase I trials with these agents are ongoing in
patients with prostate cancer and/or PIN (Sandor et al., 2002).

A mechanism to switch off tumor suppressor genes is controlled by a chemical modification
known as DNA methylation, a normal cellular process whereby cytosines in the DNA
become methylated by the enzyme DNA methyltransferase to give 5-methylcytosine (Kang
et al., 2004). However, in cancer cells, the methylation process is deregulated, and many
genes, including tumor suppressor genes, become abnormally methylated at cytosine bases.
Moreover, it seems that aberrant methylation causes recruitment of HDAC, resulting in a
more potent transcriptional inhibition of target genes (Patra et al., 2001). Many studies have
demonstrated epigenetic silencing of crucial genes, for example, AR, PTEN, and RARβ,
during prostate carcinogenesis (Yamanaka et al., 2003). Novel hypomethylating agents are
in various stages of experimental and clinical development.

7. Epilogue

Recurrent chromosomal rearrangements have not been well characterized in prostate cancer
(Papatsoris et al., 2007). *Tomlins* used a bioinformatics approach to discover candidate
oncogenic chromosomal aberrations on the basis of outlier gene expression, followed by
RNA ligase-mediated rapid amplification of cDNA ends and sequencing (Tomlins et al.,
2005). The authors identified recurrent gene fusions of the 5-prime untranslated region of
TMPRSS2 to two ETS transcription factors, ERG or ETV1, in prostate cancer tissues with
outlier expression. By using FISH, they demonstrated that 23 of 29 prostate cancer samples
harbored rearrangements in ERG or ETV1. Cell line experiments suggested that the
androgen-responsive promoter elements of TMPRSS2 mediate the overexpression of ETS
family members in prostate cancer. *Yoshimoto* demonstrated that the occurrence of these
genetic events, along with Pten haploinsufficiency, in patients with prostate cancer has a
significant clinical impact (Yoshimoto et al., 2008). Most importantly, the identification of
ERG as a cooperative initiation event in prostate tumorigenesis suggests that ERG targeted
therapies, when feasible, may be effective at preventing the transition between HGPIN and
invasive cancer, while pharmacological manipulation of the PTEN/PI3K/AKT pathway
may represent a powerful chemopreventive and chemotherapeutic tool in the future (Carver
et al., 2009). Surprisingly, the above-mentioned translocation was found in about 70–80% of
prostate cancers, but not in HGPIN. Finally, the diagnosis of prostate cancer on needle
biopsy has been refined because of the recent discovery of AMARC, which preferentially
labels prostate adenocarcinoma (Epstein, 2006). Also, in a recent peer review *Epstein*
outlined several recommendations when diagnosing PIN or atypical foci suspicious for
carcinoma in needle biopsies (Epstein & Herawi, 2006).

In conclusion, prostate cancer precursor lesions include mainly AAH and PIN (Chrisofos et al., 2007). LSC is not considered a precursor lesion of prostate cancer but shares with PIN the increased risk of diagnosing a definite cancer in subsequent biopsies. LGPIN should not be reported by pathologists due to poor inter-observer reproducibility and a relatively low risk of cancer following re-biopsy. The average incidence of HGPIN or LSC on initial needle biopsy is 6%. Following the diagnosis of HGPIN, the risk of cancer is not statistically higher compared with the risk of cancer following a benign diagnosis. Studies have shown that the risk for cancer after HGPIN diagnosis was not higher than the risk reported after diagnosis of BPH (Gallo et al., 2008). In contrast, the average risk of cancer following a diagnosis of LSC is 40%, and such patients should be re-biopsied within three to six months. Cases diagnosed as LSC have the highest likelihood of being changed upon expert review. Potential markers of prostate cancer precursor lesions include fatty acid synthetase, AMARC, and A80. However, clinical and pathological parameters do not help to stratify which men are at greater risk for a cancer diagnosis. Repeat biopsy should include increased sampling of the initial precursor lesion and adjacent ipsilateral and contralateral sites, with routine sampling of all sextant sites. Radical prostatectomy and radiotherapy are not recommended for the management of patients with HGPIN. Until the efficacy of chemopreventive agents is confirmed in well-conducted, randomized, controlled studies, there should be a reluctance to offer such agents to men with prostate cancer precursor lesion on initial biopsy.

8. References

Alexander EE, Qian J, Wollan PC, Myers R, Bostwick DG. Prostatic intraepithelial neoplasia does not appear to raise serum prostate-specific antigen concentrations. *Urology* 1996; 47: 693–698.

Amin MB, Tamboli P, Varma M, Srigley JR. Postatrophic hyperplasia of the prostate gland: a detailed analysis of its morphology in needle biopsy specimens. *Am J Surg Pathol* 1999; 23: 925–931.

Anton RC, Kattan MW, Chakraborty S, Wheeler TM. Postatrophic hyperplasia of the prostate: lack of association with prostate cancer. *Am J Surg Pathol* 1999; 23: 932–936.

Baltaci S, Orhan D, Ozer G, Tolunay O, Gogous O. Bcl-2 proto-oncogene expression in low- and high- grade prostatic intraepithelial neoplasia. *BJU Int* 2000; 85: 155–159.

Banach-Petrosky W, Ouyang X, Gao H, Nader K, Ji Y, Suh N, DiPaola RS, Abate-Shen C. Vitamin D inhibits the formation of prostatic intraepithelial neoplasia in Nkx3.1;Pten mutant mice. *Clin Cancer Res* 2006; 12: 5895–5901.

Baretton GB, Vogt T, Blasenbreu S, Lohrs U. Comparison of DNA ploidy in prostatic intraepithelial neoplasia and invasive carcinoma of the prostate: an image cytometric study. *Hum Pathol* 1994; 25: 506–513.

Berney DM, Fisher G, Kattan MW *et al.* Pitfalls in the diagnosis of prostatic cancer: retrospective review of 1791 cases with clinical outcome. *Histopathology* 2007; 51; 452–457.

Bettuzzi S, Brausi M, Rizzi F, Castagnetti G, Peracchia G, Corti A. Chemoprevention of human prostate cancer by oral administration of green tea catechins in volunteers with high-grade prostate intraepithelial neoplasia: a preliminary report from a one-year proof-of-principle study. *Cancer Res* 2006; 66: 1234–1240.

BillisA. Prostatic atrophy:an autopsy study of a histologic mimic of adenocarcinoma.*Mol Pathol* 1998;11:47-54

Boag AH, Young ID. Increased expression of the 72-kd type IV collagenase in prostatic adenocarcinoma. Demonstration by immunohistochemistry and in situ hybridization. *Am J Pathol* 1994; 144: 585-591.

Bonkhoff H. Role of the basal cells in premalignant changes of the human prostate: a stem cell concept for the development of prostate cancer. *Eur Urol* 1996; 30: 201-205.

Bostwick DG, Qian J. Effect of androgen deprivation therapy on prostatic intraepithelial neoplasia. *Urology* 582 (Suppl 1): S91-S93.

Bostwick DG, Brawer MK. Prostatic intra-epithelial neoplasia and early invasion in prostate cancer. *Cancer* 1987; 59: 788-794.

Bostwick DG. Prostatic intraepithelial neoplasia (PIN). *Urology* 1989; 34(Suppl 6): S16-S22.

Bostwick DG, Amin MB, Dundore P, Marsh W, Schultz DS. Architectural patterns of high-grade prostatic intraepithelial neoplasia. *Hum Pathol* 1993; 24: 298-310.

Bostwick DG, Dousa MK, Crawford BG, Wollan PC. Neuroendocrine differentiation in prostatic intraepithelial neoplasia and adenocarcinoma. *Am J Surg Pathol* 1994; 18: 1240-1246.

Bostwick DG, Qian J. Atypical adenomatous hyperplasia of the prostate. Relationship with carcinoma in 217 whole-mount radical prostatectomies. *Am J Surg Pathol* 1995; 19: 506-518.

Bostwick DG, Qian J, Frankel K. The incidence of high grade prostatic intraepithelial neoplasia in needle biopsies. *J Urol* 1995; 154: 1791-1794.

Bostwick DG. Prospective origins of prostate carcinoma. Prostatic intraepithelial neoplasia and atypical adenomatous hyperplasia. *Cancer* 1996; 78: 330-336.

Bostwick DG, Pacelli A, Lopez-Beltran A. Molecular biology of prostatic intraepithelial neoplasia. *Prostate* 1996; 29: 117-134

Bostwick DG, Shan A, Qian J, Darson M, Maihle NJ, Jenkins RB, Cheng L. Independent origin of multiple foci of prostatic intraepithelial neoplasia: comparison with matched foci of prostatic carcinoma. *Cancer* 1998; 83: 1995-2002.

Bostwick DG. Prostatic intraepithelial neoplasia is a risk factor for cancer. *Semin Urol Oncol* 1999; 17: 187-198.

Bostwick DG, Cheng L. Overdiagnosis of prostatic adenocarcinoma. *Semin. Urol. Oncol.* 1999; 17; 199-205

Bostwick DG. Prostatic intraepithelial neoplasia. *Curr Urol Rep* 2000; 1: 65-70.

Bostwick DG, Cheng L. *Urologic Surgical Pathology.* New York: Elsevier/Mosby, 2008.

Bourdoumis A, Papatsoris AG, Chrisofos M, Efstathiou E, Skolarikos A, Deliveliotis C.The novel prostate cancer antigen 3 (PCA3) biomarker. Int *Braz J Urol* 2010; 36: 665-9.

Brausi M, Castagnetti G, Dotti A, De Luca G, Olmi R, Cesinaro AM. Immediate radical prostatectomy in patients with atypical small acinar proliferation. Over treatment? *J Urol* 2004; 172: 906-908.

Carver BS, Tran J, Gopalan A, Chen Z, Shaikh S, Carracedo A, Alimonti A, Nardella C, Varmeh S, Scardino PT, Cordon-Cardo C, Gerald W, Pandolfi PP. Aberrant ERG expression cooperates with loss of PTEN to promote cancer progression in the prostate *Nat Genet.* 2009 May; 41: 619-624.

Chan TY, Epstein JI. Follow-up of atypical prostate needle biopsies suspicious for cancer. *Urology* 1999; 53: 351-355.

Cheng L, Shan A, Cheville JC, Qian J, Bostwick DG. Atypical adenomatous hyperplasia of the prostate: a premalignant lesion? *Cancer Res* 1998; 58: 389–391.

Cheng L, Cheville JC, Pisansky TM, Sebo TJ, Slezak J, Bergstralh EJ, Neumann RM, Singh R, Pacelli A, Zincke H, Bostwick DG. Prevalence and distribution of prostatic intraepithelial neoplasia in salvage radical prostatectomy specimens after radiation therapy. *Am J Surg Pathol* 1999; 23: 803–808.

Cheng Liang, Bostwick D.G. Atypical sclerosing adenosis of the prostate : a rare mimic of adenocarcinoma. *Histopathology* 2010; 56: 627-631.

Cheville JC, Bostwick DG. Postatrophic hyperplasia of the prostate. A histologic mimic of prostate adenocarcinoma. *Am J Surg Pathol* 1995; 19: 1068–1076.

Cohen RJ, McNeal JE, Baillie T. Patterns of differentiation and proliferation in intraductal carcinoma of the prostate: significance for cancer progression. *Prostate* 2000; 43: 11–19.

Collina G, Botticelli AR, Martinelli AM, Fano RA, Trentini GP. Sclerosing adenosis of the prostate. Report of three cases with electronmicroscopy and immunohistochemical study. *Histopathology* 1992; 20; 505–510.

Coogan C, Bostwick D, Bloom K, Gould V. Glycoprotein A-80 in the human prostate: immunolocalization in prostatic intraepithelial neoplasia, carcinoma, radiation failure, and after neoadjuvant hormonal therapy. *Urology* 2003; 61: 248–252.

Cote RJ, Skinner EC, Salem CE, Mertes SJ, Stanczyk FZ, Henderson BE, Pike MC, Ross RK. The effect of finasteride on the prostate gland in men with elevated serum prostate-specific antigen levels. *Br J Cancer* 1998; 78: 413–418.

Chrisofos M, Papatsoris AG, Lazaris A, Deliveliotis C. Precursor leasions of prostate cancer. *Crit Rev Clin Lab Sci* 2007; 44: 243-270.

Darson MF, Pacelli A, Roche P, Rittenhouse HG, Wolfert RL, Young CYF, Klee GG, Tindall DJ, Bostwick BG. Human glandular kallikrein 2 (hK2) expression in prostatic intraepithelial neoplasia and adenocarcinoma: A novel prostate cancer marker *Urology* 1997; 49: 857-862.

Darson MF, Pacelli A, Roche P, Rittenhouse HG,Wolfert RL, Saeid MS, Young CY, Klee GG, Tindall DJ, Bostwick DG. Human glandular kallikrein 2 expression in prostate adenocarcinoma and lymph node metastases. *Urology* 1999; 53: 939–944.

Davidson D, Bostwick D, Qian J, Wollan PC, Oesterling JE, Rudders RA, Siroky M, Stilmant M. Prostatic intraepithelial neoplasia is a risk factor for adenocarcinoma: predictive accuracy in needle biopsies. *J Urol* 1995; 154: 1295–1299.

DeMarzo AM, Marchi VL, Epstein JI, NelsonWG. Proliferative inflammatory atrophy of the prostate: implications for prostatic carcinogenesis. *Am J Pathol* 1999; 155: 1985–1992.

De Marzo AM, Platz EA, Epstein JI. A working group classification of focal prostate atrophy lesions. *Am J Surg Pathol.* 2006 30: 1281-1291.

Dovey Z, Corbishley CM, Kirby RS. Prostatic intraepithelial neoplasia: a risk factor for prostate cancer. *Can J Urol* 2005; 12(Suppl 1): 49–52.

Drivalos A, Papatsoris AG, Chrisofos M, Efstathiou E, Dimopoulos MA.The role of the cell adhesion molecules (integrins / cadherins) in prostate cancer. MA. *Int Braz J Urol.* 2011; 37: 302-326.

Ellis WJ, Brawer MK. Repeat biopsy: who needs it? *J Urol* 1995; 153: 1496–1498.

Epstein JI, Grignon DJ, Humphrey PA, McNeal JE, Sesterhenn IA, Troncoso P, Wheeler TM. Interobserver reproducibility in the diagnosis of prostatic intraepithelial neoplasia. *Am J Surg Pathol* 1995; 19: 873–886.

Epstein JI. Pathology of prostatic neoplasia. In Walsh PC, Retik AB, Vaughan ED, Wein AJ, Eds. *Campell's Urology, 8th Ed.* Pp 3025–3037. Philadelphia: Saunders, 2002.

Epstein JI. What's new in prostate cancer disease assessment in 2006? *Curr Opin Urol* 2006; 16: 146–151.

Epstein JI, Herawi M. Prostate needle biopsies containing prostatic intraepithelial neoplasia or atypical foci suspicious for carcinoma: implications for patient care. *J Urol* 2006; 175: 820–834.

Fadare O, Wang S, Mariappan MR. Practice patterns of clinicians following isolated diagnoses of atypical small acinar proliferation on prostate biopsy specimens. *Arch Pathol Lab Med* 2004; 128: 557–560.

Franks LM. Atrophy and hyperplasia in the prostate proper. *J Pathol Bacteriol* 1954; 68: 617–621.

Gallo F, Chiono L, Gastaldi E, Venturino E, Giberti C. Prognostic Significance of High-Grade Prostatic Intraepithelial Neoplasia (HGPIN): Risk of Prostatic Cancer on Repeat Biopsies *Urology* September 2008; 72: 628-632.

Gaudin PB, Epstein JI. Adenosis of the prostate. Histologic features in transurethral resection specimens. *Am J Surg Pathol* 1994; 18: 863–870.

Gaudin PB, Epstein JI. Adenosis of the prostate Histological features in needle biopsy specimens. *Am J Surg Pathol* 1995; 19: 737–747.

Gaudin PB, Sesterhenn IA, Wojno KJ, Mostofi FK, Epstein JI. Incidence and clinical significance of high-grade prostatic intraepithelial neoplasia in TURP specimens. *Urology* 1997; 49: 558–563.

Goeman L, Joniau S, Ponette D, Van der Aa F, Roskams T, Oyen R, Van Poppel H. Is low-grade prostatic intraepithelial neoplasia a risk factor for cancer? *Prostate Cancer Prostatic Dis* 2003; 6: 305–310.

Grignon DJ, Ro JY, Srigley JR, Troncoso P, Raymond AK, Ayala AG. Sclerosing adenosis of the prostate gland. A lesion showing myoepithelial differentiation. *Am. J. Surg. Pathol. 1992; 16; 383–391.*

Grignon DJ, Sakr WA. Atypical adenomatous hyperplasia of the prostate: a critical review. *Eur Urol* 1996; 30: 206–211.

Hailemariam S, Vosbeck J, Cathomas G, Zlobec I, Mattarelli G, Eichenberger T, Zellweger T, Bachmann A, Gasser T, Bubendorf L. Can molecular markers stratify the diagnostic value of high-grade prostatic intraepithelial neoplasia? *Human Pathology* 2011; 42: 702-709.

Haggman MJ, Macoska JA, Wojno KJ, Oesterling JE. The relationship between prostatic intraepithelial neoplasia and prostate cancer: critical issues. *J Urol* 1997; 158: 12–22.

Hellpap B, Kollermann J. Atypical acinar proliferation of the prostate. *Pathol Res Pract* 1999; 195: 795–799. Helpap B, Kollermann J, Oehler U. Limiting the diagnosis of atypical small glandular proliferations in needle biopsies of the prostate by the use of immunohistochemistry. *J Pathol* 2001; 193: 350–353.

Iczkowski KA, MacLennan GT, Bostwick DG. Atypical small acinar proliferation suspicious for malignancy in prostate needle biopsies: clinical significance in 33 cases. *Am J Surg Pathol* 1997; 21: 1489–1495.

Iczkowski KA, Bassler TJ, Schwob VS, Bassler IC, Kunnel BS, Orozco RE, Bostwick DG. Diagnosis of "suspicious for malignancy" in prostate biopsies: predictive value for cancer. *Urology* 1998; 51: 749–758. Iczkowski KA, Bostwick DG. Criteria for biopsy diagnosis of minimal volume prostatic adenocarcinoma: analytic comparison with nondiagnostic but suspicious atypical small acinar proliferation. *Arch Pathol Lab Med* 2000; 124: 98–107.

Iczkowski KA, Chen HM, Yang XJ, Beach RA. Prostate cancer diagnosed after initial biopsy with atypical small acinar proliferation suspicious for malignancy is similar to cancer found on initial biopsy. *Urology* 2002; 60: 851–854

Jones EC, Clement PB, Young RH. Sclerosing adenosis of the prostate gland. A clinicopathological and immunohistochemical study of 11 cases. *Am. J. Surg. Pathol. 1991; 15; 1171–1180.*

Joniau S, Goeman L, Pennings J, Van Poppel H. Prostatic intraepithelial neoplasia (PIN): importance and clinical management. *Eur Urol* 2005; 48: 379–385.

Joniau S, Goeman L, Roskams T, Lerut E, Oyen R, Van Poppel H. Effect of Nutritional Supplement Challenge in Patients with Isolated High-Grade Prostatic Intraepithelial Neoplasia *Urology* 2007; 69: 1102-1106.

Kang GH, Lee S, Lee HJ, Hwang KS. Aberrant CpG island hypermethylation of multiple genes in prostate cancer and prostatic intraepithelial neoplasia. *J Pathol* 2004; 202: 233–240.

Keetch DW, Humphrey P, Stahl D, Smith DS, Catalona WJ. Morphometric analysis and clinical follow up of isolated prostatic intraepithelial neoplasia in needle biopsy of the prostate. *J Urol* 1995; 154: 347–351.

Kelloff GJ, Lieberman R, Brawer MK, Crawford ED, Labrie F, Miller GJ, Kelloff GJ. Strategies for chemoprevention of prostate cancer. *Prostate Cancer Prostatic Dis* 1999; 2 (Suppl): S27–S33.

Koivisto PA, Schleutker J, Helin H, Ehren-van Eekelen C, Kallioniemi OP, Trapman J. Androgen receptor gene alterations and chromosomal gains and losses in prostate carcinomas appearing during finasteride treatment for benign prostatic hyperplasia. *Clin Cancer Res* 1999; 5: 3378–3382.

Kopelovich L, Crowell JA, Fay JR. The epigenome as a target for cancer chemoprevention. *J Nat Cancer Inst* 2003; 95: 1747–1757.

Kronz JD, Allan CH, Shaikh AA, Epstein JI. Predicting cancer following a diagnosis of high grade prostatic intraepithelial neoplasia on needle biopsy: data on men with more than one follow-up biopsy. *Am J Surg Pathol* 2001 ; 25: 1079–1085.

Langer JE, Rovner ES, Coleman BG, Yin D, Arger PH, Malkowicz SB, Nisenbaum HL, Rowling SE, Tomaszewski JE, Wein AJ, Jacobs JE. Strategy for repeat biopsy of patients with prostatic intraepithelial neoplasia detected by prostate needle biopsy. *J Urol* 1996; 155: 228-231.

Lefkowitz GK, Taneja SS, Brown J, Melamed J, Lepor H. Follow-up interval prostate biopsy 3 years after diagnosis of high grade prostatic intraepithelial neoplasia is associated with high likelihood of prostate cancer, independent of change in prostate specific antigen levels. *J Urol* 2002; 168: 1415–1418.

Lieberman R, Bermejo C, Akaza H, Greenwald P, Fair W, Thompson I. Progress in prostate cancer chemoprevention: modulators of promotion and progression. *Urology* 2001; 58: 835–842

Marks P, Rifkind RA, Richon VM, Breslow R, Miller T, Kelly WK. Histone deacetylases and cancer: causes and therapies. *Nat Rev Cancer* 2001; 1: 194-202.

McNeal JE. Morphogenesis of prostatic carcinoma. *Cancer* 1965; 18: 1659-1666.

McNeal JE, Bostwick DG. Intraductal dysplasia: a premalignant lesion of the prostate. *Hum Pathol* 1986; 17: 64-71.

McNeal JE, Bostwick DG. Intraductal dysplasia: a pre-malignant lesion of the prostate. *Hum Pathol* 1986; 17: 64-71.

Montironi R, Galluzzi CM, Diamanti L, Taborro R, Scarpelli M, Pisani E. Prostatic intra-epithelial neoplasia. Qualitative and quantitative analyses of the blood capillary architecture on thin tissue sections. *Pathol Res Pract* 1993; 189: 542-548.

Montironi R, Magi-Galluzzi C, Muzzonigro G, Prete E, Polito M, Fabris G. Effects of combination endocrine therapy on normal prostate, prostatic intraepithelial neoplasia, and prostatic adenocarcinoma. *J Clin Pathol* 1994; 47: 906-913.

Montironi R, Mazzuccelli R, Marshall JR, Bartels PH. Prostate cancer prevention: review of target populations, pathological biomarkers, and chemopreventive agents. *J Clin Pathol* 1999; 52: 793-803. Montironi R, Santinelli A, Mazzucchelli R. Prostatic intraepithelial neoplasia and prostate cancer. *Panminerva Med* 2002; 44: 213-220.

Moore CK, Karikehalli S, Nazeer T, Fisher HAG, Kaufman RP Jr, Mian BM. Prognostic significance of high grade prostatic intraepithelial neoplasia and atypical small acinar proliferation in the contemporary era. *J Urol* 2005; 173: 70-72.

Myers RB, Grizzle WE. Biomarker expression in prostatic intraepithelial neoplasia. *Eur Urol* 1996; 30: 153-166.

Nelson WG, De Marzo AM, Isaacs WB. Prostate cancer. *N Engl J Med* 2003; 349: 366-381.

Newling DW. PIN I-II: when should we interfere? *Eur Urol* 1999; 35: 504-507.

O'Dowd GJ, Miller MG, Orozco R & Veltri RW. Analysis of repeated biopsy results within 1 year after noncancer diagnosis. *Urology* 2000; 55: 553-559.

Orteil H. Involutionary changes in prostate and female breast cancer in relation to cancer development. *Can Med Assoc J* 1926; 16: 237.

O'Shaughnessy JA, Kelloff GJ, Gordon GB, Dannenberg AJ, Hong WK, Fabian CJ, Sigman CC, Bertagnolli MM, Stratton SP, Lam S, Nelson WG, Meyskens FL, Alberts DS, Follen M, Rustgi AK, Papadimitrakopoulou V, Scardino PT, Gazdar AF, Wattenberg LW, Sporn MB, Sakr WA, Lippman SM, Von Hoff DD. Treatment and prevention of intraepithelial neoplasia: an important target for accelerated new agent development. *Clin Cancer Res* 2002; 8: 314-346.

Pacelli A, Bostwick DG. Clinical significance of high-grade prostatic intraepithelial neoplasia in transurethral resection specimens. *Urology* 1997; 50: 335-359.

Papatsoris AG, Papavassiliou AG. Prostate cancer: horizons in the development of novel anti-cancer strategies. *Curr Med Chem Anticancer Agents* 2001; 1: 47-70.

Papatsoris AG, Karamouzis MV, Papavassiliou AG. Novel insights into the implication of the IGF-1 network in prostate cancer. *Trends Mol Med* 2005; 11: 52-55.

Papatsoris AG, Karamouzis MV, Papavassiliou AG. Novel biological agents for the treatment of hormone-refractory prostate cancer (HRPC). *Curr Med Chem* 2005; 12: 277-296.

Papatsoris AG, Karamouzis MV, Papavassiliou AG. The power and promise of "rewiring" the mitogen-activated protein kinase network in prostate cancer therapeutics. *Mol Cancer Ther* 2007; 6: 811-819.

Papatsoris AG, Liolitsa D, Deliveliotis C. Manipulation of the nerve growth factor network in prostate cancer. *Expert Opin Investig Drugs* 2007; 16: 303-309.

Papatsoris AG, Anagnostopoulos F. Prostate cancer screening behaviour. *Public Health* 2009; 123: 69-71.

Patra SK, Patra A, Dahiya R. Histone deacetylase and DNA methyltransferase in human prostate cancer. *Biochem Biophys Res Commun* 2001; 287: 705-713.

Powell IJ, Bock CH, Ruterbusch JJ. & Sakr W. Evidence Supports a Faster Growth Rate and/or Earlier Transformation to Clinically Significant Prostate Cancer in Black Than in White American Men, and Influences Racial Progression and Mortality Disparity. *J Urol* 2010; 183: 1792-1797.

Qian J, Bostwick DG, Takahashi S, Borell TJ, Herath JF, Lieber MM, Jenkins RB. Chromosomal anomalies in prostatic intraepithelial neoplasia and carcinoma detected by fluorescence in situ hybridization. *Cancer Res* 1995; 55: 5408-5414.

Qian J,Wollan P, Bostwick DG. The extent and multicentricity of high-grade prostatic intraepithelial neoplasia in clinically localized prostatic adenocarcinoma. *Hum Pathol* 1997; 28: 143-148.

Raghow S, Hooshdaran MZ, Katiyar S, Steiner MS. Toremifene prevents prostate cancer in the transgenic adenocarcinoma of mouse prostate model. *Cancer Res* 2002; 62: 1370-1376.

Ravery V. Towards early and more specific diagnosis of prostate cancer? Identifying a Key Target Population for Chemoprevention and Available Strategies *European Urology Supplements* 2010; 8: 103-107.

Ronnett BM, Carmichael MJ, Carter HB, Epstein JI. Does high-grade prostatic intraepithelial neoplasia result in elevated serum prostate specific antigen levels? *J Urol* 1993; 150: 386-389.

Rubin MA, Zhou M, Dhanasekaran SM, Varambally S, Barrette TR, Snada MG, Pienta KJ, Ghosh D, Chinnaiyan AM. Alpha-Methylacyl coenzyme A racemase as a tissue biomarker for prostate cancer. *JAMA* 2002; 287: 1662-1670.

Sakamoto N, Tsuneyoshi M, Enjoji M. Sclerosing adenosis of the prostate. Histopathologic and immunohistochemical analysis. *Am. J. Surg. Pathol.* 1991; 15; 660-667.

Sakr WA, Haas GP, Cassin BJ, Pontes JE, Crissman JD. Frequency of carcinoma and intraepithelial neoplasia of the prostate in young male patients. *J Urol* 1993; 150: 379-385

Sakr WA, Grignon DJ, Haas GP, Schomer KL, Heilbrun LK, Cassin BJ, Powell J, Montie JA, Pontes JE, Crissman JD. Epidemiology of high grade prostatic intraepithelial neoplasia. *Pathol Res Pract* 1995; 191: 838-841.

Sakr WA, Grignon DJ, Haas GP, Heilbrur LK, Pontes JE, Crissman JD. Age and racial distribution of prostatic intraepithelial neoplasia. *Eur Urol* 1996; 30: 138-144.

Sakr WA, Brawer MK, Moul JW, Donohue R, Schulman CG, Sakr D. Pathology and bio markers of prostate cancer. *Prostate Cancer Prostatic Dis* 1999; 2(Suppl1): S7-S14.

Sandor V, Bakke S, Robey RW, Kang MH, Blagosklonny MV, Bender J, Brooks R, Piekarz RL, Tucker E, Figg WD, Chan KK, Goldspiel B, Fojo AT, Balcerzak SP, Bates SE. Phase I trial of the histone deacetylase inhibitor, depsipeptide (FR901228, NSC 630176), in patients with refractory neoplasms. *Clin Cancer Res* 2002; 8: 718-728.

Sardana G, Dowell B, Diamandis EP. Emerging Biomarkers for the diagnosis and prognosis of prostate cancer. *Clinical Chemistry* 2008 54 : 1951-1960

Scattoni V, Roscigno M, Freschi M, Deho F, Raber M, Briganti A, Fantini G, Nava L, Montorsi F, Rigatti P. Atypical small acinar proliferation (ASAP) on extended prostatic biopsies: predictive factors of cancer detection on repeat biopsies. *Arch Ital Urol Androl* 2005; 77: 31-36.

Shah RB, Kunju LP, Shen R, LeBlanc M, Zhou M, Rubin MA. Usefulness of basal cell cocktail (34beta E12+p63) in the diagnosis of atypical prostate glandular proliferations. *Am J Clin Pathol* 2004; 122: 517-523.

Shepherd D, Keetch DW, Humphrey PA, Smith DS, Stahl D. Repeat biopsy strategy in men with isolated prostatic intraepithelial neoplasia on prostate needle biopsy. *J Urol* 1996; 156: 460-463.

Shin SS, Gould VE, Gould JE, Warren WH, Gould KA, Yaremko ML, Manderino GL, Rittenhouse HG, Tomita JT, Jansson DS. Expression of a new mucin-type glycoprotein in select epithelial dysplasias and neoplasms detected immunocytochemically with Mab A-80. *APMIS* 1989; 97: 1053-1067.

Signoretti S, Waltregny D, Dilks J, Isaac B, Lin D, Garraway L, Yang A, Montironi R, McKeonand F & Loda M. *Am J Pathol* 2000; 157: 1769-1775.

Singh PB, Nicholson CM, Ragavan N, Blades RA, Martin FL, Matanhelia SS. Risk of prostate cancer after isolated high-grade prostatic intraepithelial neoplasia (HGPIN) detected on extended core needle biopsy : a UK hospital experience. *BMC Urol* 2009; 9: 3.

Sinha AA, Quast BJ, Reddy PK, Lall V, Wilson MJ, Qian J, Bostwick DG. Microvessel density as a molecular marker for identifying high-grade prostatic intraepithelial neoplasia precursors to prostate cancer. *Exp Mol Pathol* 2004; 77: 153-159.

Sporn MB. Prevention of cancer in the next millennium: report of the chemoprevention working group to the American Association for Cancer Research. *Cancer Res* 1999; 59: 4743-4758

Steiner MS. High-grade prostatic intraepithelial neoplasia and prostate cancer risk reduction.*World J Urol* 2003; 21: 15-20.

Swinnen JV, Roskams T, Joniau S, Van Poppel H, Oyen R, Heyns W, Verhoeven G. Overexpression of fatty acid synthase is an early and common event in the development of prostate cancer. *Int J Cancer* 2002; 98: 19-22.

Tomlins SA, Rhodes DR, Perner S, Dhanasekaran SM, Mehra R, Sun XW, Varambally S, Cao X, Tchinda J, Kuefer R, Lee C, Montie JE, Shah RB, Pienta KJ, Rubin MA, Chinnaiyan AM. Recurrent fusion of TMPRSS2 and ETS transcription factor genes in prostate cancer. *Science* 2005; 310: 644-648.

Van der Kwast TH, Labrie F, Tetu B. Prostatic intraepithelial neoplasia and endocrine manipulation. *Eur Urol* 1999; 35: 508-510.

Vis AN, Van Der Kwast TH. Prostatic intraepithelial neoplasia and putative precursor lesions of prostate cancer: a clinical perspective. *BJU Int* 2001; 88: 147-157.

Vis AN, Hoedemaeker RF, Roobol M, Schroder FH, van der Kwast TH. The predictive value for prostate cancer of lesions that raise suspicion of concomitant carcinoma: an evaluation from a randomized population-based study of screening for prostate cancer. *Cancer* 2001; 92: 524-534.

Weinstein MH, Epstein JI. Significance of high-grade prostatic intraepithelial neoplasia on needle biopsy. *Hum Pathol* 1993; 24: 624-629.

Yamanaka M, Watanabe M, Yamada Y, Takagi A, Murata T, Takahashi H, Suzuki H, Ito H, Tsukino H, Katoh T, Sugimura Y, Shiraishi T. Altered methylation of multiple genes in carcinogenesis of the prostate. *Int J Cancer* 2003; 106: 382–387.

Yamauchi A, Kawai K, Tsukamoto S, Ideyama Y, Shirai T, Akaza H. Persistence of prostatic intraepithelial neoplasia after effective chemoprevention of microscopic prostate cancer with antiandrogen in a rat model. *J Urol* 2006; 175: 348–352.

Yang XJ, Lecksell K, Short K, Gottesman J, Peterson L, Bannow J, Schellhammer PF, Fitch WP, Hodge GB, Parra R, Rouse S, Waldstreicher J, Epstein JI. Does long-term finasteride therapy affect the histologic features of benign prostatic tissue and prostate cancer on needle biopsy? PLESS Study Group. Proscar Long-Term Efficacy and Safety Study. *Urology* 1999; 53: 696–700.

Yoshimoto M, Joshua AM, Cunha IW, Coudry RA, Fonseca FP, Ludkovski O, Zielenska M, Soares FA, Squire JA. Absence of TMPRSS2: ERG fusions and PTEN losses in prostate cancer is associated with a favorable outcome. *Mod Pathol* 2008; 21: 1451-1460.

Young RH, Clement PB. Sclerosing adenosis of the prostate. *Arch. Pathol. Lab Med.* 1987; 111; 363–366.

Young RH, Clement PB. 'Pseudoadenomatoid' tumour of prostate. *Histopathology* 1990; 16: 420.

Zeng L, Kyprianou N. Apoptotic regulators in prostatic intraepithelial neoplasia (PIN): value in prostate cancer detection and prevention. *Prostate Cancer Prostatic Dis* 2005; 8: 7–13.

7

Chemopreventive Target for Prostate Cancer: Prostatic Intraepithelial Neoplasia

J. Arunakaran[1], S. Banudevi[1] and A. Arunkumar[2]

[1]*Department of Endocrinology, Dr. ALM Post Graduate Institute of Basic Medical Sciences, University of Madras, Sekkizhar Campus, Taramani, Chennai Tamilnadu,*
[2]*Center of Excellence in Cancer Research, Department of Biomedical Sciences, Texas Tech University Health Sciences Center, Texas*
[1]*India*
[2]*USA*

1. Introduction

Prostate cancer is one of the most frequently diagnosed malignancies among the male population and the second common cancer-related death worldwide after lung cancer (Jemal *et al.*, 2010). It is estimated that 30% of male older than 50 years are harboring microscopic transformation of adenocarcinoma within the prostate gland. Because of the dramatic rise in the incidence of prostate cancer with the rate increasing approximately 6% per year worldwide (Eschenbach, 1996), the study of precursor lesions of prostate cancer is emerging concept in the field of prostate cancer prevention.

2. Prostatic intraepithelial neoplasia

John McNeal introduced the term *intraductal dysplasia of the prostate* in the early 1960s, postulating that carcinoma of the prostate arose from active ductal/acinar epithelium and not from atrophic acini (McNeal, 1965, 1988; Amin *et al.*, 1993). Later various terms have been rasied such as *large acinar atypical hyperplasia with malignant change* (Allam *et al.*, 1996) and *ductacinar dysplasia* (McNeal, 1988). None of these have gained popularity. The term Prostatic intraepithelial neoplasia (PIN) was first proposed by Bostwick and Brawer in 1987, and this term was accepted at the 1989 Workshop on Prostatic Dysplasia (Bethesda, Md; March 1989) as the preferred nomenclature for this preneoplastic change (Bethesda, Md; March 1989; Drago *et al.*, 1989). PIN refers to the putative precancerous end of the continuum of cellular proliferations within the lining of prostatic ducts, ductules and acini (Bostwick and Amin, 1996; Bostwick and Qian, 2004).

PIN is the most likely precursor of prostate cancer and has been described as a premalignant or pre-invasive form of prostate cancer (Sakr *et al.*, 1993; Bostwick, 1996). Although two histopathologic lesions in the prostate were proposed as being premalignant (PIN and atypical adenomatous hyperplasia, AAH), there is less evidence of a premalignant role for AAH than there is for PIN (De La Torre *et al.*, 1993; Jones and Young, 1994; Epstein, 1994; Bostwick, 1996). Within these lesions, studies have identified impaired and abnormal differentiation, increased proliferation, and abnormal DNA content and elevated *ras* protooncogene mRNA expression (Jones and Young, 1994). There are two grades of PIN (low-grade and high-grade), although the term PIN is usually used to indicate high-grade PIN (HGPIN). The high level of interobserver variability with low-grade PIN (LGPIN) limits its clinical utility, and many pathologists do not report this finding except in research studies. Low grade PIN (LGPIN) is only a very early precursor, and might even not be considered as a precancerous lesion. Moreover, the distinction between LGPIN and normal epithelium might be observer related (Lipski *et al.*, 1996; Zlotta and Schulman, 1999; Vis and Van der Kwast, 2001).

PIN lesions can only be diagnosed by histopathological examination of prostatic tissue. It is impossible to detect PIN clinically by digital rectal examination (DRE), prostate specific antigen (PSA) or ultrasound. Cytologically, LGPIN and HGPIN have clear and reproducible features (Table 1). Histologically, however, different architectural variations exist for HGPIN (Fig. 1). At least four distinct patterns can be distinguished: flat, tufting, micropapillary and cribriform (Bostwick, 1989). Less frequent are signet cell pattern, small cell neuroendocrine pattern, mucinous pattern and microvacuolar pattern. In the big majority of PIN-lesions, a tufting pattern can be found. Frequently, multiple patterns can be found at the same time.

Structural pattern	Low grade PIN (LGPIN)	High grade PIN (HGPIN)
Architecture	Crowding, stratification, irregular spacing	More changes, 4 patterns (T, MP, cribriform, flat)*
Nuclei Chromatin Nucleoli	Slight enlargement, size variation Normal Rarely prominent	Definite enlargement, less size variation Increased density and clumping Frequently prominent
Basal cell layer	Intact	May show some disruption
Basement membrane	Intact	Intact

* T indicates tufting; MP- micropapillary.

Table 1. Prostatic intraepithelial neoplasia (PIN): diagnostic criteria modified from Bostwick and Brawer (1987)

HGPIN is the most significant risk factor for prostate cancer in needle biopsy specimens. Its role as the preinvasive stage of cancer was recently confirmed conclusively in two separate mouse models (Kasper *et al.*, 1998; Garabedian *et al.*, 1998). PIN coexists with cancer in more than 85% of cases (McNeal and Bostwick, 1986; Qian *et al.*, 1997) but retains an intact or fragmented basal cell layer, unlike cancer, which lacks a basal cell layer (Bostwick and Brawer, 1987). PIN is strongly predictive of adenocarcinoma, and its identification in biopsy specimens of the prostate warrants further search for concurrent cancer. PIN alone has no apparent influence on serum PSA concentration, and it is not apparently visible by current imaging techniques.

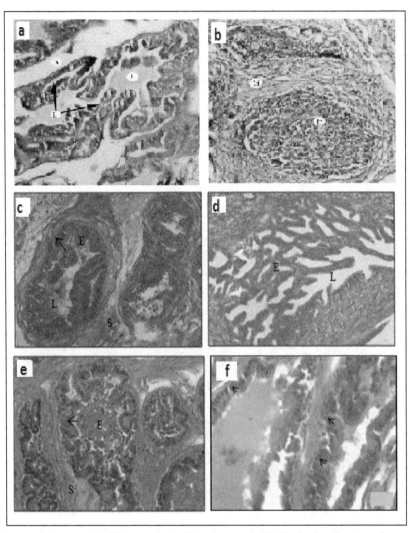

Fig. 1. **Prostatic intraepithelial neoplasia Histology. (a).** Multiple dysplastic and hyperplastic sites were seen within the same glandular epithelium (×200 magnifications; H&E) **(b).** Fully developed PIN with loss of basal epithelial cells (×200 magnifications; H&E) (Arunkumar *et al.*, 2006). Architectural pattern of high grade prostatic intraepithelial neoplasia (PIN) such as tufting **(c)**, cribriform **(d)**, and micropapillary **(e)** structures were observed (×200 magnifications; H&E) **(f)**. The cell number in PIN was markedly increased, and cell density with sparse cytoplasm (arrow headed) (×400 magnifications; H&E) (Banudevi *et al.*, 2011a & b). E- Epithelium; L- Lumen; S-Stroma

3. Identification of PIN: Histological criteria

The classification of PIN into low grade and high grade is based mainly on the cytological characteristics of the cells. The nuclei of cells composing LGPIN are enlarged, vary in size, have normal or slightly increased chromatin content, and possess small or inconspicuous nucleoli (Figs. 2 and 3). HGPIN is characterised by cells with large nuclei of relatively uniform size, an increased chromatin content, which might be irregularly distributed, and prominent nucleoli that are similar to those of carcinoma cells (Fig. 4). The basal cell layer is intact or rarely interrupted in LGPIN, but may have frequent disruptions in high grade lesions. Although the cytological features of LGPIN and HGPIN are fairly constant, the architecture shows a spectrum, varying from a flattened epithelium to a florid cribriform proliferation. Four basic patterns that often coexist have been described by Bostwick and colleagues (Bostwick *et al.*, 1993): flat, tufting, micropapillary, and cribriform. Familiarity with this diverse architectural spectrum may facilitate the histological recognition of PIN, even though these various architectural patterns have no apparent clinicopathological relevance (Bostwic *et al.*, 1993).

Neuroendocrine differentiation occurs in PIN, where it is intermediate in degree between normal prostate (which has the most cells with neuroendocrine differentiation) and carcinoma (Bostwick *et al.*, 1994a; Di Sante Agnese, 1996). Paneth cell like change of the prostatic epithelium (neuroendocrine cells with large eosinophilic granules) is considered to be a distinct form of neuroendocrine differentiation characterised by isolated cells or small groups of cells with prominent eosinophilic cytoplasmic granules.

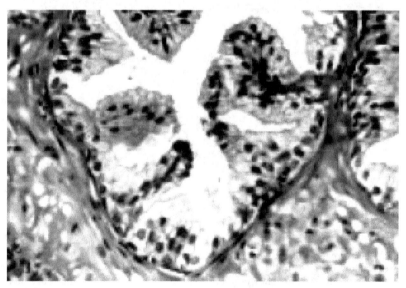

Fig. 2. **Normal prostate**. The duct is lined by a two cell layer– for example, the basal cell and the secretory or luminal cell layers.

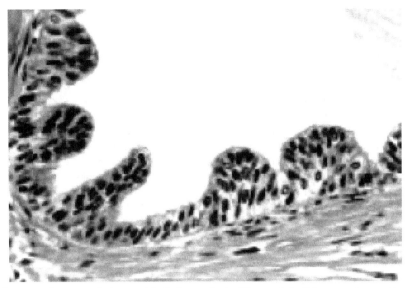

Fig. 3. Low grade prostatic intraepithelial neoplasia. The nuclei of the secretory cells are enlarged, vary in size, have normal or slightly increased chromatin content, and possess small or inconspicuous nucleoli. The basal cell layer is almost intact.

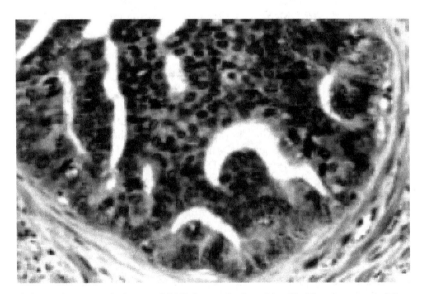

Fig. 4. **High grade prostatic intraepithelial neoplasia** with cribriform patterns. The perimeter cells show features of clearly dysplastic cells, whereas, going from the periphery towards the centre, the nuclei become smaller and the nucleoi become less apparent ("maturation phenomenon"). The basal cell layer is disrupted.

4. PIN incidence

The incidence of PIN varies according to the male population. The lowest likelihood is in men participating in PSA screening and early detection studies, with an incidence of PIN in biopsies ranging from 0.7% to 20% (Bostwick *et al.*, 1996; Langer *et al.*, 1996; Wills *et al.*, 1997; Skjorten *et al.*, 1997). Men seen by urologists in practice have PIN in 4.4% to 25% of contemporary 18-gauge needle biopsies obtained by urologists. Those undergoing transurethral resections have the highest likelihood of PIN, varying from 2.8% to 33% (Pacelli *et al.*, 1997). In such cases, all tissues should be examined, but serial sections of suspicious foci are probably not usually necessary. Select anti-keratin antibodies such as 34β-E12 (high molecular weight keratin) may be used to stain tissue sections for the presence of basal cells, recognizing that PIN retains intact or fragmented basal cell layer, whereas cancer does not (Bostwick and Brawer, 1987). By immunohistochemical analysis, Prostate tumor overexpressed-1 (PTOV1) was considered as good marker for PIN which shows strong immunoreactivity in areas of carcinoma and HGPIN (Morote *et al.*, 2008)

5. PIN distribution

Prostatic intraepithelial neoplasia is found predominantly in the peripheral zone of the prostate (75%–80%), rarely in the transition zone (10%–15%), and extremely rarely in the central zone (< 5%). This distribution parallels the frequency of the zonal predilection for prostatic carcinoma (Bostwick *et al.*, 1995; Gaudin *et al.*, 1997,; Pacelli *et al.*, 1997). The frequency of HGPIN in needle biopsy series ranges from 5% to 16% and in transurethral resection of the prostate specimens between 2.3% and 4.2% (Bostwick *et al.*, 1995; Gaudin *et al.*, 1997; Melissari *et al.*, 2006) McNeal in 1969 mentioned the multifocality of this process (McNeal, 1969; 1988); this observation has since been corroborated by others.

6. Immunohistochemistry in PIN

Numerous studies to highlight the basal cells and the secretory cells have been done (Brawer *et al.*, 1985; McNeal *et al.*, 1988a, 1988b; Perlman and Epstein, 1990; Abhrams *et al.*, 2002; Zhou *et al.*, 2003; Wu *et al.*, 2004). The basal cells and luminal cells of the prostatic glands display different keratin immunoreactivity. The high-molecular-weight cytokeratin monoclonal antibody (clone 34βE12, also referred to as CK903) recognizes keratin proteins of 49, 51, 57, and 66 kD and labels the basal cells but not the luminal/secretory cells of the prostatic glands that stain with prostate-specific antigen and prostatic acid phosphatase. α-Methylacyl CoA-racemase (AMACR) stains the cells of adenocarcinoma but usually does not stain benign prostate glands (Brawer *et al.*, 1985; McNeal *et al.*, 1988a, 1988b; Perlman and Epstein, 1990; Abhrams *et al.*, 2002; Zhou *et al.*, 2003; Wu *et al.*, 2004).

Other antibodies that also mark basal cells include p63 (McNeal, 1969) and CK5/6 (Abhrams *et al.*, 2002). p63 is a nuclear stain, whereas CK5/6 stains the cytoplasm (Abhrams *et al.*, 2002; Zhou *et al.*, 2003). The basal cell layer is present in benign epithelial proliferations, may be disrupted in HGPIN, and is absent in invasive carcinoma (Bostwick and Brawer , 1987). Bostwick and Brawer in 1987 have shown that the frequency and extent of basal cell disruption in PIN is related to the PIN grade and is greatest in HGPIN.

A numerous of Immuno histochemical studies (McNeal *et al.,* 1988a; Perlman and Epstein, 1990; Deschenes and Weidner, 1990; Nagle *et al.,* 1991; Sesterhenn *et al.,* 1991) have been done to correlate the relationship between HGPIN and invasive carcinoma, Including evaluation of monoclonal antikeratin antibody, KA4, *Ulex europaeus* lectin (UEA-l), lectin binding pattern, and vimentin. Wu *et al.* demonstrated that a significantly higher P504S (AMACR) positive rate (56.0%) was found in isolated HGPIN glands adjacent to cancer (distance less than 5 mm) compared with those away from cancer (distance more than 5 mm; 14%, $P < 0.001$) (Wu *et al.,* 2004). High-grade PIN glands adjacent to cancer also showed a higher ($P < 0.001$) P504S intensity than did those away from cancer.

Other studies including argyrophilic nucleolar organizer regions and static DNA flow cytometry suggest that HGPIN and carcinoma have similar proliferative activity and DNA content, and hence HGPIN is the most likely precursor of cancer (Sesterhenn *et al.,* 1991; Weinberg and Weidner, 1993; Amin *et al.,* 1994). Cytogenetic abnormalities (involving 7q, 8q, 10q, 16q) and numerical chromosomal changes are noted in HGPIN and carcinoma (Emmert-Buck *et al.,* 1995; Macoska *et al.,* 1995; Qian *et al.,* 1999; Al-Maghrabi *et al.,* 2002). High-grade PIN and prostate cancer share genetic and molecular markers as well, with PIN representing an intermediate stage between benign epithelium and invasive carcinoma (Brawer, 2005).

PIN offers promise as an intermediate endpoint in studies of chemoprevention of prostatic carcinoma (Bostwick *et al.,* 1994b; Aquilina *et al.,* 1997). Recognizing the slow growth rate of prostate cancer and the considerable amount of time needed in animal and human studies for adequate follow-up, the non-invasive precursor lesion PIN is a suitable intermediate histologic marker to indicate subsequent likelihood of cancer (Aquilina *et al.,* 1997).

7. Chemoprevention strategies

Chemoprevention is the administration of agents to prevent induction of cancer, or to inhibit or delay its progression. In prostatic neoplasia, the time from tumour initiation and progression to invasive carcinoma often begins in men in the fourth and fifth decades of life and extends across decades. This phenomenon represents a unique opportunity to arrest or reverse the process of carcinogenesis with the use of chemopreventive agents. Animal models in defining efficacy of chemoprevention agents against prostate cancer. Detection of inhibitory effects on de novo prostate cancer development requires a high cancer incidence and similarity of induced tumors to human prostate carcinomas. The following animal models have produced high incidences of multifocal prostate adenocarcinoma: transgenic mice with oncogenes expressed in a prostate specific fashion; Noble rats that have been treated chronically with combination of 17β-estradiol and testosterone; and Wistar or F344 rats treated sequentially with a single injection of *N*-methyl-*N*-nitrosourea (MNU) and chronic administration of testosterone.

PIN most often occurs in the first two models, and metastases are frequent in some transgenic models and the MNU-testosterone model (Shirai *et al.,* 1991; Pollard *et al.,* 1992; Kadomatsu *et al.,* 1993; Slayter *et al.,* 1994; Ingles *et al.,* 1997). The chemopreventive efficacy of a series of agents using a model in which hormone dependent prostate cancer is induced in the Wistar-Unilever rat (McCormick, 1998). This is achieved by sequential treatment with an antiandrogen (cyproterone acetate), and androgen (testosterone propionate) and a direct acting chemical carcinogen (N-methyl-N-nitrosourea), followed by chronic androgen stimulation (testosterone). This regimen reproducibly induces a high incidence (< 75%) of prostate cancer, with no gross toxicity and a low incidence of neoplasia in the seminal vesicles and other non-target tissue.

Dehydroepiandrosterone (DHEA) and 9-cis-retinoic acid (9-cis-RA) are the most active chemopreventive agents identified to date. DHEA inhibits the induction of prostate cancer when administration is started before carcinogen exposure, and when it is delayed until incipient neoplastic lesions are present. Chronic administration of 9-cis-RA starting before carcinogen exposure is highly effective in the chemoprevention of prostate cancer. Liarozole fumarate confers modest protection against induction of prostate cancer, whereas N-(4-hydroxyphenyl) retinamide (4-HPR), a-difluoromethylornithine, DL-α-tocopherol acetate (vitamin E), oltipraz, and L-selenomethionine are inactive. The differential activity of 9-cis-RA and 4-HPR suggests the ligand specificity may be a determinant of retinoid action in prostate cancer chemoprevention. Chemoprevention of MNU +T induced prostate carcinogenesis at low dose level (0.5 μg/kg body weight) of calcitriol has significant potency to inhibit prostatic hyperplasia, dysplasia and PIN, and also decreased serum PAcP activity. Calcitriol may be an effective therapy for the treatment of early prostate cancer (Senthil kumar *et al.*, 2006).

Christov *et al.* (2002) studied that PIN can be used for assessing the efficacy of chemopreventive agents on prostate carcinogenesis. Dially disulfide, an organosulfur compound of garlic, has significant potency as an inhibitor of cancer induction in Sprague Dawley rat prostate by inhibiting PIN, hyperplastic and dysplastic foci (Arunkumar *et al.*, 2006). Zinc also acts as a chemopreventive agent against prostate cancer by inducing regression in PIN (Banudevi *et al.*, 2011a). Zinc inhibits PIN an early stage of prostate cancer. This study proves the ability of zinc to restore the PIN changes particularly in the rat ventral prostate induced by carcinogen and testosterone thereby indicating its anticarcinogenic potential (Banudevi *et al.*, 2011a). As the dorsolateral prostate is most likely homologue to the human prostate (Bosland *et al.*, 1990), in our recent study we proved that zinc was found to inhibit the growth and decrease dorsolateral prostatic acid phosphatase, zinc, citrate levels, phase I drug metabolizing enzyme activities, lipid peroxide, H_2O levels, proliferating cell nuclear antigen (PCNA), Bcl2, Bcl-X_L expressions with concomitant increase in phase II enzyme activities, GSH levels, p53, Bax, caspase-3 expressions in MNU testosterone induced model of prostate carcinogenesis (Banudevi *et al.*, 2011b). Signs of dysplasia, a characteristic of prostatic intraepithelial neoplasia, were evident in the dorsolateral prostatic histoarchitecture. Thus, zinc may act as an essential trace element against MNU and testosterone induced prostatic preneoplastic progression in Sprague Dawley rats.

The most efficient strategy for developing a chemoprevention programme is to perform two clinical trials concurrently, each based on the modulation of high grade PIN but in different target populations (Nelson *et al.*, 1996; Bostwick, 1997). In patients with high grade PIN associated with prostate cancer, a prospective, double blind, placebo controlled chemoactive pilot study designed to measure the response of a potential chemopreventive agent in the period (three to six weeks) before radical prostatectomy could easily be performed.

Androgen deprivation treatment is commonly used in this population to downsize the prostate before radical prostatectomy. This study may provide information regarding the effectiveness of proposed agents on surrogate endpoint biomarkers, premalignant lesions, and cancer. In particular, such an investigation would determine the response of PIN to the agent in whole mounted radical prostatectomy specimens. In some preliminary investigations it has been shown that there is a marked decrease in the prevalence and extent of PIN in prostates after androgen deprivation treatment, as compared with untreated prostates (Ferguson et al., 1994). This is accompanied by regressive changes in the secretory epithelium. Apoptotic bodies are more often seen in the treated normal prostate, PIN, and prostate cancer than in untreated cases. This suggests that androgen ablation induces epithelial regression by enhancing apoptosis. The low proliferating cell nuclear antigen (PCNA) and Ki67 related values and the absence of mitoses in PIN as well as in normal prostate and prostate cancer in the treated cases indicates suppressed proliferation activity as a consequence of androgen deprivation treatment (Armas et al., 1993; Montironi et al., 1994). It has been reported that angiogenesis is inhibited in prostate lesions when total androgen ablation induces cell regression and activation of the apoptosis (Montironi et al., 1996). Consequently, the epithelial cells are blocked from expressing, producing, or exporting angiogenic molecules. All these findings indicate that the dysplastic prostatic epithelium is hormone dependent.

A short term prospective, double blind, placed controlled phase II chemopreventive trial with cancer as an endpoint could be done in patients with high grade PIN without cancer. Chemoprevention trials designed to reverse high grade PIN may be confounded by the presence of underlying but undetected addressed by requiring a second biopsy with negative findings for cancer before entry into the study (preferably sextant biopsies with special attention to areas of abnormality on ultrasonogram or digital rectal examination), and by including enough subjects in the study and control groups to equalise the risk of coexistent cancer between the two groups.

PIN is routinely monitored by repeat biopsy in contemporary urological practice. Periodic re-evaluation would be necessary, including physical examination, rebiopsy, and evaluation of surrogate intermediate endpoint biomarkers. If subsequent biopsy reveals prostate cancer, these patients need definitive treatment. Those with PIN or no malignancy need continued observation (Bostwick, 1997).

8. Clinical chemoprevention studies

Table 2 reports agents, the treatment periods, and the primary endpoints used in clinical chemoprevention studies sponsored or funded by the National Cancer Institute (Kellof, 1997).

Green tea catechins

Epidemiological and case–control studies have garnered support for the chemopreventive properties of green tea (Jian et al., 2004). A recent study was conducted on 60 volunteers with high-grade prostate intraepithelial neoplasia, a putative precursor of prostate adenocarcinoma (Bettuzzi et al., 2006). Patients received green tea compounds in capsule form 200 mg three times per day. Following 1 year of treatment, only 3% of patients that received the green tea polyphenols were diagnosed with cancer compared with 30% in the placebo group. Furthermore, patients that received the green tea capsules exhibited a longer latency to tumor detection and exhibited an improved quality of life. Another phase II study, in which 6 g/day of tea was administered to 42 patients with asymptomatic, androgen- independent prostate cancer, has demonstrated that a single patient achieved a PSA response of >50% that lasted for approximately 1 month. These patients suffered with side effects that include diarrhea, nausea and fatigue (Jatoi *et al.*, 2003). Another clinical study used 250 mg dose of green tea polyphenols twice daily. In this study, 6 out of 19 patients had disease control for 3 to 5 months and there was only 1 patient whose PSA rise was affected by green tea supplementation. The dose used in this study did not discernibly alter the course of hormone-refractory prostate cancer (Choan *et al.*, 2005). These results suggest that green tea possesses cancer chemopreventive properties and minimal anti-neoplastic activity against advance stage prostate cancer.

Agent	Cohort (treatment period)	Primary endpoints
Phase II		
DFMO	Scheduled for prostate cancer surgery (4–8 weeks)	Histopathology (PIN grade, nuclear polymorphism, nucleolar polymorphism, ploidy), proliferation biomarkers (PCNA, Ki-67)
	Scheduled for prostatectomy (stage A or B prostatic carcinoma or bladder cancer without prostatic carcinoma and scheduled for cystoprostatectomy) (14 days)	Drug effect measurements: ODC activity (skin and prostate), polyamine levels (prostate). Histopathology (TRUS guided biopsies). Biochemical biomarkers: PSA, PAP, testosterone
	Serum PSA 3–10 ng/ml (includes patients with prostatic carcinoma and PIN) (14 days–1 year)	Drug effect measurements: ODC activity (skin and prostate) Polyamine levels (prostate). Histopathology (TRUS-guided biopsies) Biochemical biomarkers: PSA, PAP, testosterone
DHEA	Scheduled for prostate cancer surgery (28 days)	Histopathology (PIN grade, nuclear polymorphism, nucleolar polymorphism, ploidy). Proliferation biomarkers (PCNA, Ki-67). Genetic/regulatory biomarkers (p53, bcl-2, pc-1, chromosome 8p loss)
Flutamide	Patients with high grade PIN (12 months)	PIN grade and incidence, cancer incidence, nuclear polymorphism, nucleolar size, ploidy. Other endpoints: PCNA, angiogenesis, apoptosis, LOH chromosome 8; growth factors, PSA
4-HPR	Biopsy proven non-metastatic prostate adenocarcinoma, scheduled for radical prostatectomy (4 weeks)	Genetic/regulatory biomarkers: TGFβ, c-myc, p53, plasminogen activators (tPA, uPA), apoptosis
	Scheduled for prostate cancer surgery (4–8 weeks)	Histopathology: PIN grade, nuclear polymorphism, nucleolar polymorphism, ploidy. Proliferation biomarkers: PCNA, Ki-67. Differentiation biomarkers: Lewis^y antigen. Genetic/regulatory biomarkers: p53, EGFR, TGFα
Phase III		
Finasteride	Men ⩾ 55 years of age with normal DRE and PSA < 3.0 ng/ml (7 years)	Prostate cancer incidence (grade and stage), BPH incidence and severity, overall and prostate-specific mortality, TURP, PSA levels
Selenised yeast	Skin cancer (melanoma, non-melanoma) patients, low Se areas in USA (~1 year)	PSA levels

BPH, benign prostatic hypertrophy; DFMO, difluoromethyllornithine; DHEA, dehydroepiandrosterone; DRE, digital rectal examination; EGFR, epidermal growth factor receptor; PAP, prostatic acid phosphatase; PCNA, proliferating cell nuclear antigen; PIN, prostatic intraepithelial neoplasia; PSA, prostate specific antigen; Se, selenium; TGF, transforming growth factor; TRUS, transrectal ultrasound; TURP, transurethral resection of the prostate; 4-HPR, N-(4-hydroxyphenyl)retinamide.
(Kellof, 1997).

Table 2. National Cancer Institute Chemoprevention Branch: sponsored od funded phase II/III clinical chemoprevention trials: prostate cancer

Selective estrogen receptor modulators (SERMs)

Interest in SERMs as preventive agents has been stimulated by an apparent role of estrogens in the pathogenesis of prostate cancer, through promotion of cell growth. SERMs are generally considered to be 'weak estrogens' because they possess both agonist and antagonist activities depending on the specific tissue type and on the relative ER subtype interactions. Consequently this class of agents has been called selective estrogen receptor modulators or SERMs. As with phytoestrogens, SERMs appear to possess the ability to suppress prostate carcinogenesis. Toremifene has been evaluated in a phase IIa exploratory trial in men with high-grade PIN (Steiner and Pound, 2003). After 4 months of treatment with a daily oral dose of toremifene, 18 men with high-grade PIN underwent a repeat prostate biopsy. The prostate biopsy specimens showed significantly less high-grade PIN than historical controls. This trial provided the proof-of-concept support behind a currently open 485 patient placebo controlled, randomized dose finding phase IIb/III clinical trial

(Price *et al.*, 2006). This trial is investigating the efficacy of toremifene in reducing prostate cancer incidence in men with high-grade PIN. Men with high-grade PIN are treated for 12 months with placebo or toremifene and will undergo a repeat prostate biopsy at 6 and 12 months (the trial details are available at http://www.gtxinc.com/tech/clinical.htm). In addition, the National Cancer Institute is evaluating the effects of toremifene versus placebo in men with prostate cancer prior to radical prostatectomy. The objective of this phase II clinical trial is to evaluate the effects of a toremifene on biomarkers of prostate cancer. The trial details are available at http://www.cancer.gov/search/ViewClinicalTrials.

Difluoromethyl ornithine (DFMO)

DFMO is an irreversible inhibitor of ornithine decarcoxylase involved in the synthesis of polyamines; it possesses cytostatic and cytotoxic effects (Messing *et al.*, 1999). At the clinical level, interest in exploring DFMO as a chemoprevention agent for prostate cancer has recently increased as the administration of DFMO at 0.5 g/m2 daily for 4 weeks to men scheduled for surgical interventions to treat either prostatic hyperplasia or neoplasia resulted in reduction of polyamine pools, including spermine (Simoneau *et al.*, 2001). These results confirm that DFMO reaches the target tissue and hence may warrant further study as a possible chemopreventive agent for prostate cancer.

Selective COX-2 inhibitors

Inhibition of COX-2 expression blocks its pro-inflammatory effects, reduces expression of androgen receptors and androgen inducible genes and promotes apoptosis in prostate cancer cells. Selective COX-2 expression has been observed in high-grade PIN, a putative precursor of prostate cancer, suggesting a role early in carcinogenesis (Hussain *et al.*, 2003). These results support the hypothesis that inhibition of COX-2 may be an effective preventive strategy for prostate cancers; however, an industry-sponsored large-scale trial of rofecoxib was closed after the drug was withdrawn from the market because of concerns over its cardiovascular safety.

Selenium

Many of the cellular processes and molecular markers shown to be modified by selenium play key roles in prostate cancer progression. Two clinical trials are examining the role of selenium in other groups of high-risk individuals: men who have had negative prostate biopsies and men with HGPIN. The Negative Biopsy study, which includes men who have had at least one negative sextant prostate biopsy, will be using selenium supplementation in the form of selenium yeast, SeY (200 or 400 µg) (Marshall, 2001). The high-grade PIN study will determine the incidence of prostate cancer in men with biopsy-proven high grade intraepithelial neoplasia supplemented daily with 200 µg selenomethionine (Marshall, 2001). The results of these studies will be critical in defining a role for intervention with selenium in high-risk individuals, for whom there is no current treatment.

Bicalutamide

Intermittent (weekly), low-dose bicalutamide on prostate morphology revealed a tendency to a favorable modulation of high-risk proliferative lesions such as HGPIN. Namely, HGPIN status improved in 26% of treated subjects as compared with 4% of subjects in the no-treatment arm. These findings must be interpreted with extreme caution due to the inherent limitations of prostate sampling with needle biopsy. However, our analysis including all changes in HGPIN status (i.e., both HGPIN resolution and HGPIN development after 6 months of treatment) may account, at least in part, for sampling errors, supporting the worth of our findings although they derive from a post hoc analysis. The 28% incidence of HGPIN found in our study was unusually high compared with that reported in the general population, which ranges from 0.7% to 25% in noncancerous prostate biopsies (Bubendorf *et al.*, 1998). HGPIN incidence increases with age and is higher in men seen in clinical practice compared with men participating in screening programs (Bostwick and Qian, 2004). As most subjects in our study were in their seventies and as all were urological practice outpatients, this could partly explain our findings. Moreover, whereas HGPIN by itself does not seem to increase PSA (Bostwick and Qian, 2004), the incidence of HGPIN in homogeneous cohorts of men with elevated PSA is largely unknown. Once-a-week administration of bicalutamide in men at risk for prostate cancer is feasible and reasonably safe. This finding, coupled with the encouraging signal of activity emerging from the analysis of HG-PIN changes, supports further studies of this schedule at the lowest dose for prostate cancer prevention in men at high risk despite the negative primary end-point findings on Ki-67 (Zandari *et al.*, 2009).

Nutraceuticals and Micronutrients

Nutraceutical compounds most commonly show antioxidant properties combined with other anti-neoplastic actions (Syed *et al.*, 2008). Table 3 presents the most notable nutraceutical compounds examined in prostate cancer prevention, including vitamin D, vitamin E, selenium, lycopene, soy, and green tea (Trottier *et al.*, 2010a & b).

Compound	Origin	Proposed mechanism	Strength of evidence	Outcomes	Selected references
Vitamin D (calcitriol)	Sunlight, meats, fish	Vitamin D receptor activation: cancer homeostasis, cell proliferation and differentiation	Case–control and cohort studies	Conflicting when levels are normal; modest evidence when levels are low	Trottier *et al.*, 2010 Schwartz *et al.*, 2006
Vitamin E (tocopherols and tocotrienols)	Nuts, vegetable oils, palm oil, oats, rye, wheat, rice bran	Antioxidant, proapoptotic	Cohort studies, 2 RCTS[a]	No difference from placebo in RCTS	Gaziano *et al.*, 2009 Lippman *et al.*, 2009
Lycopene	Tomatoes	Carotenoid antioxidant	Case–control and cohort studies, meta-analysis	Positive meta-analysis, but negative results from the PLCO screening trial	Etminan *et al.*, 2004 Kirsh *et al.*, 2006 Peters *et al.*, 2007
Soy and isoflavanoids	Soybeans	Phytoestrogens and tyrosine kinase inhibition causing apoptosis, limited cell growth, reduced inflammation	Case–control and cohort studies, meta-analysis	Positive effect noted with non-fermented soy and mainly in non-Western men	Yan and Spitznagel, 2009
Green tea	*Camellia sinensis* plant	EGCG is the likely active ingredient: antioxidant polyphenol and 5ARI activity	Case–control and cohort studies, 1 RCT	Conflicting for overall PCa diagnosis; possible positive effect on advanced PCa diagnosis	Kurahashi *et al.*, 2008 Brausi *et al.*, 2008

[a] SELECT (Selenium and Vitamin E Cancer Prevention Trial) and Physician's Health Study II.

RCT = randomized control trial; PLCO = Prostate, Lung, Colorectal, and Ovarian Cancer; EGCG = epigallocatechin-3-gallate; 5ARI = 5α-reductase inhibitor; PCa = prostate cancer.

Table 3. Primary prostate cancer prevention with selected nutraceutical (Trottier *et al.*, 2010)

9. Signal anomalies of human prostate cancer

Prostatic intraepithelial neoplasia (PIN) lesions could be described as low grade (LG) or high grade (HG) and it is widely perceived that HGPIN is a precursor of prostatic adenocarcinoma (Isaacs *et al.*, 2002). Since the initial growth of prostate tumor or cancer is dependent on androgens, hormone therapy in the form of medical or surgical castration constitutes a common approach for systemic treatment. However over a period of time, most cancer will develop androgen-independence, thereby making the continued androgen deprivation therapy ineffective (Taplin and Balk, 2004). While the mechanisms that drive the genesis of subclinical, microscopic PIN lesions, their progression to invasive cancer and androgen independence remain largely unknown and evidence collected in recent years point to certain molecular aberrations that pave the path of disease progression.

For example, anomalies in specific signaling molecules, including extracellular growth factors, protein tyrosine kinase cell surface receptors, intracellular transcription factors, nuclear factors and their ligands, growth suppressors, cell cycle regulators and others have indicated in some prostate carcinomas (Abate Shen and Shen, 2000; Gao and Isaacs, 2000; Roy-Burman *et al.*, 2004)

There is currently a strong focus on the genetic alterations or aberrations in gene expression that are frequently encountered in human prostate cancer in the design of mouse models. Although both men and mice harbour functionally equivalent prostate glands, there are similarities as well as differences in the anatomy and histology of the prostate in the two species. Similar epithelial cell types namely secretory, basal and neuroendocrine are found in both mouse and human prostate, although their proportions vary. While human prostate has a robust fibromuscular stroma, the mouse contains a modest stromal component. Anatomically the human prostate gland is a single, alobular structure with central, peripheral and transitional zones. In contrast, the mouse prostate is composed of four paired lobes namely anterior (AP), dorsal (DP), lateral (LP) and ventral (VP) prostate. Since DP and LP share a ductal system they are often dissected together and referred to as the dorsolateral prostate (DLP). The mouse DLP is perceived to be the most similar to the human peripheral zone in which the majority of clinically diagnosed prostate cancers are found. The mouse VP does not appear to have a human homologue and the human transitional zone does not have a murine homologue. The transitional zone constitutes a site where human nodular hyperplasia (BPH) is commonly seen. The mouse AP is analogous to the human central zone, which only infrequently represents a site of neoplasia in humans (Roy-Burman *et al.*, 2004).

10. Cell surface signaling molecules

Signaling interactions between various extracellular growth factors and the corresponding cell surface receptors converge to determine the fate of the cell with respect to proliferation, survival or death. In this context, dysregulation of several growth factors or their receptors has been implicated in prostate tumorigenesis. A number of transgenic mouse lines have been produced in which genes that are known to be overexpressed in human prostate cancer are targets. The survival factor, insulin like growth factor-I (IGF-I) which is generally overexpressed in human prostate and which may potentially be a good tumor marker in prostate cancer was a target in a transgenic line (Woodson et al., 2003). Its expression in the mouse tissues was designed by using the bovine keratin 5 promoter (Di Giovanni et al., 2000). These mice develop squamous papillomas some of which progress to carcinomas of the skin. The increased IGF I levels also lead to pathologic changes in the prostate and in other male accessory glands of these animals (Di Giovanni et al., 2000a).The severity of the lesions in the prostate ranges from PIN to carcinoma in Situ as well as tumors with neuroendocrine differentiation.

11. Fibroblast Growth Factors (FGFs)

The FGF family of heparin binding proteins is intercellular signaling molecules of which atleast 23 different members (FGF-1 – FGF-23) have been identified to date. FGF proteins are generally secreted and their effects are mediated by a complex system of FGF receptor (FGFR) tyrosine kinases, either through autocrine or paracrine mechanisms, or both (Wilkie et al., 1995; McKeehan et al., 1998). While dysregulation of several FGFs has been described in prostate development and tumorigenesis, two members, FGF 7 and FGF 8 have been further pursued through mouse modeling experiments (Djakiew, 2000; Thomson, 2001).

FGF 8b has been demonstrated to possess the most transforming and tumorigenic potential (Daphna-Iken et al., 1998). While expression of FGF 8b appears to represent the primary species in prostatic epithelium, its expression is practically undetected in the stromal component of prostate cancer (Valve et al., 2001). Increased expression of FGF 8b in prostatic lesions beginning from PIN to adenocarcinoma and its persistence in androgen independent disease has been described. The overexpression of FGF8b in prostate cancer cells has been shown to increase proliferative and invasive properties of the affected cells directly and proliferation of prostatic stromal cells indirectly (Song et al., 2000). Consistent with these results antisense down regulation of FGF8b mRNA reduces the growth rate, inhibits cologenic activity and decreases in vivo tumorigenicity of prostate tumor cells (Rudra-Ganguly et al., 1998). FGF 8 expressions in prostate cancer is regulated by the androgen receptor at the transcriptional level and that FGF 8 is angiogenic   further enhance the biological relevance of the factor in prostate cancer.

Prostatic hyperplasia appears in the LP and VP in some FGF 8b transgenic animals as early as 2 to 3 months and in DP and AP between 6 to 16 months. LGPIN lesions manifest from 5 to 7 months. 100% of the mice display multifocal prostatic epithelial hyperplasia during the first 14 months with 35% also having areas of LGPIN. In subsequent months (15 to 24 months) the profile changes to a higher incidence of LGPIN (66%) along with HGPIN (51%). Ocassionally HGPIN lesions resemble the histopathology of human prostatic carcinoma *in situ.*

12. P27kip, regulator of cyclin-cdk activity

Activation of AKT through deregulated phosphatidylinositol 3- kinase (PI3K) signaling resulting from genetic inactivation of phosphatase and tensin homolog (PTEN), mutational activation of PI3K, or the activation of upstream oncogenic tyrosine kinases is a frequent molecular event in human cancer (Brugge *et al.*, 2007; Lee *et al.*, 2007). Transgenic expression of activated AKT1 in the murine prostate induces prostatic intraepithelial neoplasia (PIN) that does not progress to invasive prostate cancer. In human epithelial cancers, reduced levels of p27^{Kip1} expression are frequently observed (Slingerland and Pagano, 2000) and are correlated with tumor progression and poor survival (Loda et al., 1997; Porter et al., 1997; Yang et al., 1998). p27^{Kip1} functions primarily as a negative regulator of cyclin-CDK activity and thus likely participates in tumor suppression by inhibiting cell-cycle progression (Chu et al., 2008). Targeted disruption of p27^{Kip1} (Cdkn1b-/-) in mice leads to prostatic hyperplasia (Cordon-Cardo et al., 1998) and development of pituitary adenomas (Fero et al., 1996, 1998) as the mice age. However, Cdkn1b-/- mice do not typically develop other spontaneous tumors (Fero et al., 1996; Kiyokawa et al., 1996; Nakayama et al., 1996).

In many human cancer cells, oncogene-induced senescence (OIS) is associated with known tumor suppressor pathways such as p53, VHL, and Rb (Serrano et al., 1997; Young et al., 2008). It has been reported that OIS occurs in many human and mouse precursors of cancer and that this phenomenon can be reversed by the inactivation of tumor suppressor pathways (Braig et al., 2005; Chen et al., 2005; Collado et al., 2005; Michaloglou et al., 2005). Majumder et al. investigated the role of p27^{Kip1} in tumor suppression in prostate cancer using both genetically engineered mice and human prostate samples (Majumder et al., 2008). They have identified a relationship among senescence induction, p27^{Kip1} expression, and PIN that supports the notion that p27^{Kip1} induction in the context of early neoplastic lesions may represent a preinvasive checkpoint linked to cellular senescence.

Importantly, is study is highly relevant to human prostate cancer (Majumder et al., 2008). Indeed, we show that p27^{Kip1} is overexpressed in human PIN not associated with invasive cancer, presumably representing the earliest phase of neoplastic transformation. In contrast, PIN adjacent to invasive cancer, where checkpoint loss may have already occurred, is associated with low levels of p27^{Kip1}. The role of p27^{Kip1} in this process is further supported by a body of data showing that loss of p27^{Kip1} is commonly found in human cancers (Chu et al., 2008) and that invasive tumor cells specifically degrade p27^{Kip1}. This in turn results in increased CDK2 activity (Loda et al., 1997).

As many as 30% of men with a diagnosis of PIN on biopsy are subsequently found to harbor an invasive prostatic adenocarcinoma on repeat biopsy (Gokden et al., 2005). Thus the CDK inhibitors might have utility in preventing cancer progression from in situ dysplasia to invasion.

13. Conclusions

High-grade PIN is the most likely precursor of prostatic adenocarcinoma, according to virtually all available evidence. PIN is associated with progressive abnormalities of phenotype and genotype that are intermediate between normal prostatic epithelium and cancer, indicating impairment of cell differentiation and regulatory control with advancing stages of prostatic carcinogenesis. There is progressive loss of some markers of secretory differentiation, whereas other markers show progressive increase. The clinical importance of recognizing PIN is based on its strong association with prostatic carcinoma. PIN has a high predictive value as a marker for adenocarcinoma, and its identification in biopsy specimens of the prostate warrants further search for concurrent invasive carcinoma. Studies to date have not determined whether PIN remains stable, regresses, or progresses, although the implication is that it can progress. Chemoprevention of early stage of prostate cancer, PIN may be useful strategy in the prostate cancer prevention.

14. Acknowledgements

The financial assistance provided by University of Madras as UWPFEP Fellowship to one of the authors Mr. A. Arunkumar and Council of Scientific and Industrial Research (CSIR), India, in the form of CSIR-SRF to Ms. S. Banudevi are gratefully acknowledged.

15. References

Abate Shen C, Shen MM. Molecular genetics of prostate cancer. *Genes Dev.* 2000; 14:2410-2434.

Abrahams NA, Ormsby AH, Brainard J. Validation of cytokeratin 5/6 as an effective substitute for keratin 903 in the differentiation of benign from malignant glands in prostate needle biopsies. *Histopathology* 2002; 41:35-41.

Allam CK, Bostwick DG, Hayes JA, *et al.* Interobserver variability in the diagnosis of high grade prostatic intraepithelial neoplasia and adenocarcinoma. *Mod Pathol.* 1996; 9:742-751.

Al-Maghrabi J, Vorobyova L, Toi A, *et al.* Identification of numerical chromosomal changes detected by interphase fluorescence in situ hybridization in high-grade prostate intraepithelial neoplasia as a predictor of carcinoma. *Arch Pathol Lab Med.* 2002; 126:165-169.

Amin MB, Ro JY, Ayala AG. Ideas in pathology: putative precursor lesions of prostatic adenocarcinoma: fact or fiction? *Mod Pathol.* 1993; 6:476-483.

Amin MB, Schultz DS, Zarbo RJ. Computerized static DNA ploidy analysis of prostatic intraepithelial neoplasia. *Arch Pathol Lab Med.* 1994; 118:260-264.

Aquilina JW, Lipsky JJ, Bostwick DG: Androgen deprivation as a strategy for prostate cancer chemoprevention. *J Natl Cancer Inst.* 1997; 89:689-696.

Armas OA, Melamed A, Aprikian A, *et al.* Effect of preoperative androgen deprivation therapy in prostatic carcinoma [abstract]. *Lab Invest.* 1993; 68:55A.

Arunkumar A, Vijayababu MR, Venkataraman P, Senthilkumar K, Arunakaran J. Chemoprevention of rat prostate carcinogenesis by diallyl disulfide, an organosulfur compound of garlic. *Biol Pharm Bull.* 2006; 29:375-379.

Banudevi S, Elumalai P, Arunkumar R, Senthilkumar K, Gunadharini DN, Sharmila G, Arunakaran J. Chemopreventive effects of zinc on N-methyl-N-nitrosourea and testosterone induced prostatic intraepithelial neoplasia in the dorsolateral prostate of male Sprague-Dawley rats. *J Cancer Res Clin Oncol.* 2011a; 137:677-686.

Banudevi S, Elumalai P, Sharmila G, Arunkumar R, Senthilkumar K, Arunakaran J. Protective effects of zinc on prostate carcinogenesis induced by N-methyl-N-nitrosourea and testosterone in adult male Sprague-Dawley rats. *Exp Biol Med.* 2011b; *in press.*

Bethesda Md. Prostatic intraepithelial neoplasia: significance and correlation with prostate-specific antigen and transrectal ultrasound. Proceedings of a workshop of the

National Prostate Cancer Detection Project; March 13, 1989; Bethesda, Md. *Urol.* 1989; 34:2–69.

Bettuzzi S, Brausi M, Rizzi F, Castagnetti G, Peracchia G, Corti A. Chemoprevention of human prostate cancer by oral administration of greentea catechins in volunteers with high-grade prostate intraepithelial neoplasia: a preliminary report from a one-year proof-of-principle study. *Cancer Res.* 2006; 6:1234–1240.

Bosland MC, Prinsen MK. Induction of dorsolateral prostate adenoma carcinomas and other accessory sex gland lesions in male Wistar rats by N-methyl-N-nitrosourea, 7,12-dimethylbenz(a) anthracene, and 3,2'-dimethyl-4-aminobiphenyl after sequential treatment with cyproterone acetate and testosterone propionate. *Cancer Res.* 1990; 50:691–699.

Bostwick DG, Amin MB, Dundore P, *et al.*: Architectural patterns of high-grade prostatic intraepithelial neoplasia. *Hum Pathol.* 1993; 24:298–310.

Bostwick DG, Amin MB. Prostate and seminal vesicles. In: Damjanov I, Linder J (eds.), Anderson's pathology, 10th edn, Vol. II, Chapter.67. St. Louis: Mosby, 1996, 2166-230.

Bostwick DG, Brawer MK. Prostatic intra-epithelial neoplasia and early invasion in prostate cancer. *Cancer* 1987; 59:788–794.

Bostwick DG, Burke HB, Wheeler TM, *et al.*: The most promising surrogate endpoint biomarkers for screening candidate chemopreventive compounds for prostatic adenocarcinoma in short-term phase II clinical trials. *J Cell Biochem.* 1994b; 19 (suppl):283–289.

Bostwick DG, Dousa M, Crawford, *et al.* Neuroendocrine differentiation in prostatic intraepithelial neoplasia and adenocarcinoma. *Am J Surg Pathol.* 1994a; 18:1240-1246.

Bostwick DG, Qian J, Frankel K. The incidence of high-grade prostatic intraepithelial neoplasia in needle biopsies. *J Urol.* 1995; 154:1791–1794.

Bostwick DG, Qian J. High-grade prostatic intraepithelial neoplasia. *Mod Pathol.* 2004; 17(3):360–379.

Bostwick DG. Phase II efficacy trials for chemoprevention in patients with PIN: strategies with androgen deprivation therapy. In: Crawford ED (ed.), Proceedings of 7th International Prostate Cancer Update. Beaver Creek, Colorado, 22-26 January 1997.Pp 485-490.

Bostwick DG. Prospective origins of prostate carcinoma. Prostatic intraepithelial neoplasia and atypical adenomatus hyperplasia. *Cancer* 1996; 78:330.

Bostwick DG. Prostatic intraepithelial Neoplasia (PIN). *Urol.* 1989; 34:16-22.

Bostwick DG. Target populations and strategies for chemoprevention trials of prostate cancer. *J Cell Biochem.* 1994; 19(suppl): 191-196.

Braig M, Lee S, Loddenkemper C, Rudolph C, Peters AH, SchlegelbergerB, Stein H, Dorken B, Jenuwein T, Schmitt CA. Oncogene- induced senescence as an initial barrier in lymphoma development. *Nature* 2005; 436:660–665

Brawer MK, Peehl DM, Stamey TA, Bostwick DG. Keratin immunoreactivity in the benign and neoplastic human prostate. *Cancer Res.* 1985; 45:3663– 3667.

Brawer MK. Prostatic intraepithelial neoplasia: an overview. *Rev Urol*. 2005; 7(suppl 3):S11–S18.

Brugge J, Hung MC, Mills GB. A new mutational AKT activation in the PI3K pathway. *Cancer Cell* 2007; 12:104–107.

Bubendorf L, Tapia C, Gasser TC, *et al.* Ki67 labeling index in core needle biopsies independently predicts tumor-specific survival in prostate cancer. *Hum Pathol.*1998; 29:949–54.

Chen Z, Trotman LC, Shaffer D, Lin HK, Dotan ZA, Niki M, Koutcher JA, Scher HI, Ludwig T, Gerald W, *et al.* Crucial role of p53-dependent cellular senescence in suppression of Pten-deficient tumorigenesis. *Nature* 2005; 436:725–730.

Choan E, Segal R, Jonker D, Malone S, Reaume N, Eapen L, Gallant V.2005. A prospective clinical trial of green tea for hormone refractory prostatecancer: an evaluation of the complementary/alternative therapy approach.*Urol Oncol*. 2005; 23:108–113.

Christov KT, Moon RC, Lantvit DD, Boone CW, Steele VE, Lubet RA, Kelloff GJ, Pezzuto JM. *Cancer Res*. 2002; 62:5178-5182.

Chu IM, Hengst L, Slingerland JM. The Cdk inhibitor p27 in human cancer: prognostic potential and relevance to anticancer therapy. *Nat Rev Cancer* 2008; 8:253–267.

Collado M, Gil J, Efeyan A, Guerra C, Schuhmacher AJ, Barradas M, Benguria A, Zaballos A, Flores JM, Barbacid, M, *et al.* Tumour biology: senescence in premalignant tumours. *Nature* 2005: 436: 642.

Cordon-Cardo C, Koff A, Drobnjak M, Capodieci P, Osman I, Millard SS, Gaudin PB, Fazzari M, Zhang ZF, Massague J, Scher HI. Distinct altered patterns of p27KIP1 gene expression in benign prostatic hyperplasia and prostatic carcinoma. *J Natl Cancer Inst*. 1998; 90:1284–1291.

Daphna-Iken D, Shankar DB, Lawshe A, Ornitz DM, Shackleford GM, MacArthur CA. FGF-8 isoforms differ in NIH3T3 cell transforming potential. *Cell Growth Differ*. 1998; 6:817-825.

De La Torre M, Haggman M, Brandstedt S, Busch C. Prostatic intraepithelial neoplasia and invasive carcinoma in total prostatectomy speciemens: distribution, volumes and DNA ploidy. *Br J Urol*. 1993; 72:207.

Deschenes J, Weidner N. Nucleolar organizer regions (NOR) in hyperplastic and neoplastic prostate disease. *Am J Surg Pathol*. 1990; 14:1148–1155.

Di Giovanni J, Boe DK, Wilker E *et al.* Constitutive expression of insulin like growth factor 1 in epidermal basal cells of transgenic mice leads to spontaneous tumor promotion. *Cancer Res*. 2000a; 63:3991-3994.

Di Giovanni J, Kiguchi K, Frijhoft A *et al.* Deregulated expression of IGF I in prostate epithelium leads to neoplasia in transgenic mice. *Cancer Res*. 2000b; 60:1561-1570.

Di Sant' Agnese PA. Neuroendocrine differentiation in the precursor of prostate cancer. *Eur Urol*. 1996; 30: 185-190.

Djakiew D. Dysregulated expression of growth factors and their receptors in the development of prostate cancer. *Prostate* 2000; 42: 150-160.

Drago JR, Mostofi FK, Lee F. Introductory remarks and workshop summary. *Urol*. 1989; 34:2-3.

Emmert-Buck MR,Vocke CD, Pozzatt RO, et al. Allelic loss of chromosome 8p12-21 in microdissected prostatic intraepithelial neoplasia. *Cancer Res.* 1995; 55:2959–2962.

Epstein JI. Adenosis (atypical adenomatous hyperplasia): histopathology and relationship to carcinoma. *Path Res Pathol.* 1994; 191:888.

Eschenbach AC. The biologic dilemma of early carcinoma of the prostate. *Cancer* 1996; 78: 326.

Ferguson J, Zincke H, Ellison E, *et al.* Decrease of prostatic intraepithelial neoplasia (PIN) following androgen deprivation therapy in patients with stage T3 carcinoma treated by radical prostatectomy. *Urol.* 1994; 44:91-95.

Fero ML, Randel E, Gurley KE, Roberts JM, Kemp CJ. The murine gene p27Kip1 is haplo-insufficient for tumour suppression. *Nature* 1998; 396:177–180.

Fero ML, Rivkin M, Tasch M, Porter P, Carow CE, Firpo E, Polyak K, Tsai LH, Broudy V, Perlmutter RM, *et al.* A syndrome of multiorgan hyperplasia with features of gigantism, tumorigenesis, and female sterility in p27(Kip1)-deficient mice. *Cell* 1996; 85: 733–744.

Gao A, Isaccs JT. The molecular basis of prostate carcinogenesis. In: Coleman WB, Tsongalis GJ (eds.), Molecular basis of human cancer, The Humana Press. Inc., Totowa NJ, 2000. Pp 365-379.

Garabedian EM, Humphrey PA, Gordon JI. A transgenic mouse model of metastatic prostate cancer originating from neuroendocrine cells. *Proc Natl Acad Sci USA* 1998; 95:15382–15387.

Gaudin PB, Sesterhenn IA, Wojno K, Mostofi FK, Epstein JI. Incidence and clinical significance of high grade prostatic intraepithelial neoplasia in TURP specimens. *Urol.* 1997; 49:558–563.

Gokden N, Roehl KA, Catalona WJ, Humphrey PA. Highgrade prostatic intraepithelial neoplasia in needle biopsy as risk factor for detection of adenocarcinoma: current level of risk in screening population. *Urol.* 2005; 65:538–542.

Hussain T, Gupta S, Mukhtar H. Cyclooxygenase-2 and prostate carcinogenesis. *Cancer Lett.* 2003; 191:125–135.

Ingles SA, Ross RK, Yu MC, *et al.* Association of prostate cancer risk with genetic polymorphisms in vitamin D receptor and androgen receptor. *J Natl Cancer Inst.* 1997; 89:166-170.

Isaacs W, De Marzo A, Nelson WG. Focus on prostate cancer. *Cancer Cell* 2002; 2:113-116.

Jatoi A, Ellison N, Burch PA, Sloan JA, Dakhil SR, Novotny P, TanW, Fitch TR, Rowland KM, Young CY, Flynn PJ. A phase IItrial of green tea in the treatment of patients with androgen independent metastatic prostate carcinoma. *Cancer* 2003; 97:1442–1446.

Jemal A, Siegel R, Ward E, Murray T, Xu J, Thun, MJ. Cancer statistics, 2010. CA. *Cancer J Clin.* 2010; 57:43-66.

Jian L, Xie LP, Lee AH, Binns CW. Protective effect of green tea against prostate cancer: a case–control study in southeast China. *Int J Cancer* 2004; 108:130–135.

Jones EC, Young RH. The differential diagnosis of prostatic carcinoma. Its distinction from premalignant and pseudocarcinomatous lesions of the prostate gland. *Am J Clin Pathol.* 1994; 101:148.

Kadomatsu K, Anzano MA, Slayter MV, *et al.* Expression of sulfated glycoprotein 2 is associated with carcinogenesis induced by N-nitroso-M-methylurea in rat prostate and seminal vesicle. *Cancer Res.* 1993; 53:1480-1483.

Kasper S, Sheppard PC, Yan Y, *et al.*: Development, preogression, and androgen-dependence of prostate tumors in probasin-large T antigen transgenic mice: A model for prostate cancer. *Lab Invest.* 1998; 78:319-333.

Kellof GJ. Chemoprevention strategies for prostate cancer. In: Crawford ED, (ed.), Proceedings of 7th International Prostate Cancer Update. Beaver Creek, Colorado, 22-26 January 1997.Pp 134-135.

Kelloff GJ, Hawk ET, Crowell JA, *et al.* Strategies for identification and clinical evaluation of promising chemoprevention agents. *Oncology* 1996; 10:1471-1481.

Kiyokawa H, Kineman RD, Manova-Todorova KO, Soares VC, Hoffman ES, Ono M, Khanam D, Hayday AC, Frohman LA, Koff A. Enhanced growth of mice lacking the cyclin-dependent kinase inhibitor function of p27(Kip1). *Cell* 1996; 85:721-732.

Langer JE, Rovner ES, Coleman BG, *et al.*: Strategy for repeat biopsy of patients with prostatic intraepithelial neoplasia detected by prostate needle biopsy. *J Urol.* 1996; 155:228-231.

Lee JY, Engelman JA, Cantley LC.Biochemistry. PI3K charges ahead. *Science* 2007; 317:206-207.

Lipski B, Garcia R, Brawer M. Prostatic intraepithelial neoplasia: significance and management. *Semin Urol Oncol.* 1996; 14:149-155.

Loda M, Cukor B, Tam SW, Lavin P, Fiorentino M, Draetta GF, Jessup JM, Pagano M. Increased proteasome-dependent degradation of the cyclin-dependent kinase inhibitor p27 in aggressive colorectal carcinomas. *Nat Med.*1997; 3:231-234.

Macoska JA, Trybus TM, Benson PD, et al. Evidence for three tumor suppressor gene loci on chromosome 8p in human prostate cancer. *Cancer Res.* 1995; 55:5390-5395.

Majumder PK, Grisanzio C, O'Connell F, Barry M, Brito JM, Xu Q, Guney I, *et al.* A Prostatic Intraepithelial Neoplasia-Dependent p27Kip1 Checkpoint Induces Senescence and Inhibits Cell Proliferation and Cancer Progression. *Cancer Cell 2008;* 14:146-155.

Mashall JR. Larry Clark's Legacy: Randomized, controlled, selenium-based, prostate cancer chemopreventiontrials. *Nutr Cancer* 2001; 40:74-77.

Mc Keehan WL, Wang F, Kan M. The heparin sulfate- fibroblast growth factor family: diversity of structure. *Prog Nucleic Acid Res Mol Biol.* 1998; 59 135-176.

McCormick DL. Chemoprevention of hormone-dependent prostate cancer in the Wistar-Unilever rat. In: Schulman C, Kelloff G (eds.), Proceedings of the International Symposium "Strategies for the chemoprevention of prostate cancer". Brussels, 30-31 October 1998. Pp 38.

McNeal JE, Alroy J, Leav I, Redwine EA, Freiha FS, Stamey TA. Immunohistochemical evidence for impaired cell differentiation in the premalignant phase of prostate carcinogenesis. *Am J Clin Pathol.* 1988b;90:23-32.

McNeal JE, Bostwick DG. Intraductal dysplasia: a premalignant lesion of the prostate. *Hum Pathol.* 1986; 17:64-71.

McNeal JE, Leav I, Alroy J. Differential lectin staining of central and peripheral zones of the prostate and alterations in dysplasia. *Am J Clin Pathol.* 1988a; 89:41–48.

McNeal JE. Morphogenesis of prostate carcinoma. *Cancer* 1965; 18:1659–66.

McNeal JE. Origin and development of carcinoma in the prostate. *Cancer.* 1969; 23:24–34.

McNeal JE. Significance of duct-acinar dysplasia in prostatic carcinogenesis. *Prostate* 1988; 13:91–102.

Melissari M, Lopez-Beltran A, Mazzucchelli R, Froio E, Bostwick DG, Montironi R. High grade prostatic intraepithelial neoplasia with squamous differentiation. *J Clin Pathol.* 2006; 59:437–439.

Messing EM, Love RR, Tutsch KD, Verma AK, Douglas J, Pomplun M, Simsiman R, Wilding G. Low-dose difluoromethylornithine and polyamine levels in human prostate tissue. *J Natl Cancer Inst.* 1999; 91:1416–1417.

Michaloglou C, Vredeveld LC, Soengas MS, Denoyelle C, Kuilman T,van der Horst CM, Majoor DM, Shay JW, Mooi WJ, Peeper DS. BRAFE600-associated senescence-like cell cycle arrest of human naevi. *Nature* 2005; 436:720–724.

Montironi R, Diamanti L, Thompson D, *et al.* Analysis of the capillary architecture in the precursors of prostate cancer: recent findings and new concepts. *Eur Urol.* 1996; 30:191-200.

Montironi R, Galluzzi CM, Scarpelli M, *et al.* Quantitative characterization of the frequency and location of cell proliferation and death in prostate pathology. *J Cell Biochem.* 1994; 19(suppl):238-45.

Morote J, Fernandez S, Alan L, Iglesias C, Planas J, Reventos J, Cajal SR, Paciucci R, and de Torres IM. PTOV1 Expression Predicts Prostate Cancer in Men with Isolated High-Grade Prostatic Intraepithelial Neoplasia in Needle Biopsy. *Clin Cancer Res.* 2008; 14: 2617-2622.

Nagle RB, Brawer MK, Kittelson J. Phenotypic relationships of prostatic intraepithelial neoplasia to invasive prostatic carcinoma. *Am J Pathol.* 1991; 138: 119–128.

Nakayama K, Ishida N, Shirane M, Inomata A, Inoue T, Shishido N, Horii I, Loh DY. Mice lacking p27(Kip1) display increased body size, multiple organ hyperplasia, retinal dysplasia, and pituitary tumors. *Cell* 1996; 85:707–720.

Nelson PS, Gleason TP, Brawer MK. Chemoprevention for prostatic intraepithelial neoplasia. *Eur Urol.* 1996; 30:269- 278.

Pacelli A, Bostwick DG. Clinical significance of high-grade prostatic intraepithelial neoplasia in transurethral resection specimens. *Urol.* 1997; 50:355–359.

Perlman EJ, Epstein JI. Blood group antigen expression in dysplasia and adenocarcinoma of the prostate. *Am J Surg Pathol.* 1990; 14:810–818.

Pollard M. The Lobund-Wistar rat model of prostate cancer. *J Cell Biochem.* 1992; 16(suppl):84-88.

Porter PL, Malone KE, Heagerty PJ, Alexander GM, Gatti LA, Firpo EJ, Daling JR, Roberts JM.. Expression of cell-cycle regulators p27Kip1 and cyclin E, alone and in combination, correlate with survival in young breast cancer patients. *Nat Med.* 1997; 3:222-225.

Price D, Stein B, Sieber P, Tutrone R, Bailen J, Goluboff E, Burzon D,Bostwick D, Steiner M. Toremifene for the prevention of prostate cancer in men with high grade prostatic intraepithelial neoplasia: results of a double-blind, placebo controlled, phase IIB clinical trial. *J Urol.* 2006; 176:965–970.

Qian J, Jenkins RB, Bostwick DG. Genetic and chromosomal alterations in prostatic intraepithelial neoplasia and carcinoma detected by fluorescence in situ hybridization. *Eur Urol.* 1999; 35:479–483.

Qian J, Wollan P, Bostwick DG. The extent and multicentricity of high grade prostatic intraepithelial neoplasia in clinically localized prostatic adenocarcinoma. *Hum Pathol.* 1997; 28:143–148.

Roy-Burman P, Wu H, Powell WC, Hagenkord J, Cohen MB. Genetically7 defined mouse models that mimic natural aspects of human prostate cancer development. *Endocrine Relat Cancer* 2004; 11:225-254.

Rudra-Garsguly N, Zheng J, Hoang AT, Roy-Burman P. Down regulation of human FGF 8 activity by antisense constructs in murine fibroblastic and human prostatic carcinoma cell systems. *Oncogene* 1998; 16:1487-1492.

Sakr WA, Haas GP, Cassin BF, Pontes JE, Crissman JD. The frequency of carcinoma and intraepithelial neoplasia of the prostate in young male patients. *Br J Urol.* 1993; 150:379.

Senthilkumar K, Arunkumar A, Sridevi N, Vijayababu MR, Kanagaraj P, Venkataraman P, Aruldhas MM, Srinivasan N, Arunakaran J. Chemoprevention of MNU and testosterone induced prostate carcinogenesis by calcitriol (vitamin D_3) in adult male albino wistar rats. *Ann Cancer Res Therap.* 2006; 14:12–18

Serrano M, Lin AW, McCurrach ME, Beach D, Lowe SW. Oncogenic ras provokes premature cell senescence associated with accumulation of p53 and p16INK4a. *Cell* 1997; 88:593–602.

Sesterhenn IA, Becker RL, Avallone FA. Image analysis of nucleoli and nucleolar organizer regions in prostatic hyperplasia, intraepithelial neoplasia, and prostatic carcinoma. *J Urogenital Pathol.* 1991; 1:61–74.

Shirai T, Yamamoto A, Iwasaki S, *et al.* Induction of invasive carcinomas of the seminal vesicles and coagulating glands of F344 rats by administration of N-methylnitrosourea or N-nitroso-bis (2-oxypropyl) amine and followed by testosterone propionate with or without high-fat diet. *Carcinogenesis* 1991;12:2169-2173.

Simoneau AR, Gerner EW, Phung M, McLaren CE, Meyskens Jr. FL. Alpha difluoromethylornithine and polyamine levels in the human prostate: results of a phase IIa trial. *J Natl Cancer Inst.* 2001; 93:57-59.

Skjorten FJ, Berner A, Harvei S, *et al.*: Prostatic intraepithelial neoplasia in surgical resections. Relationship to coexistent adenocarcinoma and atypical adenomatous hyperplasia of the prostate. *Cancer* 1997; 79:1172–1179.

Slayter MV, Anzano MA, Kadomatzu K, *et al.* Histogenesis of induced prostate and seminal vesicle carcinoma in Lobund-Wistar rats: a system for histological scoring and grading. *Cancer Res.* 1994; 54:1440-1445.

Slingerland J, Pagano M. Regulation of the cdk inhibitor p27 and its deregulation in cancer. *J Cell Physiol.* 2000; 183:10–17.

Song Z, Powell WC, Kasahara N, van Bokhoven A, Miller GJ, Roy-Burman P. The effect of fibroblast growth factor 8, isoform b, on the biology of prostate carcinoma cells and their interaction with stromal cells. *Cancer Res.* 200; 60: 6730-6736.

Steiner MS, Pound CR. Phase IIA clinical trial to test the efficacy and safety of toremifene in men with high-grade prostatic intraepithelial neoplasia. *Clin Prostate Cancer* 2003; 2:24–31.

Syed DN, Suh Y, Afaq F, Mukhtar H. Dietary agents for chemoprevention of prostate cancer. Cancer Lett 2008; 265:167–76.

Taplin ME, Balk SP. Androgen receptor: a key molecule in the progression of prostate cancer to hormone independence. *J Cell Biochem.* 2004; 91:483-490.

Thomson AA. Role of androgens and fibroblast growth factors in prostatic development. *Reproduction* 2001; 121:187-195.

Trottier G, Bostrom PJ, Lawrentschuk N, Fleshner NE. Nutraceuticals and prostate cancer prevention: a current review. *Nat Rev Urol.* 2010; 7:21–30.

Trottier G, Lawrentschuk N, Fleshner NE. Prevention strategies in prostate cancer. *Current Oncol.* 2010; 17 (Suppl 2):S4-10.

Valve EM, Nevalainenen MT, Nurmi MJ, Laato MK, MArtikainen PM, Harkonen PL. Increased expression of FGF-8 isoforms and FGF receptors in human premalignant prostatic intraepithelial neoplasia lesions and prostate cancer. *Lab Invest.* 2001; 81:815-826.

Vis AN, Van der Kwast TH. Prostatic intraepithelial neoplasia and putative precursor lesions of prostate cancer: a clinical perspective. *BJU Int.* 2001; 88:147-157.

Weinberg DS, Weidner N. Concordance of DNA content between prostatic intraepithelial neoplasia and concomitant invasive carcinoma: evidence that prostatic intraepithelial neoplasia is a precursor of invasive prostatic carcinoma. *Arch Pathol Lab Med.* 1993; 117:1132–1137.

Wilke AO, Morriss-Kay GM, Jones EY, Health JK. Functions of fibroblast growth factors and their receptors. *Curr Biol.* 1995; 5: 5000-5007.

Wills ML, Hamper UM, Partin AW, *et al.*: Incidence of highgrade prostatic intraepithelial neoplasia in sextant needle biopsy specimens. *Urol.* 1997; 49:367–373.

Woodson K, Tangrea JA, Pollak M *et al.* Models of metastatic prostate cancer: a transgenic perspective. *Prostate Cancer Prostatic Dis.* 2003; 6:204-211.

Wu CL, Yang XJ, Tretiakova M, et al. Analysis of alpha-methylacyl-CoA racemase (P504S) expression in high-grade prostatic intraepithelial neoplasia. *Hum Pathol.* 2004; 35:1008-1101.

Yang RM, Naitoh J, Murphy M, Wang HJ, Phillipson J, deKernion JB, Loda M, Reiter RE. Low p27 expression predicts poor diseasefree survival in patients with prostate cancer. *J Urol.* 1998; 159:941–945.

Young AP, Schlisio S, Minamishima YA, Zhang Q, Li L, Grisanzio C, Signoretti S, Kaelin WG. VHL loss actuates a HIF-independent senescence programme mediated by Rb and p400. *Nat Cell Biol.* 2008; 10:361–369.

Zanardi S, Puntoni M, Maffezzini M, Bandelloni R, Mori M, Argusti A, Campodonico F, Turbino L, Branchi D, Montironi R Decensi A. Phase I-II Trial of Weekly

Bicalutamide in Men with Elevated Prostate-Specific Antigen and Negative Prostate Biopsies. *Cancer Prev Res.* 2009; 2:377.

Zhou M, Shah R, Shen R, Rubin MA. Basal cell cocktail (34betaE12 _ p63) improves the detection of prostate basal cells. *Am J Surg Pathol.* 2003; 27: 365–371.

Zlotta A, Schulman C. Clinical evolution of prostatic intraepithelial neoplasia. *Eur Urol.* 1999; 35:498–503.

8

Diagnosis of Prostatic Intraepithelial Neoplasia in Luminal Cells Using Raman Spectroscopy

Suneetha Devpura[1], Jagdish Thakur[1], Seema Sethi[2],
Vaman M. Naik[3], Fazlul Sarkar[2], Wael Sakr[2] and Ratna Naik[1]

[1]*Wayne State University, Detroit, MI*
[2]*Pathology, Karmanos Cancer Institute, Detroit, MI*
[3]*Univeristy of Michigan-Dearborn, Dearborn, MI,*
USA

1. Introduction

Prostate cancer is the second most common cancer among men in worldwide based on the statistics in 2008 (Jemal, 2011). In 2008, 903,500 (14%) new cases are recorded and about 258,400 (6%) people died. The highest incidence rates are observed in the developed countries in Oceania, Europe, and North America. Since the introduction of Prostate Specific Antigen (PSA) test, which measures the level of PSA in patients' blood, for prostate cancer screening the mortality rate has decreased due to its early detection and treatment (Oesterling, 1991, Shroder, 2009). The other method for detecting the prostate cancer involves a Digital Rectal Examination (DRE) to check for growths in or enlargement of the prostate gland in men. If there is a tumor growth in the prostate, it can often be felt as a hard lump. To further confirm the tumor, a pathological examination is performed on surgically removed tissues from the suspected areas of prostate and grade of the cancer (Humphrey, 2004) is determined. According to the current policy, the age limit to obtain PSA has been lowered to 40 years. In addition, decision to perform biopsies is not followed by PSA and DRE alone, other factors such as patient age, PSA velocity, PSA density, family history, ethnicity, etc are also considered (Caroll, 2009). One of the common tissue extraction methods is needle biopsy. A recent study has shown that a Target Scan biopsy method has better accuracy over conventional practice to locate the malignant tissues (Andriole, 2007). Combined examination of PSA, DRE, and pathological tests provides a better diagnostic ability for prostate cancer (Partin, 1997). Once a patient is diagnosed with cancer, there are a few early therapeutic procedures available for patient depending on a variety of different factors, like the stage of the tumor, health of the patient and his age etc. These procedures are: radical prostatectomy, external beam radiotherapy, brachytherapy, high intensity focused ultrasound, and cryotherapy (Hricak, 2009). After radical prostatectomy, number of biopsies containing tumor and biopsy perineural invasion are found to be independent predictors of the recurrence of the disease, provided that the patients' PSA is more than 10ng/ml (Quinn, 2003). Prostate instraepethelial neoplasia (PIN) is a precursor lesion in prostate cancer which can be of high or low grade category. Usually PIN is referred to as high grade if it is capable of developing into cancer within the next 10 years (Bostwick &

Qian, 2004) or so. A study has suggested that antiactivity of certain dietary flavonoids prevents the progression of high grade PIN to cancer (Kandaswami, 2005). However, this hypothesis could not be established in a randomized double-blind study performed with 303 men in twelve Canadian centers. These men were given soy, vitamin E, and selenium on daily basis for three years. The results were not statistically significant to show their effects on decreasing the progression of cancer or eliminating it (Fleshner, 2011).

Detection and confirmation of prostate cancer is very crucial for its successful treatment and survival rate. Sometimes the standard screening programs can provide misleading results leading to wrong or over-treatments and occasionally to fatal consequences. The interpretations of histological examination of biopsies, considered as the "gold standard" for diagnosis, are often subjective and can vary significantly from one pathologist to another (Allbrook, 2001). Hence, it is imperative to detect the state of the disease with a method which is objective and capable of providing results within a very short period of time (1-2 minutes). Optical spectroscopy techniques are very well suited for these types of goals, and in addition, they are also capable of probing disease at cellular level. Raman spectroscopy is one of the optical techniques which is currently extensively investigated as a diagnostic tool for detection of different types of cancers (Laserna, 1996). This optical method can provide information about the changes in the concentrations of the constituent biomolecules of tissues and detect the progression or state of the disease. In Raman spectroscopy measurements, a laser light is incident on a sample which interacts with its molecules and gets scattered. Majority of the light scatters elastically; however a very small fraction of it scatters inelastically carrying information about the nature of the sample's molecules, their mutual interaction, and their relative concentrations in the sample (Raman, 1928). The chemical nature of molecules can be uniquely determined from a set of their vibrational energy levels and Raman spectroscopy has the capability to measure these vibrational energy levels (Gelder, 2007, Movasaghi, 2007). This powerful capability of the Raman spectroscopy provides a fundamental motivation to develop this optical technique as an objective diagnostic tool for early detection of cancers. The changes in biochemical composition of a cancerous tissue could be detected through the observed changes in Raman band intensities compared to those of normal tissue. As the biochemical compositions of a tissue or cell begin to change from its normal values, it can trigger the onset of a cancer, and Raman spectroscopy has the potential to detect those initial compositional changes. So this technique can be used to diagnose pathological condition of organs and progression of disease. Currently, the potential of this technique are being tested to detect different types of cancers: breast, prostate, bladder, cervix, skin, larynx, head and neck squamous cell carcinoma, etc. and determine their unique spectra features (Stone, 2003, Keller, 2006, Devpura, 2010, 2011). Raman spectroscopy is able to identify prostate cancer from benign with 89% accuracy in snap frozen biopsies *in vitro* (Crow, 2003). Reduction in glycogen and increment in nucleic acid contents of malignant areas are observed through Raman bands. Sensitivity of differentiating cancer grades, Gleason score 7, less than 7, and greater than 7 are more than 81%. In another study, prostate cancer was identified with 94% accuracy compared to benign and prostate intraepithelial neoplasia (Devpura, 2010). In addition, Gleason score 6, 7, and 8 were distinguished with more than 81%. However, this needs to be validated with additional studies on more tissue specimens. Bladder and prostate cancers were also investigated using a fiber optic near-infrared Raman spectrometer and an overall accuracy of 84% and 86%, respectively, was observed (Crow, 2005). An attempt to construct an integrated Raman and angular-scattering microscope was made to collect both Stokes-shifted light and elastic light (Smith & Berger, 2009) to improve the detection performance and accuracy. This will allow characterizing simultaneously the size of cell and its chemical information. Recently, Raman spectroscopy was successfully used to determine the variation of chemical composition of a cell in response to a drug treatment. A threshold concentration of a toxic amount of *Nerium Oleander* was determined (Saha, 2009). This demonstrates that this technique can be used in drug designing application.

Due to advancement in diagnostic technologies and treatment modalities, most of the cancers can be cured if detected in their early stages. Cancer is basically a disease in which abnormal cells divide without any control and these cancerous cells are able to invade other tissues, and by this process the disease spreads to other organs. Hence, it certainly will be of great advantage to detect biomolecular compositional changes at cellular level, particularly in these proliferating cells. In this study, we have focused our study on luminal cells of PIN and compare their spectral features of benign epithelia (BE) and cancerous cells of the prostate tissues. To the best of our knowledge, this is the first report of such an investigation. It is important to compare luminal cells of each pathological category since basal cells are absent in the epithelium of microacinar structures of the prostate cancer. In addition, we have investigated the stroma surrounding BE, PIN, and cancerous micro acinar clusters.

The interaction between prostatic stroma and the epithelial cells is somewhat different from the stromal cells in prostate tumors (Bowsher & Carter, 2006). The prostatic stroma, which consists of fibromuscular matrix enclosing the prostatic ducts, limits the proliferation of the epithelia unlike the stroma in the prostatic tumors which contain fibroblasts or myofibroblasts. The stroma bordering prostatic tumors is called "reactive stroma" or "carcinoma associated fibroblasts" (Bowsher & Carter, 2006). It is imperative to explore the spectral features of the stromal cells in BE, PIN, and tumor stages and understand its linkage with cancer and its progression. It appears that the reactive stroma in prostate initiates the carcinogenesis and helps its progression (Olumi, 1999, Hayward, 2001, Niu & Zia, 2009).

2. Materials and methods

In this study, we used a Renishaw RM1000 Raman microscope-spectrometer with a 785 nm laser excitation source. RM1000 is equipped with a CCD (charged-coupled device) detector, an automated *xyz* stage (ProScan II) with WIRE 2.0 software to control the recording of the Raman spectra. 50x objective was used to collect data from the tissue specimens with laser power of about 20 mW focused to a spot size of ~ 2 μm diameter on the tissue specimens. Each Raman spectrum was averaged over three scans with 20 s integration time to obtain a good signal-to-noise ratio. The scattered light was collected in a back scattering geometry and dispersed using a 1200 lines/mm grating.

2.1 Preparation of tissues for Raman spectroscopy

Tissue specimens which are embedded in paraffin wax were obtained from Karmanos Cancer Center and Harper University Hospital in Detroit, MI, USA, and were processed at the University Pathology Services at Karmanos Cancer Center. In this study, we specifically selected the tissue specimens which were purely BE, PIN or cancerous. For each specimen, two parallel adjacent sections were cut. One 5 µm thick section was stained with Haematoxylin and Eosin (H&E) and was used for pathological examination, and the second 10 µm thick layer was used for Raman spectroscopic measurements. The H&E stained slides were reviewed by three experienced pathologists and none of the cases studied here found to be in dispute. The paraffin wax was removed from the 10 µm thick tissue sections using xylene and ethanol baths following the procedures described in Devpura, 2010 and 2011. It is noted that the incomplete removal wax residues gives rise to strong and sharp Raman bands that interfere with the Raman bands of the tissue samples at 1063, 1130, 1296, 1436, and 1465 cm^{-1} (Ó Faoláin, 2005). In our spectra, we did not observe any of the sharp bands associated with wax. Assuming that the morphological features do not change across a few micrometer thick layers, the H&E stained layer was used as a guide to collect the Raman spectra from the specific sites of the adjacent unstained deparaffinized tissue section.

2.2 Raman spectroscopic measurements

A total of 34 tissue specimens obtained from 33 patients were used in this study. Out of these, 12 specimens were benign, 11 were PIN, and 11 were cancerous with a grade of 3. The Raman spectra were recorded from the unstained tissue section. The appropriate regions on the unstained tissue section were identified with the help of the adjacent stained section with regions marked by pathologist as BE, PIN, cancer, and stroma. The identification of these regions was done with an optical microscope to make sure the data were collected only from the marked regions. When collecting Raman spectra, the laser beam was focused only on the luminal cells (Figure 1). The regions marked with elliptical symbols in the Figure 1 show the regions from where the Raman spectra were collected. In addition, it was made sure that each Raman spectrum was taken from a different region of luminal cells. The Raman spectra were collected in the 500-1900 cm^{-1} region. The 600-1800 cm^{-1} region is commonly known as the "biological window" where most of the biomolecules show intense Raman excitations. An extra 100 cm^{-1} extended region were recorded at both ends of the Raman spectra to avoid any artifacts which may occur while removing the fluorescence background

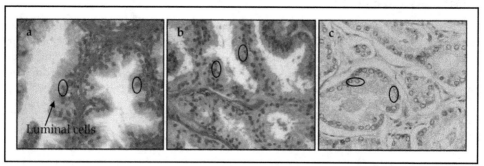

Fig. 1. Pathology pictures of (a) BE tissue, (b) PIN, and (c) Cancer (images are taken with 40x magnification). Elliptical symbols represent locations of the Raman measurements from the corresponding unstained tissue.

from each spectrum. A total of 1220 Raman spectra were collected from the tissue samples in which 207, 202, and 208 are from the luminal cells of BE, PIN, and cancer (grade 3), respectively, and the remaining (201 spectra each) from the corresponding stromal regions. The Raman measurements of stromal cells in cancer tissues were obtained from the bordering regions of the grade 3 micro acinar clusters.

2.3 Raman data processing and chemometric analysis

The collected Raman spectra were examined for non-standard noise and those with such noise were discarded from the database used for analysis. The spectra used in the analyses were cleaned from any spurious bands due to cosmic rays and noise by using wavelets method (Cao, 2007). The fluorescence background from each spectrum was removed with minmax adaptive algorithm that requires no apriori knowledge of the spectra. Finally, each spectrum was normalized with respect to the highest intensity band in the spectrum. Multivariate/chemometric statistical method, like principal component analysis, PCA, (Jolliffe, 2002) which determines correlation in the variance, was used to detect trends in the data set. The data was further analyzed using discriminant function analysis, DFA (Klecka, 1980) to classify the data. First, the data was analyzed using PCA which reduces the dimensionality of the original data set from 601 variables to 19 new variables, called the eigenvectors. These new variables captured 97% of the variance of the data. Examination of the first two eigenvectors show distinct trends in the data representing BE, PIN, and cancer. These new fewer variables carrying most of the variance of the data are useful for determining the groups in the data. To find classes in the data, we have performed DFA using 19 eigenvectors as the input variables. The classification of each pathological state is done using the leave-one-out method, where each data set is considered a new case and compared with the rest of the data pool.

3. Results and discussion

The average Raman spectral features of BE, PIN, and cancer are shown in Figure 2. We see changes in the peak intensities of most of the Raman bands (Raman band assignments are listed in the table 1) in the spectra of PIN compared to the spectra of BE and cancer tissues. These changes are fundamentally related to the changes in the concentrations of the biochemicals of BE luminal cells. Significant changes are in the region from 600 cm^{-1} to 1145 cm^{-1} which are shown in the lower panel of the Figure 2. Some of the changes are noteworthy: the band at 726 cm^{-1} (assigned to ring breathing mode of DNA/RNA bases) becomes quite intense in PIN, and in addition the Raman bands at 853 cm^{-1}, 931 cm^{-1} (v_{C-C} stretching mode of protein), 960 cm^{-1}, and 1090 cm^{-1} (symmetric phosphate stretching vibrations) also show an increase in their intensities when pathological state of cell changes from BE to PIN. While the bands at 1605 cm^{-1} and 1667 cm^{-1} (amide I) showed decrease in their intensities. When comparing the average spectral changes of PIN with cancer, we see that the peak intensities at 780 cm^{-1}, 1240 cm^{-1} (proline, tyrosine), 1330 cm^{-1} and 1605 cm^{-1} are enhanced when pathological state of luminal cells changes from PIN to cancer, while the bands at 726, 853, 931, 960, and 1090 cm^{-1} show decrease in their intensities which show similar trend like the bands of BE. The Raman bands at 780 cm^{-1} and 878 cm^{-1} showed progressive increase in their intensities when cells changes from BE to PIN and then to cancer. These Raman bands should be further investigated for their possible association with progression of prostate cancer and perhaps their use as diagnostic variables.

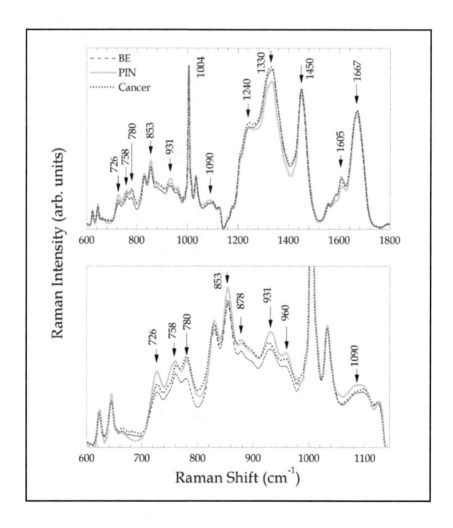

Fig. 2. Average Raman spectra of BE (blue dash line), PIN (green), and Cancer (red dotted line). The lower panel shows the Raman spectral range of 600-1145 cm-1.

Raman Shift (cm⁻¹)	Peak Assignment
726	A ring breathing mode of DNA/RNA bases
758	Symmetric breathing of tryptophan
780	DNA/uracil ring breathing mode
829	Tyrosine, phosphodiester, O-P-O stretching DNA/RNA
853	Ring breathing mode of tyrosine, C-C stretch of proline ring, glycogen
878	Tryptophan, hydroxyproline, C-O-C ring
931	C-C stretch, α-helix, protein band
960	Cholesterol, phosphate of HA
1004	Phenylalanine
1032	Phenylalanine, proline
1081	Typical phospholipids, phosphodiester groups in nucleic acids/collagen
1090	Symmetric phosphate stretching vibrations
1240	RNA, Amide III, collagen
1313	Lipid/protein
1330	Collagen, nucleic acids & phospholipids
1450	CH_2 bending mode of proteins & lipids, methylene deformation
1557	Tryptophan, tyrosine
1605	Cytocine, phenylalanine, tyrosine, C=C stretch
1667	Protein, C=C stretch, amide I

Table 1. Raman peak assignment (Gelder, 2007, Movasaghi, 2007).

3.1 Statistical analysis of the PIN, BE, and cancer data

The first three eigenvectors or the principal components (PCs) are plotted against each other in Figure 3. The left panel is the plot of PC2 vs. PC1 and the right panel represents PC3 vs. PC1. These three PCs contain 76% of the variance in the data showing different trends for each pathological state. Although some of the data seem to be overlapping with each other, the different trends are still very clear in the data. The largest variance in the data is captured by the PC1 which is shown by the spread of the BE, PIN and cancer along the PC1 axis. PC3 shows distinct trend present in the spectral data of PIN.

The average spectra of BE, PIN, and cancer showed distinct variations in the intensities of certain peaks while the spectral features of individual spectrum of each category are expected to spread about its average value and must also be different from other categories. As tissue changes from BE to PIN and then to cancer, the spectral variances in the data must exhibit some distinct classes if each of these categories is pathologically different. The classification results are shown in Figure 4 where we clearly see three distinct classes with their centroids (marked with black squares) far away from each other. It is interesting to note that PIN class is very distinct class from other classes and does not have much overlap with others. There are only three pathological states (Klecka, 1980), which means there are two discriminant functions (DFs). So the DF1 shows the maximum variance in the Raman data, and the DF2 contains the rest of the variances.

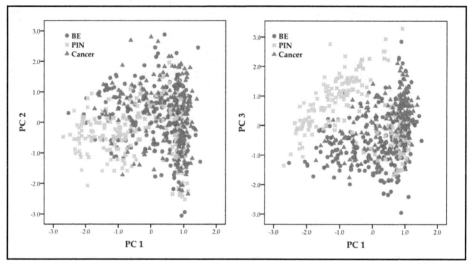

Fig. 3. PCA results of BE, PIN, and cancer. PC2 vs PC1 is on the left panel and the PC3 vs. PC1 is on to the right.

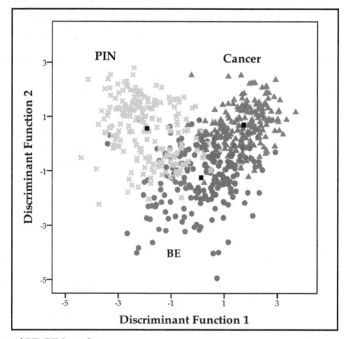

Fig. 4. DF plot of BE, PIN, and cancer.

			Predicted Group Membership			
		Group	BE	PIN	Cancer	Total
Cross-validated	Count	BE	158	20	29	207
		PIN	33	165	4	202
		Cancer	16	2	190	208
	%	BE	**76.3**	9.7	14.0	100.0
		PIN	16.3	**81.7**	2.0	100.0
		Cancer	7.7	1.0	**91.3**	100.0

Table 2. Classification results of BE, PIN, and cancer. Centre the values of "BE", "PIN" and "Cancer"

The group prediction of the Raman spectroscopy data using DFA is compared with that of pathological diagnosis: the "gold standard" for diagnosing cancer. To test the validity of our predicted classifications, we have performed leave-one-out cross-validation where the group classification for each spectrum with one of the known pathological states is determined while using the remaining data as a training set. The results of cross-validation classification results are shown in Table 2. We see that the PIN is predicted with 82% accuracy while the prediction accuracies for BE and cancer are 76% and 91 %, respectively.

3.2 Comparison of Raman spectra of stroma in BE, PIN, and cancer

It is interesting to study the stroma surrounding each of the pathological states as it could provide useful information about the onset of cancer. The nature of stroma observed was found to depend on its environment (Bowsher & Carter, 2006). Figure 5 shows the average Raman spectra of stroma surrounding BE, PIN, and cancer. The stroma surrounding cancer shows a significant enhancement in the intensity of Raman bands at 726, 758, 931, 1240, 1313, and 1330 cm^{-1} compared to stroma surrounding PIN and BE whereas bands at 1081, 1450 and 1667 cm^{-1} show a slight reduction in intensity. The Raman spectra of stromal regions in cancer and BE seem to show similarity in their biochemicals compared to that of the stroma of PIN. When comparing spectral features of the luminal cells of PIN with surrounding stroma, the Raman bands associated with DNA/RNA (726, 780, 829, and 1330 cm^{-1}) are suppressed and the bands arising from amino acids/protein/collagen (853, 931, 960, and 1240 cm^{-1}) are significantly enhanced. This can also be observed in the stroma of BE, and cancer.

The chemometric analysis was also performed on the stromal data. The PC plots of stromal investigation are shown in Figure 6. Here, the analysis generated 15 principal scores containing 98% of the variance in the Raman data. The plot is based on the first three PCs which consist of 86% of the Raman spectral information. Here, we see clear trends for stroma associated with each epithelial category: BE, PIN, and cancer.

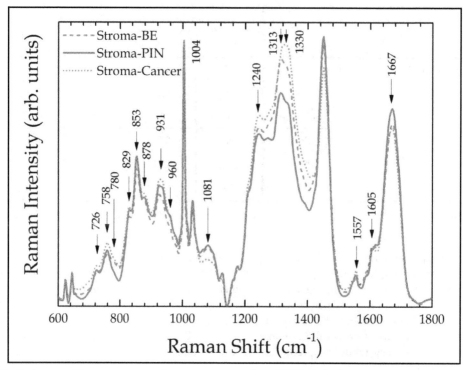

Fig. 5. Average Raman spectra of stroma surrounding BE, PIN, and cancer.

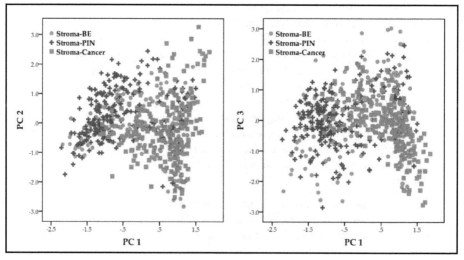

Fig. 6. PCA analysis of the Raman data of stroma.

Apparently, the stroma surrounding BE, PIN, and cancer are quite distinct as shown in the DF plot together with the overlapping average Raman spectra of stroma in BE, stroma in PIN, and stroma in cancer (see Figures 6 and 7). Here, the DFA is performed with 15 eigenvectors which contained 98% of the variance in the stromal data. Table 3 shows predicted group membership using leave-one out classification method for stromal investigation, we find that the stroma surrounding PIN is 83.6% correctly identified. The stroma in BE and cancer is classified with 81.6% and 87.1% accuracy, respectively.

It seen that when comparing stroma surrounding PIN with that of cancer, the intensities of Raman bands at 853, 931, 1240, and 1330 cm^{-1} have increased whereas the Raman bands at 1081 cm^{-1} and 1450 cm^{-1} have decreased. Thus, Raman spectroscopy can be used to distinguish easily the stroma associated with BE, PIN, and cancer accuracy. It is interesting to note when spectral data of all the pathologies are combined and analyzed statistically to find their distinct classes, we see a clear separation of stroma from BE, PIN and cancer. In addition stroma of BE, PIN, and cancer are also well separated (see Figure 8).

It should be noted that DFA constructs one less number of discriminant functions for the user-defined categories. Thus, this analysis created 5 discriminat functions due to 6 categories, and figure 8 shows only the first two discriminant functions which carry 89% of the variance in the Raman spectra. Thus, the overlapping of data in this 2-D plot may not be the same for the other dimensions. This study shows that the Raman spectroscopy can be used to distinguish the luminal cells in their normal, BE and PIN states. Further, the stroma surrounding these regions can also be distinguishes as they exhibit distinct characteristic spectral features of their own.

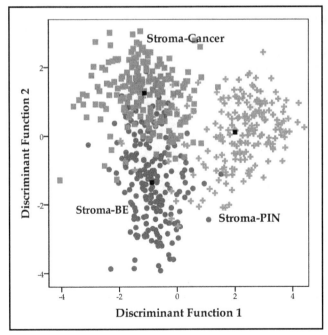

Fig. 7. DF plot of stroma surrounding BE, PIN, and cancer.

			Predicted Group Membership			
		Group	Stroma-BE	Stroma-PIN	Stroma-Cancer	Total
Cross-validated	Count	Stroma-BE	164	3	34	201
		Stroma-PIN	14	168	19	201
		Stroma-Cancer	18	8	175	201
	%	Stroma-BE	**81.6**	1.5	16.9	100.0
		Stroma-PIN	7.0	**83.6**	9.5	100.0
		Stroma-Cancer	9.0	4.0	**87.1**	100.0

Table 3. Classification results of stroma in BE, PIN, and cancer.

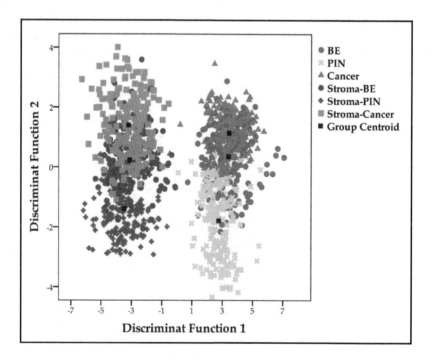

Fig. 8. DF plot of all the categories: BE, PIN, and cancer, and their surrounding stroma.

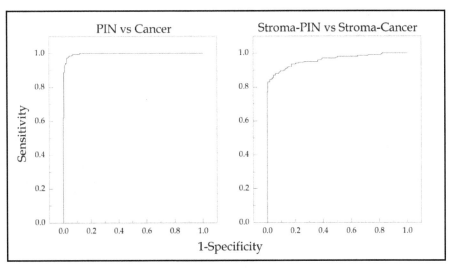

Fig. 9. ROC curves for PIN vs cancer and stromal PIN vs stromal cancer.

We also constructed Receiver Operating Characteristic (ROC) curves (Mason & Graham, 2002) for both the analyses, PIN compared to cancer and comparison between stroma around PIN and stroma associated with cancer. In an ROC graph, the sensitivity and 1-specificity are plotted against each other. Sensitivity is the ratio of true positives (cancer measurements which are correctly identified as cancer) over the total PIN data and the specificity is the ratio of true negatives (PIN measurements which are correctly identified as PIN) over the total PIN data. The sensitivity and specificity for luminal cell investigation of PIN and cancer are 98% and 99%. For the stromal investigation, the sensitivity and specificity are about 90% and 94%, respectively. The area under the curves (AUC) infers the validity of the test. An AUC = 1, indicates a perfect test whereas AUC = 0.5 implies a null result. As shown in Figure 9, both the AUC's are more than 0.96 indicating a very good test for PIN and stroma investigations.

4. Conclusions and future work

In this study, we have investigated using Raman spectroscopy, luminal cells from the tissues which are purely either BE, PIN or cancerous. We particularly focused on PIN and compared its spectral features with BE and cancer. Significant and noticeable changes in the Raman spectra of BE, PIN, and cancerous tissues are observed. As luminal cells become cancerous, the intensities of the most Raman bands in the 700-1000 cm-1 increase and the intensity changes can be interpreted in terms of changes in the biochemical composition of the tissues. In particular, the intensity of the 780 cm-1 (possibly arising from nucleic acids) Raman band increases considerably in the spectrum of cancerous tissue compared to the BE. Additionally, we also studied stromal cells surrounding each pathological state of the tissues, namely, BE, PIN and cancer and, we observed enhancement in protein contents and reduction in DNA contents when compared to the luminal cells. Chemometric analysis of the data shows that the spectral variations in the data are quite pronounced and can easily be classified with very high accuracies into distinct pathological groups. The sensitivity and specificity of luminal cells of PIN and cancer are about 98% for each and for the stroma associated with these pathologies, both the sensitivity and specificity are more than 90%. Current study suggests further investigation of different pathology grades (low and high grade) of PIN, including basal cells, are needed and to expand the Raman spectral database for better prediction capability. Since some of the PIN structures may not lead to carcinoma, by spectroscopic investigation one can explore the Raman signatures of a PIN structure which can lead to cancer.

5. Acknowledgement

We thank Dr. D. Shi, Detroit Medical Center, Detroit, for her help with prostate samples and Dr. M. D. Klein, Children's Hospital of Michigan, Detroit, for providing access to the Raman Spectroscopy facility. We express our sincere thanks to Richard A. Barber Foundation for the financial support of this work.

6. References

Allsbrook, W. C. , Mangold, K. A., Johnson, M. H., Lane, R. B., Lane, C. G. & Epstein, J. I. (2001). Interobserver reproducibility of Gleason grading of prostatic carcinoma: General Pathologists. *Human Pathology*, Vol. 32, No. 1, (January 2001), pp. (81-88)

Andriole, G. L., Bullock, T. L., Belani, J. S., Traxel, E., Yan, Y., Bostwick, D. G. & Humphrey, P. A. (2007). Is there a better way to biopsy the prostate? Prospects for a novel Transrectal systematic biopsy approach. *Urology*, Vol. 70, No. 6, Supp 6A, (December 2007), pp. (22-26).

Bostwick, D. G. & Qian, J. (2004). High-grade prostatic intraepithelial neoplasia. *Modern Pathology*, Vol. 17, (January 2004), pp. (360–379).

Bowsher, W & Carter, A. (2006). *Challenges in prostate cancer* (Second edition), Blackwell Publishing, ISBN 978-4051-0752-5, Massachusetts, USA.

Cao, A., Pandya, A. K., Serhatkulu, G. K., Weber, R. E., Dai, H., Thakur, J. S., Naik, V. M., Naik, R., Auner, G. W. & Rabah, R. J. (2007). A robust method for automated background subtraction of tissue fluorescence. *J Raman Spectrosc.*, Vol. 38, (May 2007), pp. (1199–1205)

Carroll, P., Alnerstein , P. C., Greene, K., Babalan, R. J. , Carter, H. B., Gann, P. H., Han, M., Kuban, D. A., Sartor, A. O., Stanford, J. L. & Zletman, A. (2009). In: *Prostate-Specific Antigen Best Practice Statement*, http://www.auanet.org/content/guidelines-and-quality-care/clinical-guidelines/main-reports/psa09.pdf.

Crow, P., Molckovsky, A., Stone, N., Uff, J., Wilson, B. & Wongkeesong, L.-M. (2005). Assesment of fiberoptic near-infrared Raman spectroscopy for diagnosis of bladder and prostate cancer. *Urology*, Vol. 65, (June 2005), pp. (1126-1130)

Crow, P., Stone, N., Kendall, C. A., Uff, J. S., Farmer, J. A. M., Barr, H. & Wright, M. P. J. (2003). The use of Raman spectroscopy to identify and grade prostatic adenocarcinoma in vitro. British J of Cancer, Vol. 89, pp. (106-108)

Devpura, S., Thakur, J. S., Sarkar, F. H., Sakr, W. A., Naik, V. M. & Naik, R. (2010). Detection of benign epithelia, prostatic neoplasia, and cancer regions in radical prostatectomy tissues using Raman spectroscopy. *Vibrational Spectrosc.*, Vol. 53, (July 2010), pp. (227-232)

Devpura, S., Thakur, J. S., Sethi, S, Naik, V. M. & Naik, R. (2011). Diagnosis of head and neck squamous cell carcinoma using Raman spectroscopy: tongue tissues. *J Raman Spectrosc.* (DOI: 10.1002/jrs.3070)

Fleshner, N. E., Kapusta, L., Donnelly, B., Tanguay, S., Chin, J., Hersey, K., Farley, A., Jansz, K., Siemens, R., Trpkov, K., Lacombe, L., Gleave, M., Tu D. & Parulekar, W. R. (2011). Progression From High-Grade Prostatic Intraepithelial Neoplasia to Cancer: A Randomized Trial of Combination. *J Clin. Oncol.*, Vol. 29, (June 2011), pp. (2386-2390)

Gelder, J. D., Gussem, K. D., Vandenabeele, P & Moens, L. (2007). Reference database of Raman spectra of biological molecules. *J Raman Spectrosc.* Vol. 38, (September 2007), pp. (1133-1147).

Hayward, S. W., Wang, Y., Cao, M., Hom, Y. K. & Zhang, B. (2001). Malignant transformation in a nontumorigenic human prostatic epithelial cell line. *Cancer Res.*, Vol. 61, (November 2001), pp. (8135–8142).

Hricak, H., Scardino, P. T. & Reznek, R. H. (2009). *Prostate Cancer*, Cambridge University Press, ISBN 978-0-521-88704-5, New York, USA.

Humphrey, P. A. (2004). Gleason grading and prognostic factors in carcinoma of the prostate. *Modern Pathology*, Vol. 17, (March 2004), pp. (292–306), doi:10.1038/ modpathol.3800054.

Jemal, A., Siegel, R., Bray, F., Center, M. M., Ferlay, J., Ward, E. & Forman, D. (2011). Global Cancer Statistics. *Cancer J Clin.* Vol. 5961, (March/April 2011), pp. (225-249), doi:0.3322/caac.20107 doi: 10.3322/caac.20006.

Jolliffe, I. T. (2002). *Principal Components Analysis* (Second edition), Springer-Verlag, New York, USA.

Kandaswami, C., Lee, L. T., Lee, P. P, Hwang, J. J., Ke, F. C., Huang, Y. T. & Lee, M. T. (2005). The antitumor activities of flavonoids. *In Vivo*, Vol. 19, (September-October 2005), pp. (895-909).

Keller, M. D., Kanter, E. M. & Mahadevan-Jansen A. (2006). Raman spectroscopy for cancer diagnosis. Spectroscopy, Vol. 21, No. 11, (November 2006), pp. (33-41)

Klecka, W. R. (1980). *Discriminant Analysis, Series: Quantitative applications in the social sciences*, Sage Publications Inc., Newbury Park.

Laserna, J. J. (1996). *Modern Techniques in Raman spectroscopy*, John Wiley& Sons, New York, USA.

Mason, S. J.& Graham, N. E. (2002). Areas beneath the relative operating characteristics (ROC) and relative operating levels (ROL) curves: Statistical significance and interpretation. *Q. J. R. Meteor. Soc.* Vol. 128, pp. (2145-2166).

Movasaghi, Z., Rehman, S. & Rehman, I. U. (2007). Raman spectroscopy of biological tissues. *Applied. Spectroscopy Reviews*, Vol. 42, (September 2007), pp. (493-541), doi: 10.1080/05704920701551530.

Niu, Y. -N. & Zia, S. -J. (2009). Stroma–epithelium crosstalk in prostate cancer. *Asian Journal of Andrology*, Vol. 11, (December 2008), pp. (28–35).

Oesterling, J. E. (1991). Prostate specific antigen: a critical assessment of the most useful tumor marker for adenocarcinoma of the prostate. *J Urol*, Vol. 145, No. 5, (May 1991), pp. (907-923).

Ó Faoláin, E., Hunter, M. B., Byrne, J. M., Kelehan, P., Lambkin, H. A., Byrne, H. J. & Lyng, F.M. (2005). Raman Spectroscopic Evaluation of Efficacy of Current Paraffin Wax

Section Dewaxing Agents *J Histochem Cytochem* Vol. 53(1), (March 2005), pp. (121-129).

Olumi, A. F., Grossfeld, G. D., Hayward, S. W., Carroll, P. R., Tlsty, T. D. & Cunha, G. R. (1999). Carcinoma-associated fibroblasts direct tumor progression of initiated human prostatic epithelium. *Cancer Res*, Vol. 59, (October 1999) pp. (5002-5011).

Partin A. W., Kattan, M. W., Subong, E. N., Walsh, P. C., Wojno, K. J., Oesterling, L. E., Scardino, P. T., & Pearson, J. D. (1997). Combination of prostate-specific antigen, clinical stage, and Gleason score to predict pathological state of localized prostate cancer. A multi-institutional update. *J. Am. Med. Assoc.*, Vol. 277, (May 1997), pp. (1445-1451).

Quinn, D. I., Henshall, S. M., Brenner, P. C., Kooner, R., Golovsky, D., O'Neill, G. F., Turner, J. J., Delprado, W., Grygiel, J. J., Sutherland, R. L. & Stricker, P. D. (2003). Prognostic significance of preoperative factors in localized prostate carcinoma treated with radical prostectomy. *Cancer*, Vol. 97, No. 8, (April 2003), p.p. (1884-1893).

Raman, C. V, (1928). A new type of secondary radiation. *Nature*, Vol. 121, No. 3048, (March 1928), pp. (501-502).

Saha, A. & Yakovlev, V. V. (2009). Towards a rational drug design: Raman micro-spectroscopy analysis of prostate cancer cells treated with an aqueous extract of Nerium Oleander. *J. Raman Spectrosc.*, Vol. 40, (December 2008), pp. (1459-1460).

Shroder, F. H., Hugosson, J. & Roobol, M. J. (2009). Screening and prostate-cancer mortality in a randomized European study. *New Eng J Med*, Vol. 360, (March 2009), pp. (1320-1328).

Smith, Z. J. & Berger, A. J. (2009). Construction of an integrated Raman- and angular-scattering microscope. *Review of Scientific Instruments*, Vol. 80, No. 044302, (April 2009), pp. (1-8).

Stone, N., Kendall, C., Smith, J., Crow, P. & Barr, H. (2003). Raman spectroscopy for identification of epithelial cancers. *Faraday Discuss*, Vol. 126, (September 2003), pp. (141-157).

Part 5

Intraepithelial Neoplasia of Uterus

9

Endometrial Intraepithelial Neoplasia

Nisreen Abushahin[1,4], Shuje Pang[1,2], Jie Li[1,3],
Oluwole Fadare[5] and Wenxin Zheng[1,6,7]
[1]Department of Pathology, University of Arizona College of Medicine, Tucson, AZ,
[2]Department of Pathology, Tianjin Central Hospital of Obstetrics and Gynecology, Tianjin,
[3]Department of Obstetrics and Gynecology, Qilu Hospital,
Shandong University, Jinan, Shandong,
[4]Department of Pathology, Jordan University Hospital- University of Jordan, Amman,
[5]Department of Pathology, Vanderbilt University School of Medicine, Nashville, TN
[6]Department of Obstetrics and Gynecology, University of Arizona, Tucson, AZ
[7]Arizona Cancer Center, University of Arizona, Tucson, AZ,
[1,5,6,7]USA
[2,3]China
[4]Jordan

1. Introduction

Endometrial cancers are the most common malignancies of the female genital tract in the United States, with 42,160 new cases diagnosed and 7780 cancer-associated deaths in 2009 (Jemal, Siegel et al. 2009). The histopathological classifications of endometrial cancers are numerous, but in 1983, two broad clinico-pathologic categories of endometrial carcinomas were delineated (Bokhman 1983). This conceptual classification has largely been based on light microscopic appearance, clinical behavior and epidemiology, and had been subsequently supported by molecular-cytogenetic data, which has facilitated the acceptance of the so-called *dualistic model* of endometrial carcinogenesis(Deligdisch and Holinka 1987; Lax and Kurman 1997; Sherman 2000; Matias-Guiu, Catasus et al. 2001; Lax 2004; Liu 2007), a modified and more comprehensive comparison of both types is illustrated in Table 1.

According to that model, the first and the most common type of endometrial carcinoma is called *Type I endometrial cancer*. These Type I cancers, of which the pathologic prototype is endometrioid carcinoma, represent at least 80% of newly diagnosed cases of endometrial cancer. The much less common mucinous carcinomas are also generally classified as a Type I cancer. Overall, they occur in comparatively younger age group (40-50 years)(Deligdisch and Holinka 1987; Lax and Kurman 1997; Sherman 2000; Matias-Guiu, Catasus et al. 2001; Lax 2004; Liu 2007). The tumor cells frequently express estrogen and progesterone receptors (Demopoulos, Mesia et al. 1999; Lax 2004), and their evolution appears to be driven by unopposed estrogen stimulation from either endogenous (e.g. ovarian estrogen-producing tumors) and/or exogenous sources (e.g. hormonal therapy)(Ettinger, Golditch et al. 1988; Potischman, Hoover et al. 1996; Demopoulos, Mesia et al. 1999). These tumors, therefore, mostly arise in a background of endometrial glandular hyperplasia(Lax 2004; Liu 2007). *Type*

I endometrial cancers have a relatively favorable prognostic profile compared to *type II endometrial cancer* (Creasman, Odicino et al. 2003). Several kinds of genetic alterations had been detected in *Type 1 endometrial cancers*, including PTEN inactivation(Tashiro, Blazes et al. 1997; Mutter, Ince et al. 2001), beta-catenin (CTNNB1) mutations (Konopka, Janiec-Jankowska et al. 2007), and to a lesser degree, microsatellite instability (related to inactivation of the MLH1 gene) (Esteller, Levine et al. 1998), and activational mutations of the K-ras gene (Velasco, Bussaglia et al. 2006).

Parameters	Type I	Type II
Incidence	80%	15%
Peak Age	50-60	60-70
Obesity	Common	Uncommon
Estrogen stimuli	Common	Uncommon
Precancer	EIN (classic)	EmGD (serous type & clear cell type)
Latent Precancer	PTEN null glands	P53 signature glands
Progression	Slow	Rapid
Histology	Endometrioid, mucinous	Serous, Clear cell, and Carcinosarcoma
Genetic changes	PTEN, MSI	p53, BRCA, 1pDel
Familial	HNPCC	Unknown
Prognosis	Good	Poor

Table 1. Dualistic model of endometrial cancer as modified by Zheng et al (Zheng, Xiang et al. 2011).

Type II endometrial cancer in the dualistic model are significantly less common than their Type I counterparts, and represent only10- 15% of cases. The pathologic prototype of this category is the endometrial serous carcinoma (ESC) [previously termed uterine papillary serous carcinoma (UPSC)]. *Type II endometrial cancer* typically occurs in an older age group (60-70 years) (Lax and Kurman 1997; Sherman 2000; Matias-Guiu, Catasus et al. 2001). They frequently arise in a background of inactive or resting endometrium(Lax and Kurman 1997; Sherman 2000; Matias-Guiu, Catasus et al. 2001), display a low frequency of expression of hormonal receptors, are not associated with the estrogen-associated clinical factors (such as obesity)and generally are not thought to be directly influenced by hormones (Sasano, Comerford et al. 1990; Lax, Pizer et al. 1998; Demopoulos, Mesia et al. 1999; Lax 2004; Shang 2006). Definitive risk factors for *type II endometrial cancer* are still unclear, however. In one recent study, we found that women 55 years of age or under with a personal history of breast cancer, had an increased risk of ESC as compared with controls (Liang, Pearl et al. 2010), and an earlier study by Chan et al came to comparable conclusions 2006 (Chan, Manuel et al. 2006). These Type II cancers, most notably ESC, also exhibit frequent mutation and overexpression of the p53 (Sherman, Bur et al. 1995; Nordstrom, Strang et al. 1996) and HER2/neu (Rolitsky, Theil et al. 1999) genes and proteins, respectively. They also show alterations of intercellular adhesion molecules like E-cadherin (Holcomb, Delatorre et al.

2002; Mell, Meyer et al. 2004) and claudin(Santin, Bellone et al. 2007; Konecny, Agarwal et al. 2008), and display over-expression of p16 (Chiesa-Vottero, Malpica et al. 2007; Yemelyanova, Ji et al. 2009) and IMP-3(Reid-Nicholson, Iyengar et al. 2006; Zheng, Yi et al. 2008). Overall, *type II endometrial cancer* have a relatively poor prognosis independent of other factors, and a higher mortality rate in comparison to type I cancers(Lauchlan 1981; Eifel, Ross et al. 1983; Sherman, Bitterman et al. 1992). This dualistic model has provided a valuable academic framework for the subsequent studies of myriad aspects of endometrial carcinogenesis and progression and a conceptual basis for the differential deployment of histotype-specific treatment modalities.

1.1 Endometrial Intraepithelial neoplastic lesions: The nomenclature dilemma

Endometrial cancers, especially *type II endometrial cancer*, are a significant cause of morbidity and mortality in women (Jemal, Siegel et al. 2009). This has prompted the long-standing search for optimal approaches for their prevention; one aspect of prevention is the early recognition of occult *precursor lesions* or *precancers* (Berman, Albores-Saavedra et al. 2006), along with the administration of therapeutic interventions prior to the development of overt malignancy. To establish any lesion as a precursor lesion or a precancer to one neoplasm, the putative lesion should meet some basic criteria that defines a precancer, as recognized by participants at a consensus conference on the subject sponsored by the National Cancer Institute in 2006 (Berman, Albores-Saavedra et al. 2006). This definition modifies and generalizes a definition initially proposed for endometrial intraepithelial neoplasia (Mutter 2000; Mutter, Baak et al. 2000; Mutter, Ince et al. 2001; Mutter 2002; Hecht and Mutter 2006; Mutter, Zaino et al. 2007) . The following five defining criteria must all be met: "(1) Evidence exists that the precancer is associated with an increased risk of cancer. (2) When a precancer progresses to cancer, the resulting cancer arises from cells within the precancer. (3) A precancer is different from the normal tissue from which it arises. (4) A precancer is different from the cancer into which it develops, although it has some, but not all, of the molecular and phenotypic properties that characterize the cancer. (5) There is a method by which the precancer can be diagnosed". In the last two decades, there have been significant advances made in the study of the precursors of *Type I endometrial cancer*, and this precancerous lesion is currently considered as endometrial atypical hyperplasia in the WHO classification system (that is still the most frequently used by pathologists) and the "endometrial intraepithelial neoplasia (EIN)" system that was originally proposed by Mutter et al (Mutter 2000; Mutter, Baak et al. 2000; Mutter, Ince et al. 2001; Mutter 2002; Hecht and Mutter 2006; Mutter, Zaino et al. 2007). On the other hand, studies of *Type II endometrial cancer* precursors have been relatively limited. The prototype of *Type II endometrial cancer, which is* ESC, usually arises in a background of atrophic or resting endometrium. This is in contrast to *Type I endometrial cancer*, which generally have a hyperplastic (or at least non-atrophic) background and show a strong relation to high estrogen levels. For this and a variety of other reasons, the precancers of *type I endometrial cancer* are highly unlikely to constitute the precancer lesions for *Type II endometrial cancer*. Numerous lines of evidence developed during the last decade point toward a newly recognized lesion called "endometrial glandular dysplasia (EmGD)" as the actual precursor of Type II cancers, including serous and clear cell types (Zheng, Liang et al. 2004). These lines of evidence include pathologic, genetic as well as clinical factors. Accordingly, the precancer of *type I endometrial cancer* and *type II endometrial cancer* are two distinct entities at the morphologic and molecular levels and are not related to each other. In this chapter, we explore the current state of knowledge on all types of precancerous lesions of the endometrium, based on our interpretation and modification of the dualistic model of endometrial carcinogenesis. Clinical and pathological experience in endometrial carcinogenesis had shown a significant impact of histologic subtype (endometrioid, serous, clear cell etc) on overall prognosis and survival. Considering the multiple conflicting nomenclatures that existed in studies of endometrial carcinogenesis, which lead to the inappropriate inclusion of some entities as 'precancers' (as discussed in following sections),we plan to propose a unified terminology and classification scheme for the precancerous lesions in the endometrium, which will be biology-based and clinically oriented for better patient care. Accordingly, the discussion of intraepithelial neoplastic lesions of the endometrium would embrace the precancer of *type I endometrial cancer*, which we will refer to as endometrioid EIN; as well as the precancers of *type II endometrial cancer*, that is, serous EmGD and clear cell EmGD, which will be referred to as serous EIN and clear cell EIN respectively. As summarized in Figure 1.

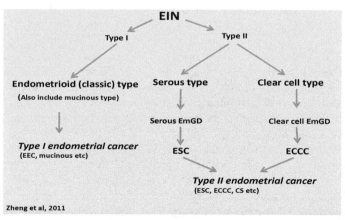

Fig. 1. A proposed unified model of endometrial carcinogenesis. Endometrial intraepithelial neoplasia (EIN) encompass a broad spectrum of morphologically and biologically distinct entities, categorized as type I, referring to the classic EIN lesion described by Mutter, and type II that include serous EmGD and clear cell EmGD as precancers of type II endometrial cancer.

2. Precancers of type I endometrial cancers

2.1 Historical backgrounds

In the past, *Type I endometrial cancer* was thought to be preceded by pan-endometrial hormonally induced changes referred to as endometrial hyperplasia. The term endometrial hyperplasia encompasses a broad spectrum of polyclonal proliferations that result from a physiological response of the endometrium to an abnormal estrogenic stimulus. The magnitude of such proliferations, reflects the quantity and duration of unopposed estrogen exposure (Trial 1996; Mutter, Zaino et al. 2007), resulting in architecturally variable glands covering a surface area that is equal or exceeds that of the stroma (i.e. gland-stroma ratio more than 1:1). The most widely observed histological features include irregular architectural remodeling of endometrial glands in functionalis layer, vascular thrombi, and stromal breakdown. A critical feature of benign hyperplasia is that no significant cytological changes are seen between the hyperplastic glands and the surrounding glands (Mutter, Zaino et al. 2007). Benign endometrial hyperplasia is most frequent around the time of the menopause, due to alterations of the normal cycle of sequentially regulated estrogen and progesterone. It may also occur following anovulatory cycles for the same reason. The most common symptoms of hyperplasia are prolonged or excessive bleeding at intervals that are initially longer than normal. Microinfarcts and estrogen withdrawal are responsible for symptomatic bleeding (Song, Rutherford et al. 2002; Ferenczy 2003). Other patients may complain of intermittent spotting, commonly attributed to patchy stromal breakdown secondary to estrogen-induced microthrombi. A rapid decline in the prolonged estrogen stimulation causes massive apoptosis of the endometrial glands and stroma resulting in heavy shedding. Occasionally, the decrease in estrogen levels is sufficiently gradual that generalized apoptosis and shedding fail to take place as regular menstruation.

The World Health Organization (WHO) 1994 classification system subdivided endometrial hyperplasia according to architectural complexity and cytological atypia into 4 subgroups: simple, simple hyperplasia without atypia, complex, and atypical complex hyperplasia (Scully RE 1994), as illustrated in Figure 2. The WHO 1994 endometrial hyperplasia schema confines most precancers of *type I endometrial cancer* in the atypical hyperplasia subgroup, but in the opinion of many pathologists and investigators, there are several problems associated with this classification. First, this classification system is poorly reproducible among pathologists (Hunter, Tritz et al. 1994; Skov, Broholm et al. 1997; Zaino 2000). Second, this system is missing diagnostic elements that have only become clear in recent years. Of these elements, the localizing topographic distribution of a clonally expanding precancer and the need to establish size thresholds for diagnosis. Third, it is a purely morphology-based system without any supporting molecular and morphometric studies that precisely quantifies diagnostic architectural changes. The search for an alternative classification system for endometrial carcinomas had lead to the introduction of the Endometrial Intraepithelial Neoplasia (EIN) entity.

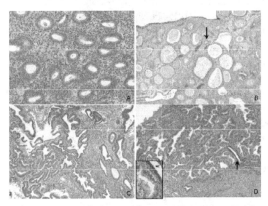

Fig. 2. A: simple hyperplasia, B: simple hyperplasia without atypia, C: complex hyperplasia, D: atypical complex hyperplasia. Arrows indicate residual uninvolved glands. Inset: magnified area with arrow, a hyperplastic gland shows atypical distinct morphology compared to adjacent resting endometrial gland (RE).

2.2 EIN: Mutter's model for *type I endometrial cancer*

The concept of EIN and the diagnostic schema was introduced by Mutter and the Endometrial Collaborative Group in 2000 (Mutter 2000); and later launched at Brigham and Women's Hospital in 2002 (Mutter 2002; Hecht, Ince et al. 2005), to replace the older hyperplasia-based nomenclature, the currently used terminology by WHO 1994 classification system(Scully RE 1994), which implies endometrial hyperplasia as the precancerous lesion of type I cancers. This concept was the result of cautious correlation of genetically ascertained pre-malignant lesions with histopathologic feature and clinical outcomes. A better vision of the carcinogenesis of *type I endometrial cancer* was achievable with the advent of polymerase chain reaction–based clonal assays and relevant biomarkers that facilitated a molecular, rather than purely morphologic approach to precancer diagnosis. The molecular entity of EIN is thought to be a clinically pertinent lesion that can be reproducibly diagnosed by pathologists and targeted for therapeutic intervention. According to this model, the premalignant lesions are referred to as EIN to distinguish them from the diffuse estrogen associated changes of benign endometrial hyperplasia. EIN is a" histologically recognizable localized lesion composed of a clonal proliferation of glands and that usually carry one or several of the genetic abnormalities associated with endometrioid carcinoma" (Mutter, Boynton et al. 1996; Mutter, Baak et al. 2000) . This model had been supported by molecular and morphometric studies. First, the monoclonality of EIN lesions was proven utilizing nonrandom X-chromosome inactivation (HUMARA assay) and clonal propagation of altered microsatellites (Mutter, Chaponot et al. 1995; Jovanovic, Boynton et al. 1996). Second, the identification of lineage continuity with subsequent carcinomas that occur in the same patient, fulfilling a vital standard for molecular definition of precancers(Mutter, Baak et al. 2000). Third, the application of computer based morphometry has been successful at further improving diagnostic reproducibility of precursor diagnosis, and have a better correlation between morphologic features and patient actual clinical outcome(Baak, Nauta et al. 1988; Baak, Wisse-Brekelmans et al. 1992). The applied morphometric measures were combined into a threshold D-score (detailed below). In 2005, a meta-analysis study of the cumulative outcome prediction experience of the D-score (Baak, Mutter et al. 2005), showed that patients with a D-score less than 1 have an overall 89-fold increased cancer risk than those with D-score more than 1. Even if one excludes concurrent cancers, those diagnosed within 12 months of EIN, cancer risk over the next two decades is 45-fold that of controls. Comparison of the WHO 94 and the EIN systems, with correlation of the clinical outcome reveals a degree of overlapping. Mutter et al (Hecht, Ince et al. 2005) had found that, for the simple non-atypical hyperplasia, only a minimal risk for endometrial cancer is believed to be present (only 5% are re-diagnosed as EIN upon review). Complex atypical hyperplasia has the highest risk of cancer and 80% of cases are rediagnosed as EIN (the greatest overlap). Therefore, majority of EIN lesions are actually equivalent to most of the WHO atypical complex hyperplasia category. Detailed relationship is illustrated in Figure 3.

2.2.1 Diagnostic features of EIN

As defined by Mutter et al, EIN is 'the premalignant clone of an endometrial lesion that is characteristically offset from the background endometrium by its altered cytology and crowded architecture'. This definition implies the use of an internal control for cytologic atypia, which is the benign resting endometrium, combined with the distinctive topography of a clonal process. The average age of women with EIN is 52 years (Baak, Mutter et al.

2005), almost a decade younger than the average age for cases of endometrioid endometrial carcinoma in patients with concurrent endometrioid carcinoma (Baak, Mutter et al. 2005; Hecht and Mutter 2006), and when those patients are excluded, the average time following EIN detection to carcinoma diagnosis is 4 years(Baak, Mutter et al. 2005). The clinical significance of EIN lesions is that they represent a long-term cancer risk that is 45-fold greater than that of their benign endometrial hyperplasia counterparts (Hecht and Mutter 2006; Mutter, Zaino et al. 2007). This distinctive clinical profile, is further supported by morphometric measures, summed up under the term D-score, which includes the volume percent stroma (a measure of gland crowding); standard deviation of the shortest nuclear axis (a measure of nuclear pleomorphism); and gland outer surface density (a measure of branching and folding). The morphometric techniques were effective at discriminating those endometrial lesions which progress to adenocarcinoma from those that do not (Baak, Nauta et al. 1988). However, we would like to point out that the morphometric techniques are so far mainly limited to research applications rather than in general practice.

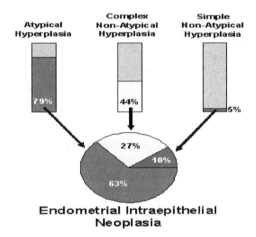

Fig. 3. Mutter's diagram (Hecht, Ince et al. 2005) for overlaps between WHO and EIN classification systems.

Based on the model of EIN, 5 strict morphologic features are applied, and all 5 criteria must be met in each case to maintain a high level of diagnostic specificity and predictive value. The diagnostic criteria are summarized in Table 2.

Architecture

A feature that makes EIN lesions readily visible under low magnification. Area of glands exceeds that of stroma, thus, the surface area of glands (combined epithelium and lumen) is greater than that of the stroma that contains them, and the tissue proportion occupied by stroma is less than 50%. However, this ratio is also used as a diagnostic feature for benign endometrial hyperplasias. To overcome the potential source of diagnostic confusion, strict search for the other 4 criteria is critical to confirm the lesion in question is EIN and not a focus of endometrial hyperplasia. Another important point to mention is the condition of the endometrial stroma within the area of question.

EIN feature	Definitions
Architecture	Area of glands exceeds that of stroma (glands/stroma > 1). Lesion composed of individual glands, which may branch slightly and vary in shape.
Cytology	Nuclear and/or cytoplasmic features of epithelial cells differ between glands with abnormal architecture and those with normal background. May include change in nuclear polarity, nuclear pleomorphism, or altered cytoplasmic differentiation state. Highly abnormal cytology if no normal comparison glands are present
Size	Maximum linear dimension exceeds 1 mm
Exclude benign mimics	Benign conditions with overlapping criteria :disordered proliferative, basalis, secretory, polyps, repair, etc.
Exclude cancer	Carcinoma if mazelike glands, solid areas, or significant cribriform growth.

Table 2. Essential diagnostic criteria of EIN as outlined by Mutter et al (Mutter, Zaino et al. 2007).

Cytologic Changes

This must be judged individually in each case using the native background endometrium as the internal control (Figure 4 A). No unified diagnostic cytologic features are settled for EIN, this is due to several factors. First, the variability of the cytological characteristics of the endometrial glandular epithelium among specimens, according to fixation, processing, and staining. Second, this inconsistency also depends largely on the fluctuating hormonal environment. Third, not all EIN lesions maintain endometrioid differentiation (Mutter, Zaino et al. 2007), and commonly acquire metaplastic changes, including mucinous,

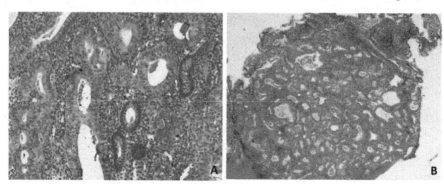

Fig. 4. Diagnostic features to look for in EIN lesions. EIN glands are cytologically distinct from the surrounding normal endometrial glands (A, 200x). The lesion should be at least 1mm in dimension (B, 40x).

squamous morular, tubal, eosinophilic, or micropapillary changes. The practicing pathologist should be careful, however, not to confuse the relatively mild cytologic atypia of EIN lesions with the striking atypia with possibly hobnailed nuclei seen in the precancer of *type II endometrial cancer* (serous EmGD and clear cell EmGD).

Size

The lesion must be at least 1 mm in dimension in a single tissue fragment (Figure 4B). This "golden" number needs to be present in only one dimension of the lesion. Separate foci cannot be added to achieve this minimum size, it must be met in a single focus(Mutter, Zaino et al. 2007). The reason why a size parameter is needed in such a lesion, is probably to confer reproducibility and predictive value in pathological and the clinical sides, respectively. It may also significantly reduce the risk of EIN overdiagnosis in minute randomly detected foci of glandular crowding. One problem of the size limit is that about 20% of EIN lesions are diffuse and non-localized by the time they are detected(Mutter, Zaino et al. 2007), such a diagnostic difficulty might be overcome by largely depending on the other diagnostic criteria. Lesions that have most of the diagnostic criteria for EIN but are <1mm in dimension are still of unknown clinical significance, but are thought to be a good indication for subsequent follow-up endometrial sampling. However, in a recent study by Huang et al(Huang, Mutter et al. 2010), 71 579 consecutive gynecological pathology reports were retrieved, of which, 206 (0.3%) cases with 'gland crowding' were identified, in which 69% (143/206) had follow-up sampling. Of these, 33 (23%) had an outcome diagnosis of EIN (27 cases; 19%) or carcinoma (6 cases; 4%). Included were 18 cases (55%) diagnosed within the first year and presumed concurrent, and an additional 15 (45%) discovered after 1 year and interpreted as a later phase of disease or new events (Huang, Mutter et al. 2010). The authors suggested that such "gland crowding" is significant and deserves mention in pathologic reports.

Exclusion of benign mimics

Many innocent conditions are frequently encountered during routine examination of endometrial specimens, and these may be the source of diagnosticdifficulty in the exclusion of a potential EIN lesion. These may include (but are not limited to) artifactually pushed together or telescoped endometrial glands; or crowding related to the late secretory endometrium, in which the gland density may be very high in the deep functionalis where predecidual change is minimal (Figure 5A). Some portions of specialized but otherwise normal endometrium such as lower uterine segment or uterine basalis may also cause confusion, these are usually identified by their fibrous stromal context and quiescent epithelium. Another more serious misinterpretation comes when dealing with endometria under the influence of estrogen withdrawal, either during the normal menstruation or as a result of hormonal imbalances, the resulting microscopic picture is collapsed glands and stromal condensation. This frequently results in irregular glands lacking much stromal separation, giving an EIN-like picture (Figure 5B). Overall, the most commonly overdiagnosed lesions as EIN are probably endometrial polyps (as well as with endometrial hyperplasia), yet, their characteristic altered stroma and thick vessels, are readily distinct from the stromal features of EIN.

Exclusion of cancer

EIN lesions are composed of clusters of individually recognizable glands, whereas endometroid carcinoma show more complex growth patterns not seen in EIN, such as solid, cribriform, or complex interlacing mazelike growth (Figure 6A &B). The presence of myoinvasion is also diagnostic of carcinoma (Figure 6C), but this is better applied to hysterectomy specimens where intact myometrial wall is present. Overly malignant cytologic features beyond that seen in EIN are also present (Figure 4D).

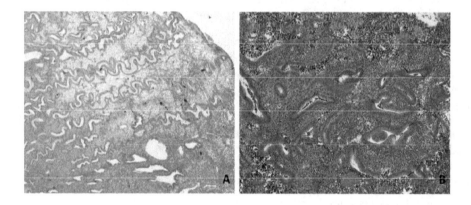

Fig. 5. Benign mimics of EIN. Late secretory endometrium displays prominent tortuous glands and crowding (A, 40x). Breakdown changes with artificially crowded irregular glands due to stromal collapse (B, 100x).

In our current proposal for a unified and simplified nomenclature for the precancers of endometrial cancers, we prefer to refer to this classic EIN entity described above as 'endometrioid EIN'.

2.2.2 Differential diagnosis

Many of the important mimics of EIN lesions have been discussed in the preceding sections (exclusion of benign mimics and exclusion of cancer). Another differential diagnostic consideration that is worth mention is the precancer of *type II endometrial cancer* (in our opinion, serous EmGD and clear cell EmGD). Unlike *type II endometrial precancers*, endometrioid EIN cells lack high-grade cytologic features, including hobnail nuclei. Also endometrioid EIN usually has a high-level estrogen stimulation, yet this is not the usual scenario in serous or clear cell EIN. Immunohistochemical studies with p53 and IMP3 are useful as these markers are positive in serous EmGD but not in endometrioid EIN.

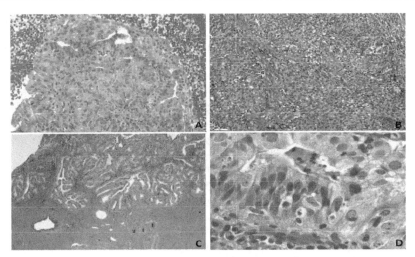

Fig. 6. Features of endometrioid carcinoma, and not EIN. Cribriform glands (A,100x), solid growth pattern (B, 40x), presence of muscular invasion (C, 20x), or frankly malignant cytologic features (D, 200x).

2.3 Molecular insights

2.3.1 *Type I endometrial cancer* and sex hormones

The normal endometrial epithelial cells are highly responsive to sex hormones, namely, estrogens and progesterone; and the morphology of the endometrium at any point is the sum of these responses and interactions. Consequently, the risk for *type I endometrial cancer* is significantly affected by the reciprocal and "opposing"actions of estrogens and progesterones, and is a dynamic process that depends on temporal changes in tissue responsiveness. The same is true for the resultant cancer in which the neoplastic cells retain this feature of hormone-responsiveness. Studies of the expression profiling of *type 1 endometrial cancer* cells had shown resemblance to the expression profile seen in estrogen-driven proliferative endometrium, it also lacks expression of some genes induced by progestins (Hecht and Mutter 2006).

On one hand, estrogens are promoters of cell proliferation and inhibitors of apoptosis, the effect of which is a manifestation of a complex downstream sequence of transcription changes that may involve modulation of tumor suppressor genes. These changes may include alterations in PTEN, PAX2, and HOXB13 among others. PTEN expression in normal endometrial glands is greatly elevated by estrogens and reduced by progestins during hormonal fluctuations of the normal menstrual cycle(Mutter, Lin et al. 2000). As promoters of proliferative activity, estrogens may also increase probability of arbitrary mutations(Cairns 1998), as well as increase the rate of mutagenesis through free radical formation (Burcham 1999), although the magnitude of this effect is minimal. Overall, large-scale population studies had shown that women exposed to "unopposed" estrogens have 2 to 10 folds increased risk for *type I endometrial cancer*, and this wide range is influenced by the dose and duration of exposure (Parazzini, La Vecchia et al. 1991; Potischman, Hoover et al. 1996; Trial 1996; Zeleniuch-Jacquotte, Akhmedkhanov et al. 2001). Moreover, EIN lesions are thought to attain high levels of nuclear estrogen receptors (Mutter, Ince et al. 2001), thus, estrogens may also act as growth positive selectors of the previously mutated cells, allowing their clonal expansion.

On the other hand, progestins have the ability to "oppose" the biologic effects of coexisting estrogens through down-regulation of the estrogen receptor itself in the endometrial epithelial cells. This is actually the basis of combined estrogen/progesterone therapies, in which the net effect is dominated by the progestin component. It is also known to be the reason behind the protective influence of combined oral contraceptives, women who uses these drugs are said to have 0.5 to 0.7 fold risk of *type I endometrial cancer* compared to controls (Grimes and Economy 1995; Weiderpass, Adami et al. 1999). Other protective effects of circulating progestins are due to downregulation of proliferative promoters like PTEN, thus it was found that PTEN mutant clones have a tendency to involute under the influence of progestins (Zheng, Baker et al. 2004). The anti-cancer role of progestins is further mediated by induction of apoptosis through the increased expression of Bcl-2 and BAX (Vereide, Kaino et al. 2005).

2.3.2 Molecular alterations in *type I endometrial cancer* and endometrioid EIN

Type I endometrial cancer demonstrate large numbers of genetic changes in which the sequential order of mutation, and the final combination of defects differ considerably between individual examples (Hecht and Mutter 2006). Common genetic changes in endometrioid carcinoma include, but are not limited to, microsatellite instability (MSI) (Risinger, Berchuck et al. 1993; Duggan, Felix et al. 1994; Kobayashi, Matsushima et al. 1996; Mutter, Boynton et al. 1996; Catasus, Bussaglia et al. 2004), or specific mutation of *PTEN* (Risinger, Hayes et al. 1997; Tashiro, Blazes et al. 1997; Levine, Cargile et al. 1998; Maxwell, Risinger et al. 1998; Gurin, Federici et al. 1999; Mutter, Lin et al. 2000), *K-ras* (Enomoto, Inoue et al. 1991; Fujimoto, Shimizu et al. 1993; Duggan, Felix et al. 1994; Sakamoto, Murase et al. 1998; Mutter, Wada et al. 1999; Swisher, Peiffer-Schneider et al. 1999; Lax, Kendall et al. 2000; Lagarda, Catasus et al. 2001), and β-catenin genes (Kobayashi, Matsushima et al. 1996; Fukuchi, Sakamoto et al. 1998; Mirabelli-Primdahl, Gryfe et al. 1999; Schlosshauer, Pirog et al. 2000). As previously described, it is the clonal origin of EIN that supports its definition as a precancer. Moreover, studies by Mutter et al gave considerable evidence that comparison of the type and magnitude of genomic damage between endometrioid EIN and *type I endometrial cancer* (Mutter, Boynton et al. 1996; Esteller, Catasus et al. 1999; Mutter, Baak et al. 2000; Mutter, Lin et al. 2000), indicates a greater cumulative mutational burden in the later , a feature considered one milestone in the definition of precancer (Berman, Albores-Saavedra et al. 2006). Below is a discussion of the most commonly encountered molecular alterations.

PTEN

Inactivation of the *PTEN* tumor-suppressor gene (formerly known as *MMAC1*) is the most common genetic defect in endometrioid carcinoma and is seen in up to 83% of tumors that are preceded by a histologically discrete premalignant phase (Mutter, Lin et al. 2000). It acts as tumor suppressor genes because their proteins may counteract the effect of the proteins encoded by the protein kinase group of protooncogenes (Matias-Guiu, Catasus et al. 2001).

The most frequently encountered alterations of PTEN in endometrial cancers are LOH at chromosome 10q23 (40% of EC) (Jones, Koi et al. 1994; Peiffer, Herzog et al. 1995); and somatic mutation, which are almost exclusively found in *type I endometrial cancer* (37 to 61%) (Kong, Suzuki et al. 1997; Tashiro, Blazes et al. 1997; Maxwell, Risinger et al. 1998; Bussaglia, del Rio et al. 2000). A number of investigators have found a concordance between microsatellite instability status and PTEN mutations; the mutations occur in 60% to 86% of MSI (+) endometrioid carcinoma, but in only 24% to 35% of the MSI (-) tumors (Matias-Guiu, Catasus et al. 2001). Such results have led to the assumption that PTEN could be a likely target for mutations in MSI (+) EC. PTEN inactivation was detected frequently in EIN lesions (63% of cases) (Mutter, Ince et al. 2001). However, the routine application of PTEN as an informative marker of premalignant lesions is still questionable due to several facts. First, PTEN mutations have been detected in endometrial hyperplasia with and without atypia (19% and 21% respectively) (Levine, Cargile et al. 1998; Maxwell, Risinger et al. 1998; Bussaglia, del Rio et al. 2000). Second, the lack of PTEN inactivation in about one third of studied EIN lesions(Mutter, Zaino et al. 2007). And finally, the finding of somatic inactivation of PTEN in scattered benign endometrial gland in 43% of cases.(Mutter, Ince et al. 2001)

B-catenin (CTNNB1)

Gain of function mutations in exon 3 of *CTNNB1* gene at 3p21 are seen in 25% to 38% of *type I endometrial cancer*s (Fukuchi, Sakamoto et al. 1998; Schlosshauer, Pirog et al. 2000). B-catenin is a component of the E-cadherin-catenin unit essential for cell differentiation and maintenance of normal tissue architecture, and plays an important role in signal transduction (Hecht and Mutter 2006). B-catenin mutation may represent a pathway to endometrial carcinogenesis characterized by squamous differentiation and independent of PTEN (Su, Vogelstein et al. 1993). B-catenin expression change is usually a diffuse process seen in all tumor cells, and is present in some premalignant lesions (Hecht and Mutter 2006). This suggests that B-catenin mutation is an early step of endometrial tumorigenesis that is clonally represented in all tumor cells (Matias-Guiu, Catasus et al. 2001; Saegusa, Hashimura et al. 2001; Hecht and Mutter 2006). Furthermore, B-catenin might regulate the expression of the matrix metalloprotease-7, which would have a role in the establishment of the microenvironment necessary for the initiation and maintenance of growth of the primary tumors and their metastasis. (Brabletz, Jung et al. 1999). Further studies are needed to explore the role of B-catenin in *type I endometrial cancer* carcinogenesis.

K-RAS

K-RAS mutations have been identified in 10% to 30% of *type I endometrial cancer* (Sasaki, Nishii et al. 1993; Swisher, Peiffer-Schneider et al. 1999; Lax, Kendall et al. 2000). There is a higher frequency of *K-ras* mutations in cancers with microsatellite instability (MSI), and many of these are characterized by methylation related GC3AT transitions(Lagarda, Catasus et al. 2001). Several investigators had previously found associations between k-RAS mutations and the presence of coexistent endometrial hyperplasia, (Tsuda, Jiko et al. 1995) lymph node metastases, and clinical outcome in postmenopausal patients over 60 years of age (Ito, Watanabe et al. 1996). In addition, some investigators have reported an almost complete absence of k-RAS mutations in serous and clear cell carcinomas of the endometrium (Caduff, Johnston et al. 1995). In the study by Mutter et al (Lagarda, Catasus et al. 2001), the authors reported k-RAS mutations in 18.9% of 58 endometrial cancers, all of them were of endometrioid type. They also described a higher frequency of k-RAS mutations in MSI (+) carcinomas (6 of 14, 42.8%) than in MSI (-) tumors (5 of 44, 11.3%), which lead to the assumption that k-RAS mutations are common in endometrial cancer with the microsatellite mutator phenotype. In the same series, k-RAS mutations were detected in only one of 22 endometrial hyperplasia cases. In this case, atypical hyperplasia coexisted with carcinoma; interestingly, both lesions exhibited MLH-1 promoter hypermethylation, MSI, and identical PTEN mutations, but they had different k-RAS mutations; of the remaining 21 endometrial hyperplasias, 6 had shown MLH-1 promoter hypermethylation and one had both MLH 1 methylation and MSI. Accordingly, the authors hypothesized that "both k-RAS and MSI are closely related phenomena that may occur simultaneously before and during clonal expansion".

Microsatelite instability (MSI)

Among sporadic *type I endometrial cancer* of all grades, approximately 20% demonstrate a molecular phenotype referred to as Microsatelite instability (MSI) (Risinger, Berchuck et al. 1993; Burks, Kessis et al. 1994; Duggan, Felix et al. 1994; Mutter, Boynton et al. 1996). MSI is rare (< 5%) in *type II endometrial cancer* (Faquin, Fitzgerald et al. 2000; Goodfellow, Buttin et al. 2003). Microsatellites are short segments of repetitive DNA bases that are scattered throughout the genome; they are found predominantly in noncoding DNA. Due to DNA repair errors made during replication, the tendency to develop changes in the number of repeat elements as compared with normal tissue is termed MSI. MLH1 inactivation, a component of the mismatch repair system, is the most common mechanism in endometrial carcinoma and is accomplished by hypermethylation of CpG islands in the gene promoter, a process known as epigenetic silencing.(Esteller, Levine et al. 1998) Inherited or somatically acquired mutations of *MSH6*, another mismatch repair element, are also common in patients with MSI endometrial cancers.(Goodfellow, Buttin et al. 2003) MSI in general, and abnormal methylation of MLH1 in particular, is an early event in endometrial carcinogenesis that has been described in precancerous lesions (Mutter, Boynton et al. 1996; Levine, Cargile et al. 1998; Esteller, Catasus et al. 1999). The significance of MSI may also be a result of its ability to specifically inactivate other genes which contain susceptible repeat elements, such as transforming growth factor receptor type II, (TGF-âRII), BAX, insulin-like growth factor II receptor (IGFIIR), and hMSH3, resulting in secondary tumor sub-clones with an increased capacity to invade and metastasize (Ouyang, Shiwaku et al. 1997; Catasus, Matias-Guiu et al. 2000).

p53

p53 is a nuclear phosphoprotein that provoke cell cycle arrest or apoptosis through induction of P21Wafl/Cip1 and hMdm2 in response to cellular stress (Hecht and Mutter 2006). Mutations involving p53 are among the most commonly encountered molecular abnormalities in *type II endometrial cancer* (detailed in subsequent sections), and are usually due to p53 truncation mutations (Alkushi, Lim et al. 2004). On the other hand, only 5% of *type I endometrial cancers* show aberrant accumulation of inactivated p53 protein (Lax, Kendall et al. 2000), may be secondary to changes in its upstream regulatory proteins (Soslow, Shen et al. 1998) rather than truncation mutations. Examples of such upstream regulatory molecules include *MDM2* and *p14 ARF*, that regulate p53 levels and their dysregulation had been shown to cause detectable levels of p53 in the absence of p53 mutation, and may be associated with adverse clinical outcomes(Soslow, Shen et al. 1998; Schmitz, Hendricks et al. 2000; Pijnenborg, van de Broek et al. 2006). p53 overexpression and high protein levels are thought to be associated with high grade and stage, but is also an independent prognostic factor (Alkushi, Lim et al. 2004). Other possible causes for p53 accumulation in *type I endometrial cancer*s may be nonspecific DNA damage such as that induced by irradiation which is known to induce accumulation of wild-type *p53* (MacCallum and Hupp 1999).

3. Precancers of type II endometrial cancer

3.1 Precursor of endometrial serous carcinoma

3.1.1 Historical backgrounds

Endometrial carcinoma with papillary features and psammoma bodies had been described in the literature as early as 1960s (Karpas and Bridge 1963; Hameed and Morgan 1972; Factor 1974). Nevertheless, the concept of "serous" differentiation and the distinguished aggressive behavior of such cancers were recognized 2 decades later by Lauchlan in 1981 (Lauchlan 1981), and shortly followed by Kempson and Hendrickson (Hendrickson, Ross et al. 1982). These concepts were further established by subsequent studies and case series focusing on morphologic features and patient survival relative to *type I endometrial cancer* (Lauchlan 1981; Eifel, Ross et al. 1983; Sherman, Bitterman et al. 1992). In 1992, Sherman et al illustrated 32 cases of endometrial cancer with a serous component (13 pure and 19 mixed histotypes), the author noted the presence of "cytologically malignant cells closely resembling the invasive serous carcinoma in the surface endometrium adjacent to the tumor" in 28 out of the 32 studied cases, and were entitled "intraepithelial carcinoma" (Sherman, Bitterman et al. 1992). Spiegel et al described a similar lesion in 1995,and designated it as "endometrial carcinoma in-situ" (Spiegel 1995).Within the same year, Ambros et al introduced the designationof endometrial intraepithelial carcinoma (EIC), as a lesion that was repeatedly and distinctively associated with endometrial carcinoma with a serous differentiation (Ambros, Sherman et al. 1995). Main histologic patterns illustrated in Figure 7.

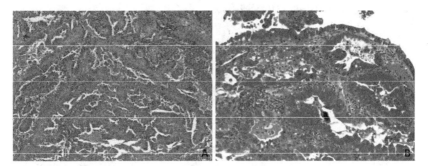

Fig. 7. Morphology of endometrial serous carcinoma. Papillary pattern (A) and glandular pattern (B).

However, in 1998, Zheng et al used the designation "uterine surface carcinoma" instead to describe that lesion, with emphasis on the multicentricity and relatively worse behavior in comparison to carcinoma in-situ per se, questioning the appropriateness of such a lesion as a precancer (Zheng, Khurana et al. 1998). A similar approach had been published in 2000 by Wheeler et al, who proposed the term "minimal uterine serous carcinoma", adding the size parameter (<1cm) to the definition of that lesion (Wheeler, Bell et al. 2000).

3.1.2 Zheng's model for precursors of type II endometrial cancer

Serous EIC is still used in the most recent (2003)WHO classification as the precancerous lesion for serous endometrial carcinoma (Tavassoli FA 2003). However, in our opinion, the fact that stage 1A non-myoinvasive serous carcinomas are known to display extrauterine disease in 17-67% of studied cases (Carcangiu, Tan et al. 1997; Gehrig, Groben et al. 2001; Zheng and Schwartz 2005), strongly argues against the designation of serous EIC as a true "precancer". Many years of gynecological surgical experience and studies of patient outcome have show that many patients diagnosed with serous EIC and treated with simple hysterectomy without surgical staging, had recurrences or intra-abdominal carcinomatosis (Carcangiu, Tan et al. 1997; Gehrig, Groben et al. 2001; Chan, Loizzi et al. 2003; Zheng and Schwartz 2005). Consequently, serous EIC is better recognized as an endometrial serous carcinoma with an early, non-myoinvasive growth pattern (Zheng and Schwartz 2005).

Careful examination of the definition of a precancer established in the National Cancer Institute Consensus in 2006, resulted in the conclusion that EmGD fulfills most of the defined criteria as a precancer of *Type II endometrial cancer*. The diagnostic criteria of serous EmGD were established by Zheng et al in 2004 (Zheng, Liang et al. 2004). Using morphological as well as immunohistochemical features, the EmGD lesions display changes that bridge the gap between benign endometrium and frankly malignant epithelium of serous EIC (Figure 8); the dysplastic epithelium of EmGD has cytologic features that are more atypical than resting endometrium but fall short of serous EIC (Figure 9), as discussed in Table 3.

Macroscopic features

Grossly, no visible lesions could be identified in the corresponding areas of EmGD (Zheng, Liang et al. 2004).

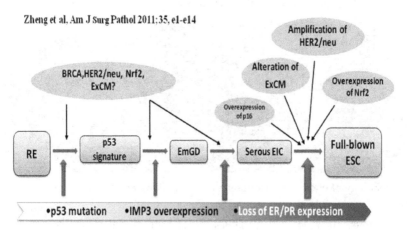

Fig. 8. Proposed model for endometrial serous carcinogenesis by Zheng et al (Zheng, Xiang et al. 2011).

3.1.3 Diagnostic criteria of serous EmGD

Serous EmGD Criterion	Comments
Patient age	Postmenopausal women, classically elder than 55 years old
Architecture & Cytology	Atypical endometrial glandular epithelium. The degree of atypia falls short of serous EIC. Many in endometrial polyp
Size limit and background	No size limit. Background is often atrophic or weakly proliferative, could be proliferative and rarely hyperplastic.
Exclude mimics	Benign conditions with overlapping features: bleeding or curettage associated atypia, repair, polyp with metaplastic changes
Exclude cancer	Serous EIC has frankly malignant cells same as in ESC/UPSC

Table 3. Serous EIN (serous EmGD) fact sheet

Microscopic features

The EmGD lesions are frequently multifocal (86% of cases) (Fadare and Zheng 2008). Classically, EmGD is characterized by glands and/or surface endometrial epithelium with atypical cytologic features. The cells of EmGD shows oval or round nuclei with a 2-3 folds nuclear enlargement compared with the benign resting endometrium. The nuclei are either hyperchromatic or with open chromatin patterns. When hyperchromasia is present, the degree of hyperchromasia is less than that of frankly malignant cells seen in serous EIC. Nucleoli are usually conspicuous instead of prominent. Partial loss of cell polarity is seen when nuclear stratification is present. A few stratifications may be seen. Mitotic figures and apoptotic bodies are appreciable, but not easily identified. Small papillary structures can be identified in EmGD glands, the thin fibrovascular cores of the EmGD papillae are also lined by dysplastic cells instead of malignant cells as in serous EIC or ESC. Occasional mitotic figures are present, but no abnormal mitoses are seen in EmGD lesions. Apoptotic bodies are scarce and in one of our series ranged from 0 to 5 per gland with an average of 1.5/gland (Zheng, Liang et al. 2004). The most common microscopic patterns include glandular involvement, either as a single or a group of EmGD glands within the endometrium or within an endometrial polyp. Another pattern is surface epithelial involvment of the endometrium or lining a polyp. EmGD foci are usually smaller than 1 mm in size. This may be related to the fact that they often presented as a single or a group of a few glands. However, occasionally, potential serous EmGD glands form clusters. When in endometrial polyps, the overall size of serous EmGD lesions may reach several millimeters. The stroma around the serous EmGD glands is usually fibrotic, but desmoplastic reactions are not seen. Background endometrium is often atrophic or weakly proliferative endometrium, but it could also be proliferative or rarely hyperplastic. This is actually a significant point to keep in mind, since nowadays; a considerable number of post-menopausal women are using hormonal replacement therapy compared to those who did 2 or 3 decades ago. In clinical practice, this is translated to the fact that about 40% of women with serous EIC or ESC have a non-atrophic endometrium as a background (Zheng, Liang et al. 2004) (34% proliferative,

6% hyperplastic endometrium). The significance of these findings is, in one hand, it provides further evidence that hormones are not risk factors for *type II endometrial cancer*; and on the other hand, pathologists should keep endometrial serous carcinoma as a differential diagnosis even in the existence of endometrial hyperplasia, in order to avoid the misdiagnosis of *type I endometrial cancer*, and the substantial consequences on patient management and outcome.

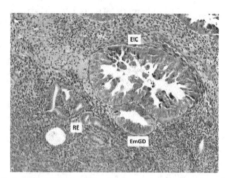

Fig. 9. Endometrial glandular dysplasia (EmGD) morphology. EmGD bridges the morphologic gap between benign resting endometrium (RE) and endometrial intraepithelial carcinoma (EIC).

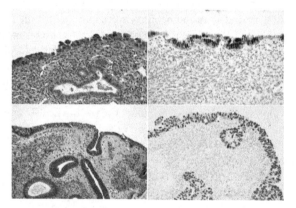

Fig. 10. p53 immunohistochemical stain in EmGD lesions. EmGD may involve the surface epithelium (upper left) or endometrial glands (lower left). The right panel shows diffuse nuclear positivity for p53 stain in areas corresponding to those in the left panel.

3.1.4 Molecular alterations of serous EIN

p53 Mutations. The p53 tumor suppressor gene, located on chromosome 17p 13.1, is probably the most commonly altered gene in human cancers (Harris 1993; Pietsch, Sykes et al. 2008), with the mutations commonly resulting in p53 protein over-expression (Darvishian, Hummer et al. 2004; Liang, Chambers et al. 2004; Jia, Liu et al. 2008). An extremely high rate of p53 alteration and over-expression (90% of our studied cases) had been detected in endometrial serous carcinoma, as evaluated by immunohistochemical staining (Figure 10) (Zheng, Cao et al. 1996; Zheng, Khurana et al. 1998). In 2008, our group studied the frequency of TP53 gene mutations in exons 5 and 8 from laser-captured microdissected endometrial samples (Jia, Liu et al. 2008). In that specific context, the TP53 gene mutations had shown a successive increment from p53 signature glands (42%) to EmGD (43%), to serous EIC (63 to 72%), and to ESC (96%) (Jia, Liu et al. 2008). The benign endometria from the control group, in contrast, showed no mutation in non-signature glands. Analogous findings were found in a later study by Zhang et al in 2009 (Zhang, Liang et al. 2009). It is concluded that p53 gene mutation is a critical and early step in endometrial serous carcinogenesis, and that p53 is an important diagnostic immunohistochemical tool in this situation (Liang, Chambers et al. 2004; Jia, Liu et al. 2008; Zheng, Xiang et al. 2011).

BRCA Mutations

A subset of relatively younger women with hereditary breast cancers are also at increased risk for the development of subsequent ovarian/tubal serous malignancies as a manifestation of hereditary breast-ovarian syndrome (Hall, Jamison et al. 2001; Arai, Utsunomiya et al. 2004), and in patients with BRCA mutations (Lavie, Hornreich et al. 2004). An earlier paper by Curtis et al in 1973, had described other malignancies that may follow breast cancer, including endometrial cancer (Inskip and Curtis 2007). The exact nature of this link between breast and endometrial cancer is still unclear; however, a few foundational studies had shed some light on evidences for such a relationship. In 1999, Hornreich et al reported a case of "uterine serous papillary carcinoma" in 2 siblings with both endometrial and ovarian serous carcinomas who were carrying identical mutation in BRCA1 gene (Hornreich, Beller et al. 1999). Genetic analysis showed loss of the wild-type allele, suggesting a link between germline BRCA1 mutation and serous cancer. A following study of Ashkenazi Jewish patients with endometrial serous carcinoma confirmed a high incidence of BRCA1 mutation and LOH (75% of tumor samples) (Lavie, Hornreich et al. 2004). A contradictory result was found in a study by the same group on a population of germline-BRCA mutation carriers, showing no relationship to increased risk of endometrial cancer. Another disagreement was reported by Liu et al, who found no significant increase in BRCA1 mutation in sporadic endometrial cancers (Liu, Ho et al. 1997).

Based on epidemiologic data, Chan et al had reported an association between breast cancer and endometrial cancers with aggressive histological types (Chan, Manuel et al. 2006). In 2007, a Swedish study found that 7.28% of patients undergoing genetic counseling for an increased risk of breast cancer, had family histories of both endometrial and breast cancers (von Wachenfeldt, Lindblom et al. 2007). We recently published a large cohort study that had found a history of a prior breast cancer in 20% of women with ESC, with the incidence being significantly higher in patients who were 55 years old or younger (41.5%) in comparison to those older than 55 years (16%) (Liang, Pearl et al. 2010).

In the light of the current controversy, further studies are absolutely needed to clarify the possible role of BRCA mutations in ESC. Contemporary data regarding BRCA mutations in serous EIC and EmGD lesions are still lacking.

Alteration of extracellular adhesion molecules

Studies of the role of extracellular adhesion molecules in the development of ESC are limited relative to studies of tumor supressor genes and oncogens. As aforementioned, ESC has the unusual capacity to metastasize outside the uterus even in the absence of myometrial invasion. This might be linked to alterations of the extracellular adhesion molecules of the neoplastic serous epithelium. Such alterations likely assist the transtubal spread of ESC into the peritoneum, and consequently result in the advanced stage of disease at time of diagnosis, even with limited uterine disease. The phenomenon of transtubal spread of serous carcinoma cells had been emphasized by our study of serous EIC lesions in 2005 (Zheng and Schwartz 2005). In that study, 67% (6 out of 9) serous EIC cases had peritoneal carcinomatosis, and among these cases, 50% showed free-floating serous malignant cells and cell clusters in the tubal lumena (Zheng and Schwartz 2005). Our suggestion was further supported by the former findings of identical clones of cells in serous EIC and serous carcinomas at extrauterine sites (Kupryjanczyk, Thor et al. 1996; Baergen, Warren et al. 2001).

Of these extracellular molecules, *E-cadherin* and *claudins* had been described as potential contributors to the biological aggressiveness of ESC. E-cadherin downregulation had been previously reported to be associated with the progression of endometrial cancers (Sakuragi, Nishiya et al. 1994; Holcomb, Delatorre et al. 2002; Mell, Meyer et al. 2004). Holocomb et al described a reduction of E-cadherin expression in 62% and 87% of their studied serous carcinoma and clear cell carcinoma, respectively (Holcomb, Delatorre et al. 2002). A recent study showed that loss of E-cadherin may be attributed to L1CAM upregulation in the aggressive subtype of endometrial cancer (Huszar, Pfeifer et al.). Claudins are a family of extracellular tight junction proteins that are said to be up regulated in ovarian cancers (Rangel, Agarwal et al. 2003), and possibly related to cancer progression (Santin, Bellone et al. 2007). Expression of claudins, especially claudin-3 and claudin-4, is also higher in *type II endometrial cancers* relative to *type I endometrial cancers* (Konecny, Agarwal et al. 2008). *CD44* is a protein involved in cell adhesion and leukocyte homing, CD44v6 is one of its isoforms that may be related to lymphovascular space invasion and metastasis. A significant loss of CD44 and that particular isoform CD44v6 had been detected in ESC compared to EEC (Soslow, Shen et al. 1998). *β-catenin* is a transcriptional activator downstream of the Wnt signaling pathway. Many types of human cancers harbor mutations of β-catenin, including endometrial cancers. Fukuchi et al (Fukuchi, Sakamoto et al. 1998)detected B-catenin mutations in 13% (10 out of 76) of their ESC cases. Whether or not serous EmGD show such alterations of extracellular adhesion molecules is still unclear and is the subject of future studies.

Amplification of HER2/neu

HER2/neu, also known as c-erb B2, is a protoncogen that encodes the human epidermal growth factor receptor (Gehrig, Groben et al. 2001). Amplification of HER2/neu and the overexpression of its protein had been shown in many human malignancies, including ESC (Santin, Bellone et al. 2002; Casalini, Iorio et al. 2004), some studies even described that in association with advanced stage and poor prognosis in ESC (Santin, Bellone et al. 2002; Villella, Cohen et al. 2006). Although similar overexpression of HER2/neu by immunohistochemistry had been shown by the authors in serous EmGD and serous EIC in the studied cases, no data regarding the gene amplification is available so far.

Overexpression of IMP-3

Insulin-like growth factor m-RNA binding protein 3, or IMP-3 is a protoncogen expressed predominantly in embryonic tissues and rarely in adult tissues except for the placenta and gonads (Nielsen, Christiansen et al. 1999; Yaniv and Yisraeli 2002). Some studies have revealed that IMP-3 is associated with cell migration and tumor invasion (Yaniv and Yisraeli 2002; Vikesaa, Hansen et al. 2006), and it could predict metastasis and prognosis in renal cell carcinoma(Jiang, Chu et al. 2006). In 2008, our group studied the expression of this oncofetal protein in serous endometrial carcinoma and its proposed precursor lesions using immunohistochemical staining (Zheng, Yi et al. 2008), we found that IMP-3 was overexpressed in 14% (3 of 21) EmGD lesions, 89% (16 of 18) serous EIC, and 94% (48 of 51) ESC cases. This was significantly higher than the expression detected in only 5 out of 70 (7%) EEC cases, and was not identified in its precancer lesion (EIN) (0 of 35 cases). These findings imply that IMP-3 overexpression may contribute in the early steps of ESC development and aggressive behavior (Zheng, Yi et al. 2008).

Overexpression of Nrf2

NF-E2-related factor 2, or for simplicity, Nrf2, is a newly described transcription factor that is thought to boost the chemo-resistance of cancer cells (Wang, Sun et al. 2008). Nrf2 has been the subject of multiple studies by our group. In one of these studies in 2010 (Jiang, Chen et al.), Nrf2 showed high expression in 89% (41 of 46) of ESC cases, compared to marginal expression of Nrf2 in 28% (14 of 51) of EEC cases, while none of the studied benign endometria showed such an expression (0 of 20). Transient silencing of endogenous Nrf2 enhanced the sensitivity to chemotherputic agents in SPEC-2 cells derived from ESC (Zheng, Xiang et al. 2011). In addition, Overexpression of Keap1, a negative regulatory gene of Nrf2, significantly sensitized those ESC-derived SPEC-2 cells and its xenografts to chemotherapeutic drugs. More recently, we have also examined Nrf2 expression in precursor lesions of ESC, and found that Nrf2 was expressed in 40% of EmGD lesions, and 44% of serous EIC lesions in the studied cases (Chen, Yi et al.); it also showed a lesser degree of expression in clear cell carcinoma (13%) and its proposed precursor lesions, clear cell EmGD and EIC (10% and 25%), respectively. In the same study, only 6% of EEC (3 out of 50) and none of the atypical endometrial hyperplasia/EIN showed overexpression of Nrf2 (Chen, Yi et al.). The relationship between Nrf2 and early steps of ESC carcinogenesis and p53 mutations is currently under exploration in our laboratory.

Overexpression of p16

p16, also known as CDKN2A, is a tumor supressor gene that had been extensively studied in the context of HPV-related cancers and their precursors (Keating, Cviko et al. 2001; Keating, Ince et al. 2001; Negri, Egarter-Vigl et al. 2003). In cervical HPV-related cancers, the mechanism of p16 overexpression may be mediated by HPV E7 viral protein. More recently, p16 has been also shown to be overexpressed in the cells of ESC in multiple studies (Reid-Nicholson, Iyengar et al. 2006; Chiesa-Vottero, Malpica et al. 2007; Yemelyanova, Ji et al. 2009). The mechanism of this overexpression, however, is probably different from that described in viral- related malignancies, since HPV DNA in situ hybridization has been negative in all studied cases (Chiesa-Vottero, Malpica et al. 2007). ESC, it is rather linked to the inactivation of RB gene through dysregulation of the p16INK4a/cyclin D-CDK/pRb-E2F pathway (Reid-Nicholson, Iyengar et al. 2006). Reid- Nicholson et al, (Reid-Nicholson, Iyengar et al. 2006) reported that p16 overexpression was detected in 92% (22 of 24) of ESC cases, compared with 7% (3 of 42) FIGO grade 1and 2 EEC, and 25% (10 of 40) of FIGO grade 3 EEC cases. Unpublished data from our laboratory also show p16 overexpression in lesions of EmGD and serous EIC (Zheng w et al, unpublished), however, it was also diffusely present in benign endometrial samples, raising the question of the practical relevance of this biomarker in serous carcinogenesis.

3.1.5 Differential diagnosis

The diagnosis of serous EIN can be difficult because it does not present as a mass. It may be a focal finding in an otherwise unremarkable endometrial polyp. This is particularly true when a biopsy sample is encountered. The overall clinico-pathologic picture is significant to avoid misdiagnosis (Table 3). The recognition of serous EmGD in an endometrial biopsy or a curettage specimen may aid the pathologist to diagnose serous EIC or to raise concerns for the presence of concurrent ESC before a hysterectomy is undertaken. Attention should be paid in the interpretation of endometrial specimens not to confuse EmGD with any of the following pathologic entities:

Reparative epithelial changes in benign endometrium

These benign changes are frequently encountered post –endometrial curettage or biopsy and rarely show the architectural patterns seen in EmGD. The cytologic atypia is minimal. Numerous mitotic figures are lacking as well. In difficult cases, the use of immunohistochemical stains for p53, IMP-3 and ki-67 can be very useful (Figure 11 A &11B).

Serous EIC

The most useful criterion here is nuclear atypia, which is marked in serous EIC, even identical to that of invasive ESC. Mitotic figures are also more frequent in EIC.

Benign endometrial metaplasias

These may include hobnail metaplasia, papillary metaplasia and also some cases of Arias-Stella reaction. The cytologic atypia in these various types of metaplasia are minimal and they are usually associated with bleeding or breakdown changes in adjacent endometrium. Mitotic activity is rarely seen. Other characteristic cytologic features, such as hobnail nuclei

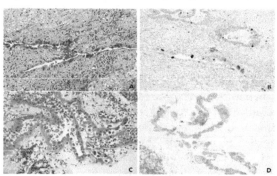

Fig. 11. Reactive endometrial changes. Post-abortive decidulized endometrium may show atypical glandular cells (A), but immunostaining with p53 is weak and focal (B). Hobnail metaplasia in endometrial curretings (C), lack of mitoses and negative p53 stain (D) helps to rule out malignancy.

or cytoplasmic clearing or eosinophilia can further help in the distinction. Pathologic examination should always keep pace with the clinical data and presentation, as a history of a preceeding conception or dilation and curettage will help minimize the misdiagnosis of such benign changes with serous EmGD. An example is illustrated in Figure 11C & 11D.

Endometrial hyperplasia

As previously mentioned, it is of clinical significance to accurately differentiate between the precursors of type I and *type II endometrial cancer*, due to the influence on management and patient outcome. In most of the cases, this should be straightforward, bearing in mind that *type I endometrial cancer* precursors usually lack the highly atypical nuclear features seen in type II precancers, including hobnail appearance, round large nuclei and prominent nucleoli. However, in case of doubt, correlation with positive immunohistochemical stains for p53, IMP-3 would help diagnose *type II endometrial cancer* precancers.

3.1.6 Clinical significance and future management

At present, there are no standard management guidelines for patients with EmGD. The approach at our institute is based on our consideration of these patients to be at a significantly higher risk for the development of endometrial malignancy than their counterparts without EmGD, and that this risk is accentuated by factors such as a personal history of breast cancer or BRCA mutations(Chan, Manuel et al. 2006; Liang, Pearl et al. 2010). For patients without breast cancer history and/or BRCA mutations, if the diagnosis of EmGD is confirmed in a biopsy, we recommend complete dilation and curettage (D&C) for larger sampling. If the diagnosis was made on an endometrial curettage, we recommend periodic follow-up (no more than every 6 months) with transvaginal ultrasound and pelvic examinations. The presence of any abnormalities during this period that may be a harbinger for neoplasia, such as persistent abnormal uterine bleeding, abnormal glandular cells on Papanicolaou tests, palpable pelvic masses, or ultrasound abnormalities, should warrant a complete D and C. For those patients with BRCA mutations or a personal history of breast cancer, our gynecologic oncologists typically offer the option for a hysterectomy. Whether or not complete staging is performed would then be dependent on the intraoperative frozen section findings. If a serous cancer is identified, irrespective of its size in representative sections of the uterus, a complete staging, including omentectomy is performed. If no such focus is identified, the procedure is limited to the hysterectomy with or without the salpingoophorectomies. It should be emphasized, however, that additional studies are required to more clearly define the clinical significance of the lesion in everyday practice. This would entail a larger systematic study of endometrial biopsies to establish the time frame between the development of EmGD and ESC, the proportion of EmGD cases that evolve to ESC, and follow-up of prospectively diagnosed cases to confirm that they are never associated with extrauterine disease in a short term.

3.2 Clear cell EmGD

3.2.1 Historical background

Endometrial clear cell carcinoma (ECCC) is a rare variant of endometrial type II cancer, accounting for 1% to 6% of all endometrial carcinomas cases (Webb and Lagios 1987; Abeler and Kjorstad 1991). It is now established that precursor lesions exist for the more common and more thoroughly studied types of endometrial cancers, including the spectrum of atypical hyperplasia and classic (endometrioid) EIN for *type I endometrial cancer* (Kurman, Kaminski et al. 1985; Mutter 2000; Mutter 2002; Mutter, Zaino et al. 2007; Scully RE 1994); and serous endometrial glandular dysplasia (EmGD) for ESC (Zheng, Liang et al. 2004; Zheng, Liang et al. 2007; Fadare and Zheng 2009; Zheng, Xiang et al. 2011). However, the other rarer and accordingly less studied variants of endometrial carcinoma, including ECCC, has not been the focus of similar searches. A few pioneer studies are mentioned in the following sections.

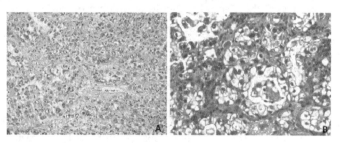

Fig. 12. Clear cell carcinoma of the endometrium.

3.2.2 Putative precursor for endometrial clear cell carcinoma: Clear cell EmGD

In 2004, Moid and Berezowski (Moid and Berezowski 2004) described a distinctive lesion in a hysterectomy specimen from a 70-year-old woman which they designated endometrial intraepithelial carcinoma, clear cell type (EIC, clear cell type). The lesion comprised surface epithelium and glands that were lined by cells with "clear cytoplasm, marked nuclear pleomorphism, coarse chromatin, irregular nuclear membranes, and prominent eosinophilic nucleoli" and an occasional hobnail appearance. No mitotic figures were recognized. There was no evidence of stromal or myometrial invasion. The lesions showed "focal" staining for p53, a "moderate to high proliferative index," and no evidence of extrauterine extension. In 2006, our group studied the characteristic clinicopathologic features of these putative precursor lesions (Fadare, Liang et al. 2006). 14 cases of pure ECCC and 16 endometrial carcinomas with a greater than 10% clear cell component were evaluated, the adjacent benign endometria were searched for lesions that were morphologically distinct from the background benign endometrium and which were not clearly classifiable as a non-neoplastic process. A total of 38 benign uteri and 30 uteri with EEC served as the control groups. In 90% of cases, we identified a spectrum of atypical endometrial glandular and surface changes that were distinct from both the background benign endometrium and the adjacent ECCC. These changes were not identified in any of the control group cases. Transition from resting endometrium to clear cell EmGD, or from clear cell EmGD to clear cell EIC, was detected in 11 (41%) of 27 cases(Fadare, Liang et al. 2006). These morphological changes were also maintained by immunohistochemical stains, which showed that the clear cell EmGD lesions had p53 staining scores and MIB1 proliferative indices that were intermediate between the resting endometrium in which they were identified and the adjacent ECCC. The lesions also showed markedly reduced frequency of ER and PR expression compared with the background endometrium. According to our findings, we hypothesized that these lesions represent precancerous lesions of ECCC. There has been an inadequate number of cases described to know if clear cell EIC in isolation, like serous EIC, has any capacity or propensity for extrauterine extension. Additional studies, as have previously been carried out on serous EmGD(Zheng, Liang et al. 2007; Zheng, Xiang et al. 2011), are required to conclusively establish the precancerous nature of these per National Cancer Institute criteria(Berman, Albores-Saavedra et al. 2006).

3.2.3 Morphologic features of clear cell EmGD

The features of clear cell EmGD are a spectrum of morphological changes involving a single gland, a few glandular clusters or surface epithelium lined by cells with cytoplasmic clarity or eosinophilia, or hobnail nuclei, and varying degrees of nuclear atypia. These changes were graded on a scale of 1 to 3 (Fadare, Liang et al. 2006), primarily depending on the level of cytologic atypia of the constituent cells. A lesion is grade 1 if there is nuclear enlargement (2- to 3-fold compared with resting endometrium) (Figure 13). Grade 3 nuclei show marked pleomorphism and prominent nucleoli comparable to frank ECCC (Figure 12). Grade 2 changes display intermediate features. Mitotic figures were rare in grade 1 and 2 lesions but were easily seen in grade 3 lesions. Morphologically and conceptually, grade 3 lesions were classifiable as clear cell EIC, whereas grade 1 and 2 lesions were designated clear cell endometrial glandular dysplasia (clear cell EmGD).

3.2.4 Molecular alterations of clear cell EmGD

The genetic aspects of ECCC are not fully understood, and further studies are required to establish the exact pathogenesis of this unusual tumor. The information assembled from previous efforts suggests that the molecular pathogenesis of ECCC is different from that of EEC and ESC, and that the molecular alterations frequently detected in EEC and ESC, including PTEN, K-ras, and TP53 mutations, are not commonly seen in ECCC. Lax et al (Lax, Pizer et al. 1998); noted that a division of ECCC cases display morphologic features suggestive of ESC (ECCC with serous features) and that the latter showed a higher Ki-67 proliferative index than did typical ECCC. Furthermore, ECCC with serous features were associated with serous endometrial intraepithelial carcinoma (EIC) in 50% of cases. In 2004, An et al (An, Logani et al. 2004), studied 16 ECCCs (including 11 pure and 5 mixed cases) for mutations in the PTEN and p53 genes, and for microsatellite instability. These alterations were detected in only a minority of the pure cases, but they were present in the mixed tumors. In addition, in the 2 cases of mixed ECCC/ESC, identical p53 mutations were identified in the 2 histologically distinct parts of the tumor. In one case of a mixed ECCC/EEC, identical p53 and PTEN mutations, as well as microsatellite instability, were identified in the 2 components. The authors concluded that ECCC "represent a heterogeneous group of tumors that arise via different pathogenetic pathways."

As previously noted, molecular alterations that are characteristic of ESC, such as p53 mutations or down-regulation of E-cadherin, may also be seen in ECCC but at a significantly lower frequency (Lax, Pizer et al. 1998; Holcomb, Delatorre et al. 2002; Yalta, Atay et al. 2009). On the other hand, expression of ER and PR is seen at a considerably lower rate in ECCC than is typical of EEC. Other alterations that have been reported in ECCC include decreased expression of the metastasis suppressor CD82 (Kangai-1) and frequent hypermethylation of the stem cell-related transcription factor (SOX2), and up-regulation of the oncogenesis-related protein HNF-1A. The first 2 of these alterations are considered to be linked to type II cancers in general, rather than ECCC in particular(Wong, Huo et al.). The precise molecular alterations in clear cell EmGD and clear cell EIC are still uncertain and further molecular and genetic studies are necessary to elucidate them.

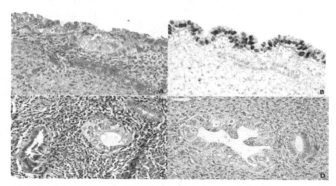

Fig. 13. Clear cell endometrial glandular dysplasia (clear cell EmGD). It can be seen in the surface epithelium (A), or single glands (C&D). Imunohistochemical stain for p53 is positive (B).

3.2.5 Clinical significance and future management

Due to limited number of clear cell EmGD cases that have been studied,the practical clinical impact of this diagnosis, especially if it is found in isolation in an endometrial biopsy sample, is simply unclear. Guidelines on how to manage such cases will not be available until more retrospective and prospective studies are done. Clear cell carcinoma is a rare type of endometrial type II carcinoma. Studies of precursor lesions are so far scarse. We previously proposed clear cell EmGD as a putative precursor due to similarities in morphologic and immunophenotypic features of clear cell carcinoma. However, follow-up and molecular studies are required to establish an ancestry connection between the clear cell EmGD, clear cell EIC, and ECCC and to illuminate the genetic pathways involved in the development and progression of these putative precursor lesions.

4. Conclusion

Endometrial carcinomas encompass a wide spectrum of morphologically and biologically distinct entities. These can be categorized into 2 major pathways (*type I* and *type II endometrial cancer*) according to the dualistic model of endometrial carcinogenesis. Both types have histologic subtypes, and are distinct in their risk factors, molecular background, precancerous lesions and overall patient outcome. The histologic subtype of endometrial cancer has been demonstrated as a significant prognostic factor. The previously used contradicting nomenclature systems for endometrial precancers had been a basis for confusion and low reproducibility among pathologists. They also resulted in the inappropriate inclusion of certain lesions as precancer lesions that did not qualify as such (e.g. simple hyperplasia without atypia for *type I endometrial cancer*, and serous EIC for *type II endometrial cancer*); which in our opinion makes it essential to search for a more simple and unified nomenclature system in this context. Based on the previously detailed dualistic model of endometrial carcinogenesis, and with emphasis on the strict criteria of a precancer as defined by the 2006 National Cancer Institute consensus; we believe that endometrioid EIN (as defined by Mutter et al) is the precancer lesion for *type I endometrial cancer*. For *type II endometrial cancer*, on the other hand, our recent studies confirmed that serous EIN (serous EmGD as previously defined) is the precancerous lesion for serous carcinoma. Similarly, clear cell EIN (previously defined as clear cell EmGD) as a putative precancer for clear cell carcinoma. The precancers of type I and type II endometrial cancer are morphologically and biologically distinct entities, and to the best of our knowledge do not overlap or function as precancer of their cancer counterparts. Much is still to be explored regarding the nature, clinical significance, and appropriate management of those precancer lesions. The newly proposed unified endometrial intraepithelial neoplasia classification system, hopefully, will reduce the confusion in clinic and ultimately benefit patients.

5. References

Abeler, V. M. and K. E. Kjorstad (1991). "Clear cell carcinoma of the endometrium: a histopathological and clinical study of 97 cases." *Gynecol Oncol* 40(3): 207-17.

Alkushi, A., P. Lim, et al. (2004). "Interpretation of p53 immunoreactivity in endometrial carcinoma: establishing a clinically relevant cut-off level." *Int J Gynecol Pathol* 23(2): 129-37.

Ambros, R. A., M. E. Sherman, et al. (1995). "Endometrial intraepithelial carcinoma: a distinctive lesion specifically associated with tumors displaying serous differentiation." *Hum Pathol* 26(11): 1260-7.

An, H. J., S. Logani, et al. (2004). "Molecular characterization of uterine clear cell carcinoma." *Mod Pathol* 17(5): 530-7.

Arai, M., J. Utsunomiya, et al. (2004). "Familial breast and ovarian cancers." *Int J Clin Oncol* 9(4): 270-82.

Baak, J. P., G. L. Mutter, et al. (2005). "The molecular genetics and morphometry-based endometrial intraepithelial neoplasia classification system predicts disease progression in endometrial hyperplasia more accurately than the 1994 World Health Organization classification system." *Cancer* 103(11): 2304-12.

Baak, J. P., J. J. Nauta, et al. (1988). "Architectural and nuclear morphometrical features together are more important prognosticators in endometrial hyperplasias than nuclear morphometrical features alone." *J Pathol* 154(4): 335-41.

Baak, J. P., E. C. Wisse-Brekelmans, et al. (1992). "Assessment of the risk on endometrial cancer in hyperplasia, by means of morphological and morphometrical features." *Pathol Res Pract* 188(7): 856-9.

Baergen, R. N., C. D. Warren, et al. (2001). "Early uterine serous carcinoma: clonal origin of extrauterine disease." *Int J Gynecol Pathol* 20(3): 214-9.

Berman, J. J., J. Albores-Saavedra, et al. (2006). "Precancer: a conceptual working definition -- results of a Consensus Conference." *Cancer Detect Prev* 30(5): 387-94.

Bokhman, J. V. (1983). "Two pathogenetic types of endometrial carcinoma." *Gynecol Oncol* 15(1): 10-7.

Brabletz, T., A. Jung, et al. (1999). "beta-catenin regulates the expression of the matrix metalloproteinase-7 in human colorectal cancer." *Am J Pathol* 155(4): 1033-8.

Burcham, P. C. (1999). "Internal hazards: baseline DNA damage by endogenous products of normal metabolism." *Mutat Res* 443(1-2): 11-36.

Burks, R. T., T. D. Kessis, et al. (1994). "Microsatellite instability in endometrial carcinoma." *Oncogene* 9(4): 1163-6.

Bussaglia, E., E. del Rio, et al. (2000). "PTEN mutations in endometrial carcinomas: a molecular and clinicopathologic analysis of 38 cases." *Hum Pathol* 31(3): 312-7.

Caduff, R. F., C. M. Johnston, et al. (1995). "Mutations of the Ki-ras oncogene in carcinoma of the endometrium." *Am J Pathol* 146(1): 182-8.

Cairns, J. (1998). "Mutation and cancer: the antecedents to our studies of adaptive mutation." *Genetics* 148(4): 1433-40.

Carcangiu, M. L., L. K. Tan, et al. (1997). "Stage IA Uterine Serous Carcinoma: A Study of 13 Cases." *The American Journal of Surgical Pathology* 21(12): 1507-1514.

Casalini, P., M. V. Iorio, et al. (2004). "Role of HER receptors family in development and differentiation." *J Cell Physiol* 200(3): 343-50.

Catasus, L., E. Bussaglia, et al. (2004). "Molecular genetic alterations in endometrioid carcinomas of the ovary: similar frequency of beta-catenin abnormalities but lower rate of microsatellite instability and PTEN alterations than in uterine endometrioid carcinomas." *Hum Pathol* 35(11): 1360-8.

Catasus, L., X. Matias-Guiu, et al. (2000). "Frameshift mutations at coding mononucleotide repeat microsatellites in endometrial carcinoma with microsatellite instability." *Cancer* 88(10): 2290-7.

Chan, J. K., V. Loizzi, et al. (2003). "Significance of comprehensive surgical staging in noninvasive papillary serous carcinoma of the endometrium." *Gynecol Oncol* 90(1): 181-5.

Chan, J. K., M. R. Manuel, et al. (2006). "Breast cancer followed by corpus cancer: is there a higher risk for aggressive histologic subtypes?" *Gynecol Oncol* 102(3): 508-12.

Chen, N., X. Yi, et al. "Nrf2 expression in endometrial serous carcinomas and its precancers." *Int J Clin Exp Pathol* 4(1): 85-96.

Chiesa-Vottero, A. G., A. Malpica, et al. (2007). "Immunohistochemical overexpression of p16 and p53 in uterine serous carcinoma and ovarian high-grade serous carcinoma." *Int J Gynecol Pathol* 26(3): 328-33.

Creasman, W. T., F. Odicino, et al. (2003). "Carcinoma of the corpus uteri." *Int J Gynaecol Obstet* 83 Suppl 1: 79-118.

Darvishian, F., A. J. Hummer, et al. (2004). "Serous endometrial cancers that mimic endometrioid adenocarcinomas: a clinicopathologic and immunohistochemical study of a group of problematic cases." *Am J Surg Pathol* 28(12): 1568-78.

Deligdisch, L. and C. F. Holinka (1987). "Endometrial carcinoma: two diseases?" *Cancer Detect Prev* 10(3-4): 237-46.

Demopoulos, R. I., A. F. Mesia, et al. (1999). "Immunohistochemical comparison of uterine papillary serous and papillary endometrioid carcinoma: clues to pathogenesis." *Int J Gynecol Pathol* 18(3): 233-7.

Duggan, B. D., J. C. Felix, et al. (1994). "Microsatellite instability in sporadic endometrial carcinoma." *J Natl Cancer Inst* 86(16): 1216-21.

Eifel, P. J., J. Ross, et al. (1983). "Adenocarcinoma of the endometrium. Analysis of 256 cases with disease limited to the uterine corpus: treatment comparisons." *Cancer* 52(6): 1026-31.

Enomoto, T., M. Inoue, et al. (1991). "K-ras activation in premalignant and malignant epithelial lesions of the human uterus." *Cancer Res* 51(19): 5308-14.

Esteller, M., L. Catasus, et al. (1999). "hMLH1 promoter hypermethylation is an early event in human endometrial tumorigenesis." *Am J Pathol* 155(5): 1767-72.

Esteller, M., R. Levine, et al. (1998). "MLH1 promoter hypermethylation is associated with the microsatellite instability phenotype in sporadic endometrial carcinomas." *Oncogene* 17(18): 2413-7.

Ettinger, B., I. M. Golditch, et al. (1988). "Gynecologic consequences of long-term, unopposed estrogen replacement therapy." *Maturitas* 10(4): 271-82.

Factor, S. M. (1974). "Papillary adenocarcinoma of the endometrium with psammoma bodies." *Arch Pathol* 98(3): 201-5.

Fadare, O., S. X. Liang, et al. (2006). "Precursors of endometrial clear cell carcinoma." *Am J Surg Pathol* 30(12): 1519-30.

Fadare, O. and W. Zheng (2008). "Endometrial Glandular Dysplasia (EmGD): morphologically and biologically distinctive putative precursor lesions of Type II endometrial cancers." *Diagn Pathol* 3: 6.

Fadare, O. and W. Zheng (2009). "Insights into endometrial serous carcinogenesis and progression." *Int J Clin Exp Pathol* 2(5): 411-32.

Faquin, W. C., J. T. Fitzgerald, et al. (2000). "Sporadic microsatellite instability is specific to neoplastic and preneoplastic endometrial tissues." *Am J Clin Pathol* 113(4): 576-82.

Ferenczy, A. (2003). "Pathophysiology of endometrial bleeding." *Maturitas* 45(1): 1-14.

Fujimoto, I., Y. Shimizu, et al. (1993). "Studies on ras oncogene activation in endometrial carcinoma." *Gynecol Oncol* 48(2): 196-202.

Fukuchi, T., M. Sakamoto, et al. (1998). "Beta-catenin mutation in carcinoma of the uterine endometrium." *Cancer Res* 58(16): 3526-8.

Gehrig, P. A., P. A. Groben, et al. (2001). "Noninvasive papillary serous carcinoma of the endometrium." *Obstet Gynecol* 97(1): 153-7.

Goodfellow, P. J., B. M. Buttin, et al. (2003). "Prevalence of defective DNA mismatch repair and MSH6 mutation in an unselected series of endometrial cancers." *Proc Natl Acad Sci U S A* 100(10): 5908-13.

Grimes, D. A. and K. E. Economy (1995). "Primary prevention of gynecologic cancers." *Am J Obstet Gynecol* 172(1 Pt 1): 227-35.

Gurin, C. C., M. G. Federici, et al. (1999). "Causes and consequences of microsatellite instability in endometrial carcinoma." *Cancer Res* 59(2): 462-6.

Hall, H. I., P. Jamison, et al. (2001). "Second primary ovarian cancer among women diagnosed previously with cancer." *Cancer Epidemiol Biomarkers Prev* 10(9): 995-9.

Hameed, K. and D. A. Morgan (1972). "Papillary adenocarcinoma of endometrium with psammoma bodies. Histology and fine structure." *Cancer* 29(5): 1326-35.

Harris, C. C. (1993). "p53: at the crossroads of molecular carcinogenesis and risk assessment." *Science* 262(5142): 1980-1.

Hecht, J. L., T. A. Ince, et al. (2005). "Prediction of endometrial carcinoma by subjective endometrial intraepithelial neoplasia diagnosis." *Mod Pathol* 18(3): 324-30.

Hecht, J. L. and G. L. Mutter (2006). "Molecular and pathologic aspects of endometrial carcinogenesis." *J Clin Oncol* 24(29): 4783-91.

Hendrickson, M., J. Ross, et al. (1982). "Uterine papillary serous carcinoma: a highly malignant form of endometrial adenocarcinoma." *Am J Surg Pathol* 6(2): 93-108.

Holcomb, K., R. Delatorre, et al. (2002). "E-cadherin expression in endometrioid, papillary serous, and clear cell carcinoma of the endometrium." *Obstet Gynecol* 100(6): 1290-5.

Hornreich, G., U. Beller, et al. (1999). "Is Uterine Serous Papillary Carcinoma a BRCA1-Related Disease? Case Report and Review of the Literature." *Gynecologic Oncology* 75(2): 300-304.

Huang, E. C., G. L. Mutter, et al. (2010). "Clinical outcome in diagnostically ambiguous foci of /`gland crowding/' in the endometrium." *Mod Pathol* 23(11): 1486-1491.

Hunter, J. E., D. E. Tritz, et al. (1994). "The prognostic and therapeutic implications of cytologic atypia in patients with endometrial hyperplasia." *Gynecol Oncol* 55(1): 66-71.

Huszar, M., M. Pfeifer, et al. "Up-regulation of L1CAM is linked to loss of hormone receptors and E-cadherin in aggressive subtypes of endometrial carcinomas." *J Pathol* 220(5): 551-61.

Inskip, P. D. and R. E. Curtis (2007). "New malignancies following childhood cancer in the United States, 1973-2002." *Int J Cancer* 121(10): 2233-40.

Ito, K., K. Watanabe, et al. (1996). "K-ras point mutations in endometrial carcinoma: effect on outcome is dependent on age of patient." *Gynecol Oncol* 63(2): 238-46.

Jemal, A., R. Siegel, et al. (2009). "Cancer statistics, 2009." *CA Cancer J Clin* 59(4): 225-49.

Jia, L., Y. Liu, et al. (2008). "Endometrial glandular dysplasia with frequent p53 gene mutation: a genetic evidence supporting its precancer nature for endometrial serous carcinoma." *Clin Cancer Res* 14(8): 2263-9.

Jiang, T., N. Chen, et al. "High levels of Nrf2 determine chemoresistance in type II endometrial cancer." *Cancer Res* 70(13): 5486-96.

Jiang, Z., P. G. Chu, et al. (2006). "Analysis of RNA-binding protein IMP3 to predict metastasis and prognosis of renal-cell carcinoma: a retrospective study." *Lancet Oncol* 7(7): 556-64.

Jones, M. H., S. Koi, et al. (1994). "Allelotype of uterine cancer by analysis of RFLP and microsatellite polymorphisms: frequent loss of heterozygosity on chromosome arms 3p, 9q, 10q, and 17p." *Genes Chromosomes Cancer* 9(2): 119-23.

Jovanovic, A. S., K. A. Boynton, et al. (1996). "Uteri of women with endometrial carcinoma contain a histopathological spectrum of monoclonal putative precancers, some with microsatellite instability." *Cancer Res* 56(8): 1917-21.

Karpas, C. M. and M. F. Bridge (1963). "Endometrial Adenocarcinoma with Psammomatous Bodies." *Am J Obstet Gynecol* 87: 935-41.

Keating, J. T., A. Cviko, et al. (2001). "Ki-67, cyclin E, and p16INK4 are complimentary surrogate biomarkers for human papilloma virus-related cervical neoplasia." *Am J Surg Pathol* 25(7): 884-91.

Keating, J. T., T. Ince, et al. (2001). "Surrogate biomarkers of HPV infection in cervical neoplasia screening and diagnosis." *Adv Anat Pathol* 8(2): 83-92.

Kobayashi, K., M. Matsushima, et al. (1996). "Mutational analysis of mismatch repair genes, hMLH1 and hMSH2, in sporadic endometrial carcinomas with microsatellite instability." *Jpn J Cancer Res* 87(2): 141-5.

Konecny, G. E., R. Agarwal, et al. (2008). "Claudin-3 and claudin-4 expression in serous papillary, clear-cell, and endometrioid endometrial cancer." *Gynecol Oncol* 109(2): 263-9.

Kong, D., A. Suzuki, et al. (1997). "PTEN1 is frequently mutated in primary endometrial carcinomas." *Nat Genet* 17(2): 143-4.

Konopka, B., A. Janiec-Jankowska, et al. (2007). "Molecular genetic defects in endometrial carcinomas: microsatellite instability, PTEN and beta-catenin (CTNNB1) genes mutations." *J Cancer Res Clin Oncol* 133(6): 361-71.

Kupryjanczyk, J., A. D. Thor, et al. (1996). "Ovarian, peritoneal, and endometrial serous carcinoma: clonal origin of multifocal disease." *Mod Pathol* 9(3): 166-73.

Kurman, R. J., P. F. Kaminski, et al. (1985). "The behavior of endometrial hyperplasia. A long-term study of "untreated" hyperplasia in 170 patients." *Cancer* 56(2): 403-12.

Lagarda, H., L. Catasus, et al. (2001). "K-ras mutations in endometrial carcinomas with microsatellite instability." *J Pathol* 193(2): 193-9.

Lauchlan, S. C. (1981). "Tubal (serous) carcinoma of the endometrium." *Arch Pathol Lab Med* 105(11): 615-8.

Lavie, O., G. Hornreich, et al. (2004). "BRCA germline mutations in Jewish women with uterine serous papillary carcinoma." *Gynecol Oncol* 92(2): 521-4.

Lax, S. F. (2004). "Molecular genetic pathways in various types of endometrial carcinoma: from a phenotypical to a molecular-based classification." *Virchows Arch* 444(3): 213-23.

Lax, S. F., B. Kendall, et al. (2000). "The frequency of p53, K-ras mutations, and microsatellite instability differs in uterine endometrioid and serous carcinoma: evidence of distinct molecular genetic pathways." *Cancer* 88(4): 814-24.

Lax, S. F. and R. J. Kurman (1997). "A dualistic model for endometrial carcinogenesis based on immunohistochemical and molecular genetic analyses." *Verh Dtsch Ges Pathol* 81: 228-32.

Lax, S. F., E. S. Pizer, et al. (1998). "Clear cell carcinoma of the endometrium is characterized by a distinctive profile of p53, Ki-67, estrogen, and progesterone receptor expression." *Hum Pathol* 29(6): 551-8.

Levine, R. L., C. B. Cargile, et al. (1998). "PTEN mutations and microsatellite instability in complex atypical hyperplasia, a precursor lesion to uterine endometrioid carcinoma." *Cancer Res* 58(15): 3254-8.

Liang, S. X., S. K. Chambers, et al. (2004). "Endometrial glandular dysplasia: a putative precursor lesion of uterine papillary serous carcinoma. Part II: molecular features." *Int J Surg Pathol* 12(4): 319-31.

Liang, S. X., M. Pearl, et al. (2010). "Personal history of breast cancer as a significant risk factor for endometrial serous carcinoma in women aged 55 years old or younger." *Int J Cancer* 128(4): 763-70.

Liu, F. S. (2007). "Molecular carcinogenesis of endometrial cancer." *Taiwan J Obstet Gynecol* 46(1): 26-32.

Liu, F. S., E. S. Ho, et al. (1997). "Mutational analysis of the BRCA1 tumor suppressor gene in endometrial carcinoma." *Gynecol Oncol* 66(3): 449-53.

MacCallum, D. E. and T. R. Hupp (1999). "Induction of p53 protein as a marker for ionizing radiation exposure in vivo." *Methods Mol Biol* 113: 583-9.

Matias-Guiu, X., L. Catasus, et al. (2001). "Molecular pathology of endometrial hyperplasia and carcinoma." *Hum Pathol* 32(6): 569-77.

Maxwell, G. L., J. I. Risinger, et al. (1998). "Mutation of the PTEN tumor suppressor gene in endometrial hyperplasias." *Cancer Res* 58(12): 2500-3.

Mell, L. K., J. J. Meyer, et al. (2004). "Prognostic significance of E-cadherin protein expression in pathological stage I-III endometrial cancer." *Clin Cancer Res* 10(16): 5546-53.

Mirabelli-Primdahl, L., R. Gryfe, et al. (1999). "Beta-catenin mutations are specific for colorectal carcinomas with microsatellite instability but occur in endometrial carcinomas irrespective of mutator pathway." *Cancer Res* 59(14): 3346-51.

Moid, F. and K. Berezowski (2004). "Pathologic quiz case: a 70-year-old woman with postmenopausal bleeding. Endometrial intraepithelial carcinoma, clear cell type." *Arch Pathol Lab Med* 128(11): e157-8.

Mutter, G. L. (2000). "Endometrial intraepithelial neoplasia (EIN): will it bring order to chaos? The Endometrial Collaborative Group." *Gynecol Oncol* 76(3): 287-90.

Mutter, G. L. (2002). "Diagnosis of premalignant endometrial disease." *J Clin Pathol* 55(5): 326-31.

Mutter, G. L., J. P. Baak, et al. (2000). "Endometrial precancer diagnosis by histopathology, clonal analysis, and computerized morphometry." *J Pathol* 190(4): 462-9.

Mutter, G. L., K. A. Boynton, et al. (1996). "Allelotype mapping of unstable microsatellites establishes direct lineage continuity between endometrial precancers and cancer." *Cancer Res* 56(19): 4483-6.

Mutter, G. L., M. L. Chaponot, et al. (1995). "A polymerase chain reaction assay for non-random X chromosome inactivation identifies monoclonal endometrial cancers and precancers." *Am J Pathol* 146(2): 501-8.

Mutter, G. L., T. A. Ince, et al. (2001). "Molecular identification of latent precancers in histologically normal endometrium." *Cancer Res* 61(11): 4311-4.

Mutter, G. L., M. C. Lin, et al. (2000). "Altered PTEN expression as a diagnostic marker for the earliest endometrial precancers." *J Natl Cancer Inst* 92(11): 924-30.

Mutter, G. L., M. C. Lin, et al. (2000). "Changes in endometrial PTEN expression throughout the human menstrual cycle." *J Clin Endocrinol Metab* 85(6): 2334-8.

Mutter, G. L., H. Wada, et al. (1999). "K-ras mutations appear in the premalignant phase of both microsatellite stable and unstable endometrial carcinogenesis." *Mol Pathol* 52(5): 257-62.

Mutter, G. L., R. J. Zaino, et al. (2007). "Benign endometrial hyperplasia sequence and endometrial intraepithelial neoplasia." *Int J Gynecol Pathol* 26(2): 103-14.

Negri, G., E. Egarter-Vigl, et al. (2003). "p16INK4a is a useful marker for the diagnosis of adenocarcinoma of the cervix uteri and its precursors: an immunohistochemical study with immunocytochemical correlations." *Am J Surg Pathol* 27(2): 187-93.

Nielsen, J., J. Christiansen, et al. (1999). "A family of insulin-like growth factor II mRNA-binding proteins represses translation in late development." *Mol Cell Biol* 19(2): 1262-70.

Nordstrom, B., P. Strang, et al. (1996). "Endometrial carcinoma: the prognostic impact of papillary serous carcinoma (UPSC) in relation to nuclear grade, DNA ploidy and p53 expression." *Anticancer Res* 16(2): 899-904.

Ouyang, H., H. O. Shiwaku, et al. (1997). "The insulin-like growth factor II receptor gene is mutated in genetically unstable cancers of the endometrium, stomach, and colorectum." *Cancer Res* 57(10): 1851-4.

Parazzini, F., C. La Vecchia, et al. (1991). "The epidemiology of endometrial cancer." *Gynecol Oncol* 41(1): 1-16.

Peiffer, S. L., T. J. Herzog, et al. (1995). "Allelic loss of sequences from the long arm of chromosome 10 and replication errors in endometrial cancers." *Cancer Res* 55(9): 1922-6.

Pietsch, E. C., S. M. Sykes, et al. (2008). "The p53 family and programmed cell death." *Oncogene* 27(50): 6507-21.

Pijnenborg, J. M., L. van de Broek, et al. (2006). "TP53 overexpression in recurrent endometrial carcinoma." *Gynecol Oncol* 100(2): 397-404.

Potischman, N., R. N. Hoover, et al. (1996). "Case-control study of endogenous steroid hormones and endometrial cancer." *J Natl Cancer Inst* 88(16): 1127-35.

Rangel, L. B., R. Agarwal, et al. (2003). "Tight junction proteins claudin-3 and claudin-4 are frequently overexpressed in ovarian cancer but not in ovarian cystadenomas." *Clin Cancer Res* 9(7): 2567-75.

Reid-Nicholson, M., P. Iyengar, et al. (2006). "Immunophenotypic diversity of endometrial adenocarcinomas: implications for differential diagnosis." *Mod Pathol* 19(8): 1091-100.

Risinger, J. I., A. Berchuck, et al. (1993). "Genetic instability of microsatellites in endometrial carcinoma." *Cancer Res* 53(21): 5100-3.

Risinger, J. I., A. K. Hayes, et al. (1997). "PTEN/MMAC1 mutations in endometrial cancers." *Cancer Res* 57(21): 4736-8.

Rolitsky, C. D., K. S. Theil, et al. (1999). "HER-2/neu amplification and overexpression in endometrial carcinoma." *Int J Gynecol Pathol* 18(2): 138-43.

Saegusa, M., M. Hashimura, et al. (2001). "beta- Catenin mutations and aberrant nuclear expression during endometrial tumorigenesis." *Br J Cancer* 84(2): 209-17.

Sakamoto, T., T. Murase, et al. (1998). "Microsatellite instability and somatic mutations in endometrial carcinomas." *Gynecol Oncol* 71(1): 53-8.

Sakuragi, N., M. Nishiya, et al. (1994). "Decreased E-cadherin expression in endometrial carcinoma is associated with tumor dedifferentiation and deep myometrial invasion." *Gynecol Oncol* 53(2): 183-9.

Santin, A. D., S. Bellone, et al. (2002). "Overexpression of HER-2/neu in uterine serous papillary cancer." *Clin Cancer Res* 8(5): 1271-9.

Santin, A. D., S. Bellone, et al. (2007). "Overexpression of claudin-3 and claudin-4 receptors in uterine serous papillary carcinoma: novel targets for a type-specific therapy using Clostridium perfringens enterotoxin (CPE)." *Cancer* 109(7): 1312-22.

Sasaki, H., H. Nishii, et al. (1993). "Mutation of the Ki-ras protooncogene in human endometrial hyperplasia and carcinoma." *Cancer Res* 53(8): 1906-10.

Sasano, H., J. Comerford, et al. (1990). "Serous papillary adenocarcinoma of the endometrium. Analysis of proto-oncogene amplification, flow cytometry, estrogen and progesterone receptors, and immunohistochemistry." *Cancer* 65(7): 1545-51.

Schlosshauer, P. W., E. C. Pirog, et al. (2000). "Mutational analysis of the CTNNB1 and APC genes in uterine endometrioid carcinoma." *Mod Pathol* 13(10): 1066-71.

Schmitz, M. J., D. T. Hendricks, et al. (2000). "p27 and cyclin D1 abnormalities in uterine papillary serous carcinoma." *Gynecol Oncol* 77(3): 439-45.

Scully RE, B. T., et al., Ed. (1994). *Histological Typing of Female Genital Tract Tumors* Uterine corpus. New York, NY, Springer Verlag.

Shang, Y. (2006). "Molecular mechanisms of oestrogen and SERMs in endometrial carcinogenesis." *Nat Rev Cancer* 6(5): 360-8.

Sherman, M. E. (2000). "Theories of endometrial carcinogenesis: a multidisciplinary approach." *Mod Pathol* 13(3): 295-308.

Sherman, M. E., P. Bitterman, et al. (1992). "Uterine serous carcinoma. A morphologically diverse neoplasm with unifying clinicopathologic features." *Am J Surg Pathol* 16(6): 600-10.

Sherman, M. E., M. E. Bur, et al. (1995). "p53 in endometrial cancer and its putative precursors: evidence for diverse pathways of tumorigenesis." *Hum Pathol* 26(11): 1268-74.

Skov, B. G., H. Broholm, et al. (1997). "Comparison of the reproducibility of the WHO classifications of 1975 and 1994 of endometrial hyperplasia." *Int J Gynecol Pathol* 16(1): 33-7.

Song, J., T. Rutherford, et al. (2002). "Hormonal regulation of apoptosis and the Fas and Fas ligand system in human endometrial cells." *Mol Hum Reprod* 8(5): 447-55.

Soslow, R. A., P. U. Shen, et al. (1998). "Distinctive p53 and mdm2 immunohistochemical expression profiles suggest different pathogenetic pathways in poorly differentiated endometrial carcinoma." *Int J Gynecol Pathol* 17(2): 129-34.

Soslow, R. A., P. U. Shen, et al. (1998). "The CD44v6-negative phenotype in high-grade uterine carcinomas correlates with serous histologic subtype." *Mod Pathol* 11(2): 194-9.

Spiegel, G. W. (1995). "Endometrial carcinoma in situ in postmenopausal women." *Am J Surg Pathol* 19(4): 417-32.

Su, L. K., B. Vogelstein, et al. (1993). "Association of the APC tumor suppressor protein with catenins." *Science* 262(5140): 1734-7.

Swisher, E. M., S. Peiffer-Schneider, et al. (1999). "Differences in patterns of TP53 and KRAS2 mutations in a large series of endometrial carcinomas with or without microsatellite instability." *Cancer* 85(1): 119-26.

Tashiro, H., M. S. Blazes, et al. (1997). "Mutations in PTEN are frequent in endometrial carcinoma but rare in other common gynecological malignancies." *Cancer Res* 57(18): 3935-40.

Tavassoli FA, D. P., (Eds), Ed. (2003). *World Health Organization Classification of Tumors. Pathology and Genetics of Tumors of the Breast and Female genital Organs.* Lyon, IARC Press.

Trial, T. W. G. f. t. P. (1996). "Effects of hormone replacement therapy on endometrial histology in postmenopausal women. The Postmenopausal Estrogen/Progestin Interventions (PEPI) Trial. The Writing Group for the PEPI Trial." *JAMA* 275(5): 370-5.

Tsuda, H., K. Jiko, et al. (1995). "Frequent occurrence of c-Ki-ras gene mutations in well differentiated endometrial adenocarcinoma showing infiltrative local growth with fibrosing stromal response." *Int J Gynecol Pathol* 14(3): 255-9.

Velasco, A., E. Bussaglia, et al. (2006). "PIK3CA gene mutations in endometrial carcinoma: correlation with PTEN and K-RAS alterations." *Hum Pathol* 37(11): 1465-72.

Vereide, A. B., T. Kaino, et al. (2005). "Bcl-2, BAX, and apoptosis in endometrial hyperplasia after high dose gestagen therapy: a comparison of responses in patients treated with intrauterine levonorgestrel and systemic medroxyprogesterone." *Gynecol Oncol* 97(3): 740-50.

Vikesaa, J., T. V. Hansen, et al. (2006). "RNA-binding IMPs promote cell adhesion and invadopodia formation." *EMBO J* 25(7): 1456-68.

Villella, J. A., S. Cohen, et al. (2006). "HER-2/neu overexpression in uterine papillary serous cancers and its possible therapeutic implications." *Int J Gynecol Cancer* 16(5): 1897-902.

von Wachenfeldt, A., A. Lindblom, et al. (2007). "A hypothesis-generating search for new genetic breast cancer syndromes--a national study in 803 Swedish families." *Hered Cancer Clin Pract* 5(1): 17-24.

Wang, X. J., Z. Sun, et al. (2008). "Nrf2 enhances resistance of cancer cells to chemotherapeutic drugs, the dark side of Nrf2." *Carcinogenesis* 29(6): 1235-43.

Webb, G. A. and M. D. Lagios (1987). "Clear cell carcinoma of the endometrium." *Am J Obstet Gynecol* 156(6): 1486-91.

Weiderpass, E., H. O. Adami, et al. (1999). "Use of oral contraceptives and endometrial cancer risk (Sweden)." *Cancer Causes Control* 10(4): 277-84.

Wheeler, D. T., K. A. Bell, et al. (2000). "Minimal uterine serous carcinoma: diagnosis and clinicopathologic correlation." *Am J Surg Pathol* 24(6): 797-806.

Wong, O. G., Z. Huo, et al. "Hypermethylation of SOX2 Promoter in Endometrial Carcinogenesis." *Obstet Gynecol Int* 2010.

Yalta, T., L. Atay, et al. (2009). "E-cadherin expression in endometrial malignancies: comparison between endometrioid and non-endometrioid carcinomas." *J Int Med Res* 37(1): 163-8.

Yaniv, K. and J. K. Yisraeli (2002). "The involvement of a conserved family of RNA binding proteins in embryonic development and carcinogenesis." *Gene* 287(1-2): 49-54.

Yemelyanova, A., H. Ji, et al. (2009). "Utility of p16 expression for distinction of uterine serous carcinomas from endometrial endometrioid and endocervical adenocarcinomas: immunohistochemical analysis of 201 cases." *Am J Surg Pathol* 33(10): 1504-14.

Zaino, R. J. (2000). "Endometrial hyperplasia: is it time for a quantum leap to a new classification?" *Int J Gynecol Pathol* 19(4): 314-21.

Zeleniuch-Jacquotte, A., A. Akhmedkhanov, et al. (2001). "Postmenopausal endogenous oestrogens and risk of endometrial cancer: results of a prospective study." *Br J Cancer* 84(7): 975-81.

Zhang, X., S. X. Liang, et al. (2009). "Molecular identification of "latent precancers" for endometrial serous carcinoma in benign-appearing endometrium." *Am J Pathol* 174(6): 2000-6.

Zheng, W., H. E. Baker, et al. (2004). "Involution of PTEN-null endometrial glands with progestin therapy." *Gynecol Oncol* 92(3): 1008-13.

Zheng, W., P. Cao, et al. (1996). "p53 overexpression and bcl-2 persistence in endometrial carcinoma: comparison of papillary serous and endometrioid subtypes." *Gynecol Oncol* 61(2): 167-74.

Zheng, W., R. Khurana, et al. (1998). "p53 immunostaining as a significant adjunct diagnostic method for uterine surface carcinoma: precursor of uterine papillary serous carcinoma." *Am J Surg Pathol* 22(12): 1463-73.

Zheng, W., S. X. Liang, et al. (2007). "Occurrence of endometrial glandular dysplasia precedes uterine papillary serous carcinoma." *Int J Gynecol Pathol* 26(1): 38-52.

Zheng, W., S. X. Liang, et al. (2004). "Endometrial glandular dysplasia: a newly defined precursor lesion of uterine papillary serous carcinoma. Part I: morphologic features." *Int J Surg Pathol* 12(3): 207-23.

Zheng, W. and P. E. Schwartz (2005). "Serous EIC as an early form of uterine papillary serous carcinoma: recent progress in understanding its pathogenesis and current opinions regarding pathologic and clinical management." *Gynecol Oncol* 96(3): 579-82.

Zheng, W., L. Xiang, et al. (2011). "A proposed model for endometrial serous carcinogenesis." *Am J Surg Pathol* 35(1): e1-e14.

Zheng, W., X. Yi, et al. (2008). "The oncofetal protein IMP3: a novel biomarker for endometrial serous carcinoma." *Am J Surg Pathol* 32(2): 304-15.

Permissions

The contributors of this book come from diverse backgrounds, making this book a truly international effort. This book will bring forth new frontiers with its revolutionizing research information and detailed analysis of the nascent developments around the world.

We would like to thank Supriya Srivastava, for lending her expertise to make the book truly unique. She has played a crucial role in the development of this book. Without her invaluable contribution this book wouldn't have been possible. She has made vital efforts to compile up to date information on the varied aspects of this subject to make this book a valuable addition to the collection of many professionals and students.

This book was conceptualized with the vision of imparting up-to-date information and advanced data in this field. To ensure the same, a matchless editorial board was set up. Every individual on the board went through rigorous rounds of assessment to prove their worth. After which they invested a large part of their time researching and compiling the most relevant data for our readers. Conferences and sessions were held from time to time between the editorial board and the contributing authors to present the data in the most comprehensible form. The editorial team has worked tirelessly to provide valuable and valid information to help people across the globe.

Every chapter published in this book has been scrutinized by our experts. Their significance has been extensively debated. The topics covered herein carry significant findings which will fuel the growth of the discipline. They may even be implemented as practical applications or may be referred to as a beginning point for another development. Chapters in this book were first published by InTech; hereby published with permission under the Creative Commons Attribution License or equivalent.

The editorial board has been involved in producing this book since its inception. They have spent rigorous hours researching and exploring the diverse topics which have resulted in the successful publishing of this book. They have passed on their knowledge of decades through this book. To expedite this challenging task, the publisher supported the team at every step. A small team of assistant editors was also appointed to further simplify the editing procedure and attain best results for the readers.

Our editorial team has been hand-picked from every corner of the world. Their multi-ethnicity adds dynamic inputs to the discussions which result in innovative outcomes. These outcomes are then further discussed with the researchers and contributors who give their valuable feedback and opinion regarding the same. The feedback is then collaborated with the researches and they are edited in a comprehensive manner to aid the understanding of the subject.

Apart from the editorial board, the designing team has also invested a significant amount of their time in understanding the subject and creating the most relevant covers. They scrutinized every image to scout for the most suitable representation of the subject and create an appropriate cover for the book.

The publishing team has been involved in this book since its early stages. They were actively engaged in every process, be it collecting the data, connecting with the contributors or procuring relevant information. The team has been an ardent support to the editorial, designing and production team. Their endless efforts to recruit the best for this project, has resulted in the accomplishment of this book. They are a veteran in the field of academics and their pool of knowledge is as vast as their experience in printing. Their expertise and guidance has proved useful at every step. Their uncompromising quality standards have made this book an exceptional effort. Their encouragement from time to time has been an inspiration for everyone.

The publisher and the editorial board hope that this book will prove to be a valuable piece of knowledge for researchers, students, practitioners and scholars across the globe.

List of Contributors

Angela Celetti and Chiara Luise
Istituto di Endocrinologia e Oncologia Sperimentale, CNR, c/o Dipartimento di Biologia e Patologia Cellulare e Molecolare, Università Federico II, Napoli, Italy

Maria Siano and Stefania Staibano
Dipartimento di Scienze Biomorfologiche e Funzionali, Università Federico II, Napoli, Italy

Francesco Merolla
Istituto di Endocrinologia e Oncologia Sperimentale, CNR, c/o Dipartimento di Biologia e Patologia Cellulare e Molecolare, Università Federico II, Napoli, Italy
Dipartimento di Scienze Biomorfologiche e Funzionali, Università Federico II, Napoli, Italy

Napaporn Tananuvat
Departments of Ophthalmology and Pathology, Faculty of Medicine, Chiang Mai University, Thailand

Nirush Lertprasertsuke
Departments of Pathology, Faculty of Medicine, Chiang Mai University, Thailand

Valentín Huerva
Department of Ophthalmology, Universitary Hospital"Arnau de Vilanova" and IRB, Lleida

Francisco J. Ascaso
Department of Ophthalmology,"Lozano Blesa" University Clinic Hospital, Zaragoza, Spain

Chia-Yang Liu and Winston W.Y. Kao
Edith J. Crawley Vision Research Center/Department of Ophthalmology, College of Medicine/University of Cincinnati, Cincinnati OH, USA

Simonetta Monti and Andres Del Castillo
Senology Division, European Institute of Oncology, Milan, Italy

A.G. Papatsoris, C. Kostopoulos, V. Migdalis and M. Chrisofos
2nd Department of Urology, School of Medicine, University of Athens, Sismanoglio General Hospital, Athens, Greece

J. Arunakaran and S. Banudevi
Department of Endocrinology, Dr. ALM Post Graduate Institute of Basic Medical Sciences, University of Madras, Sekkizhar Campus, Taramani, Chennai, Tamilnadu, India

A. Arunkumar
Center of Excellence in Cancer Research, Department of Biomedical Sciences, Texas Tech University Health Sciences Center, Texas, USA

Suneetha Devpura, Jagdish Thakur and Ratna Naik
Wayne State University, Detroit, MI, USA

Seema Sethi, Fazlul Sarkar and Wael Sakr
Pathology, Karmanos Cancer Institute, Detroit, MI, USA

Vaman M. Naik
Univeristy of Michigan-Dearborn, Dearborn, MI, USA

Shuje Pang
Department of Pathology, University of Arizona College of Medicine, Tucson, AZ, USA
Department of Pathology, Tianjin Central Hospital of Obstetrics and Gynecology, Tianjin, China

Jie Li
Department of Pathology, University of Arizona College of Medicine, Tucson, AZ, USA
Department of Obstetrics and Gynecology, Qilu Hospital, Shandong University, Jinan, Shandong, China

Nisreen Abushahin
Department of Pathology, University of Arizona College of Medicine, Tucson, AZ, USA
Department of Pathology, Jordan University Hospital- University of Jordan, Amman, Jordan

Oluwole Fadare
Department of Pathology, Vanderbilt University School of Medicine, Nashville, TN, USA

Wenxin Zheng
Department of Pathology, University of Arizona College of Medicine, Tucson, AZ, USA
Department of Obstetrics and Gynecology, University of Arizona, Tucson, AZ, USA
Arizona Cancer Center, University of Arizona, Tucson, AZ, USA